Parametric Sensitivity in Chemical Systems

The behavior of a chemical system is affected by many physicochemical parameters. The sensitivity of the system's behavior to changes in parameters is known as parametric sensitivity. When a system operates in a parametrically sensitive region, its performance becomes unreliable and changes sharply with small variations in parameters. Thus, it is of great value to those who design and operate chemical reactors and systems to be able to predict sensitivity behavior.

This book is the first to provide a thorough treatment of the concept of parametric sensitivity and the mathematical tool it generated, sensitivity analysis. The emphasis is on applications to real situations. The book begins with definitions of various sensitivity indices and describes the numerical techniques used for their evaluation. Extensively illustrated chapters discuss sensitivity analysis in a variety of chemical reactors – batch, tubular, continuous-flow, fixed-bed – and in combustion systems, air pollution processes, and metabolic processes. In addition, various plots and simple formulas are provided to readily evaluate the operational behavior of reactors. Chemical engineers, graduate students, researchers, chemists and other practitioners will welcome this valuable resource.

Arvind Varma is the Arthur J. Schmitt Professor of Chemical Engineering at the University of Notre Dame.

Massimo Morbidelli is Professor of Chemical Reaction Engineering at ETH Zentrum, Switzerland.

Hua Wu is Senior Chemical Engineer at Ausimont Research & Development Center, Milano, Italy.

T0211196

Parametric Sensitivity in Chemical Systems

A. Varma
M. Morbidelli
H. Wu

CAMBRIDGE
UNIVERSITY PRESS

CAMBRIDGE UNIVERSITY PRESS
Cambridge, New York, Melbourne, Madrid, Cape Town, Singapore, São Paulo

Cambridge University Press
The Edinburgh Building, Cambridge CB2 2RU, UK

Published in the United States of America by Cambridge University Press, New York

www.cambridge.org
Information on this title: www.cambridge.org/9780521621717

First published 1999
This digitally printed first paperback version 2005

A catalogue record for this publication is available from the British Library

Library of Congress Cataloguing in Publication data

Varma, Arvind.
    Parametric sensitivity in chemical systems / cubic equations of
state and their mixing rules / A. Varma. M. Morbidelli, H. Wu.
        p.   cm. – (Cambridge series in chemical engineering)
    Includes bibliographical references and index.
    ISBN 0-521-62171-2 (hb)
    1. Chemical processes – Mathematical models.   I. Morbidelli,
Massimo.   II. Wu, H. (Hua)   III. Title.   IV. Series.
TP155.7.V37   1999
660′.281′015118 – dc21   98-45450
                                                                        CIP

ISBN-13 978-0-521-62171-7 hardback
ISBN-10 0-521-62171-2 hardback

ISBN-13 978-0-521-01984-2 paperback
ISBN-10 0-521-01984-2 paperback

To our parents

Contents

Contents

Contents

Preface

The behavior of physical and chemical systems depends on values of the parameters that characterize the system. The analysis of how a system responds to changes in the parameters is called *parametric sensitivity*. For the purposes of reliable design and control, this analysis is important in virtually all areas of science and engineering. While similar concepts and techniques can be applied in different types of systems, we focus on chemical systems where chemical reactions occur.

In many cases, when one or more parameters are varied slightly, while holding the remaining parameters fixed, the response of a chemical system also changes slightly. However, under other sets of parameter combinations, the chemical system may respond with an enormous change, even if one or more parameters are varied only slightly. In this case, we say that the system behaves in a *parametrically sensitive* manner. Clearly, it becomes difficult to control the chemical system when it operates in a parametrically sensitive region, and sometimes this leads to so-called *runaway* behavior that ends up with catastrophic results. This book is concerned with parametric sensitivity and parametrically sensitive behavior of chemical systems, analyzed with a unified conceptual and theoretical framework.

In Chapter 2, we define various sensitivity indices and illustrate numerical techniques that are commonly used for their evaluation. Then, in Chapters 3 to 4, sensitivity analysis is used to identify the parametrically sensitive regions in various types of reactors, such as batch, tubular, continuous-flow stirred tank, and fixed-bed, where either a single or complex reactions occur. In Chapter 7, we use explosions in hydrogen–oxygen mixtures as an example to show that the same analysis can be used to quantify critical ignition conditions in combustion systems. Chapters 8–10 comprise the second part of the book, where sensitivity analysis is employed as an effective mathematical tool to analyze various chemical systems. These include mechanistic studies and model reduction in chemical kinetics, air pollution, and metabolic processes.

This book should appeal to all who are interested in the behavior of chemical systems, including chemists and chemical, mechanical, aerospace, and environmental engineers. Also, the applied mathematicians should find here a rich source of interesting mathematical problems. Finally, we hope that industrial practitioners will find the concepts and results described in this book to be useful for their work.

This book can be used either as a text for a senior graduate-level specialized course, or as a supplementary text for existing courses in reaction engineering, applied mathematics, design, and control. In this context, although we do not provide unsolved problems at the end of chapters, there are a relatively large number of examples illustrating the concepts and results. The book can also be used as a reference for industrial applications in reactor design, operation and control.

It is a pleasure to acknowledge here our debt of gratitude to Professor John H. Seinfeld of the California Institute of Technology. He encouraged our writing from the beginning, and looked over drafts of Chapters 2 and 9, providing valuable suggestions for improvements. In addition, Dr. Vassily Hatzimanikatis of du Pont Central Research Department kindly provided a keen evaluation of our draft of Chapter 10.

The last thought goes to our families. Our wives (Karen, Luisella, and Guixian) and children (Anita and Sophia; Melissa and Oreste; Xian and Dino) deeply support us and our work, even as they suffer some neglect during the course of writing projects such as this. We cherish their love and affection.

Arvind Varma
Massimo Morbidelli
Hua Wu

1

Introduction

1.1   The Concept of Sensitivity

The behavior of a chemical *system* is affected by many physicochemical *parameters*. Changing these parameters, we can alter the characteristics of the system to realize desired behavior or to avoid undesired behavior. In general, different parameters affect a system to different extents, and for the same parameter, its effect may depend on the range over which it is varied. By *parametric sensitivity*, we mean the sensitivity of the system behavior with respect to changes in parameters.

Let us illustrate the concept of sensitivity using some examples. Figure 1.1 shows the effect of changes in the initial temperature on the temperature evolution in a batch reactor for acetic anhydride hydrolysis, measured experimentally by Haldar and Rao (1992). There is a critical change in the temperature profile as the initial temperature increases from 319.0 to 319.5 K. In particular, an increase in the initial temperature by 0.5 K leads to a change in the temperature maximum by about 31 K. This experimental observation indicates that the system temperature becomes *sensitive to small variations* in the initial temperature in a specific region, called the *parametrically sensitive region*.

Figure 1.2 shows similar sensitivity phenomena in a tubular reactor obtained by numerical computations, given by Bilous and Amundson (1956) in their pioneering work on parametric sensitivity in the context of chemical reactors. In this example, the ambient temperature of a tubular reactor, where an exothermic reaction occurs, is changed. It is seen in Fig. 1.2a that when the ambient temperature increases by 2.5 K from 335 to 337.5 K, the temperature maximum (hot spot) along the reactor length changes by about 70 K. Moreover, such a variation also causes a sharp change in the corresponding concentration profile along the reactor, as shown in Fig. 1.2b. Thus, *when a system operates in the parametrically sensitive region, its performance becomes unreliable and changes sharply with small variations in parameters.*

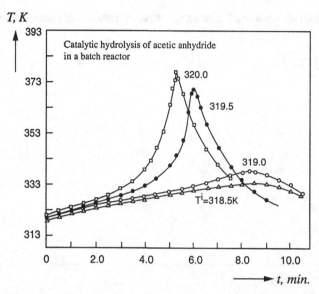

Figure 1.1. Catalytic hydrolysis of acetic anhydride in a batch reactor. Temperature profiles as a function of time for various values of the initial temperature, measured experimentally by Haldar and Rao (1992).

For a chemical system to operate in a reliable and safe manner, it is often required to identify the sensitive regions in the system *parameter space*. An example is shown in Fig. 1.3a, where for a fixed-bed catalytic reactor in which vinyl acetate synthesis occurs, the sensitive region in the cooling versus heat of reaction parameter plane was identified by Emig *et al.* (1980) through a large number of experiments. The symbols o and ● denote low- and high-temperature operating conditions, respectively. These data clearly define a boundary (broken curve) separating the low-temperature from high-temperature operating conditions. In particular, let us consider two operating conditions in Fig. 1.3a near the boundary, indicated by points 1 and 2. The corresponding temperature profiles are shown in Fig. 1.3b. As may be seen, although the two conditions are close in terms of parameters, their temperature profiles are substantially different, indicating that the reactor is operating in the parametrically sensitive region.

Sensitive regions have also been investigated experimentally for other reacting systems, especially for combustion processes. An example is the sensitive region in the initial pressure-temperature plane for hydrogen oxidation in a closed vessel, identified by Lewis and von Elbe (1961), as shown in Fig. 1.4. In particular, the boundary representing the sensitive region divides the parameter plane into two parts. For a fixed initial pressure, as the initial temperature increases, the system undergoes a sharp transition near the boundary, from non-explosion on the left-hand side to explosion on the right-hand side.

It should be noted that although the sensitive region for each parameter can, in principle, be identified experimentally, only a few experimental studies on parametric

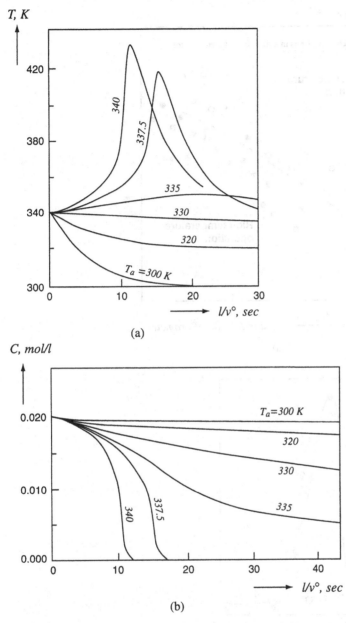

Figure 1.2. Numerical calculated (a) temperature and (b) concentration profiles along the length, l, of a tubular reactor; v°, represents the reaction mixture velocity. From Bilous and Amundson (1956).

sensitivity have been reported to date in the literature. This is because each system involves many physicochemical parameters, so that detailed experimental investigation becomes cumbersome. Thus, it is of great interest to predict theoretically the sensitivity behavior of a chemical system, through appropriate model simulations. The

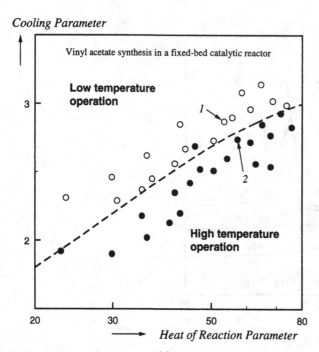

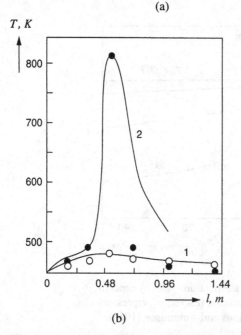

Figure 1.3. Vinyl acetate synthesis in a fixed-bed catalytic reactor. (a) Sensitive operation region in the cooling versus heat of reaction parameter plane, measured experimentally by Emig *et al.* (1980), where ○ = low temperature operation and ● = high temperature operation. (b) Temperature profiles along the reactor length corresponding to the two operation conditions indicated by points *1* and *2* in (a).

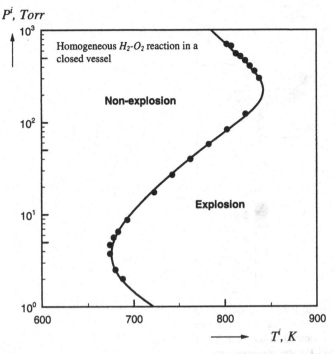

Figure 1.4. Stoichiometric H_2–O_2 mixtures in a closed vessel. The boundary representing the sensitive region in the initial pressure-initial temperature plane, which separates the non-explosion from explosion regions, measured experimentally by Lewis and von Elbe (1961).

essential problems in theoretical predictions are how to describe the sensitive region rigorously and quantitatively, from a mathematical point of view, and how to give a measure for the system sensitivity. Much work has been reported in the literature and comprises the first part of this book.

1.2   Uses of the Sensitivity Concept

Although the phenomenon of sensitivity has been introduced above as an essentially undesired system behavior that needs to be avoided in practice, the concept of sensitivity has now generated an effective mathematical tool, called *sensitivity analysis*, which is widely applied in a variety of fields in science and engineering. Examples include systems control, process optimization, chemical and nuclear reactor design, cell biology, and ecology. The wide applications arise because the sensitivity concept determines a relation between system behavior and a parameter, and the *sensitivity value* quantifies this relationship.

A typical example is the application of sensitivity analysis in kinetic model reduction of a complex reacting system. A complex reacting process (*e.g.*, combustion)

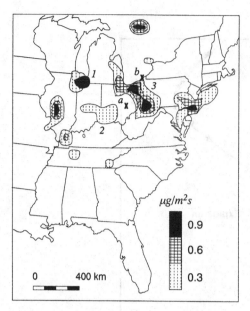

Figure 1.5. Emissions of the ground-level SO$_2$ for July 4, 1974. The numbers $1, 2$, and 3 indicate the emission sources located in the vicinity of Gary, Cincinnati, and Pittsburgh, respectively, while a and b are two receptors of interest. From Hidy and Mueller (1976).

generally involves a few hundred elementary reactions and includes several hundred kinetic parameters. In practical applications, it is often required to reduce such a complex kinetic model to its minimum size, which can give the same description as the complex one for a specific system behavior of interest. The general procedure for kinetic model reduction using sensitivity analysis can be briefly illustrated as follows. The first step is to compute the sensitivity value of the selected system behavior with respect to variations in each elementary reaction in the complex kinetic model, by solving the sensitivity equations derived from basic equations of the system (see Chapter 2). Then, based on the obtained sensitivity values, we classify the reactions. In particular, if a reaction leads to a high sensitivity value, indicating that the system behavior is sensitive to this reaction, then it needs to be included in the reduced kinetic model. On the contrary, if a reaction gives a low sensitivity value, it can be safely excluded from the reduced kinetic model, since the system behavior is insensitive to its presence. Thus the complex kinetic model is reduced.

A second example about applications of sensitivity analysis is related to air pollution control. In this context, an important problem is to determine the influence of specific pollutant sources on specific target locations (receptors). Figure 1.5 shows SO$_2$ emissions in the eastern United States during some particular meteorological conditions. A question, for example, is how the SO$_2$ emission sources in the vicinities of Gary, Cincinnati, and Pittsburgh (indicated by 1, 2, and 3, respectively) and their variations affect the ground-level SO_3 concentrations at the two receptors indicated by a and b. The answer can be obtained through a proper sensitivity analysis.

A final example that we mention is the wide application of sensitivity analysis in metabolic systems, for an understanding of the sensitivities of metabolite concentrations and production fluxes with respect to variations in enzyme concentrations. Additionally, an important goal for molecular biologists is to reduce a complex biochemical system to its elemental units, to explain it at the molecular level, and then to use this knowledge to reconstruct it. However, cellular components exhibit large interactions, about which relatively little is known at present. Since sensitivity analysis, as discussed above, gives the relationships between the system behavior and parameters, we can expect it to have potential applications in understanding how a living cell will respond to variations in novel environments.

1.3   Overview of the Book Contents

As noted above, the literature on parametric sensitivity and sensitivity analysis is abundant, involving a variety of fields in science and engineering. It is beyond our expertise to write a book that covers all the fields. Thus we focus on chemical systems where reactions occur. We provide a theoretical background for parametric sensitivity, examine uses of sensitivity concept, and illustrate the theories using examples.

Chapter 2 gives the definitions of various sensitivity indices and illustrates the numerical techniques that are most commonly used for their evaluation.

We then discuss, in Chapters 3–7, the use of sensitivity analysis to identify the sensitive regions for various types of reactors, such as batch, tubular, continuous-flow stirred tank, and fixed-bed, where either a single or complex reactions occur. Chapter 3 involves the simple case of an irreversible exothermic reaction occurring in a batch reactor. This represents the classical Semenov (1928) problem, *i.e.*, thermal explosion in a closed vessel. In this case, there is a vast literature in developing criteria for identifying the critical conditions for thermal explosion. Detailed descriptions and comparisons of these criteria are given through various examples. Chapters 4, 5, and 6 analyze the parametric sensitivity of homogeneous tubular, continuous-flow stirred tank, and fixed-bed reactors, respectively. The influence of a second reaction and interactions between chemical and various transport phenomena on the sensitive regions are examined. An effort is made to supply for practitioners useful plots where the sensitive regions are indicated for each type of reactor. Moreover, in many cases, explicit expressions are given, which allow for estimating the sensitive regions analytically. In Chapter 7, we return to analyze explosions in mixtures of hydrogen or hydrocarbons with oxygen, in closed vessels. Although these mixtures may appear simple, they involve a large number of elementary reactions. It will be seen that one can use the sensitivity concept to describe the various explosion phenomena observed experimentally.

Chapters 8, 9, and 10 comprise the second part of this book and illustrate applications of sensitivity analysis as an effective mathematical tool to various chemical systems. In Chapter 8, we discuss its applications to chemical reaction kinetics. This is one of the fields where sensitivity analysis has been most widely used. In this context, two important applications are discussed: (1) to understand the main reaction path or mechanism in a detailed kinetic model consisting of a large number of elementary reactions, and (2) to extract important (or to eliminate unimportant) elementary reactions from a complex kinetic model so as to obtain a reduced, minimum kinetic model that provides essentially equivalent predictions. The latter is particularly crucial when the reaction model needs to be coupled with a complex model for transport processes. The illustration examples are oxidation of wet carbon monoxide, Belousov-Zhabotinsky oscillations, hydrogen–oxygen explosions, and combustion of methane–ethane–air mixtures.

In Chapter 9, the applications of sensitivity analysis to air pollution control are discussed. Air pollution processes involve large simulation models (simultaneous reactions and three-dimensional pollutant transport), characterized by many physicochemical, kinetic, and meteorological parameters with large uncertainties. We discuss the use of sensitivity analysis to evaluate the influence of parameter variations on model predictions, through *functional* and *global* sensitivity analyses. Two examples are given: (1) studies of relations between pollutant emission sources and regional air quality, and (2) evaluation of pollutant variations with large changes in system parameters.

Chapter 10 is dedicated to metabolic systems. We discuss two specific approaches to sensitivity analysis and illustrate them by two examples: (1) anaerobic fermentation of the yeast *Saccharomyces cerevisiae* with glucose under dynamic conditions, and (2) gluconeogenesis from lactate under steady-state conditions. The basic idea is to provide a flavor of the potential of sensitivity analysis in this field.

References

Bilous, O., and Amundson, N. R. 1956. Chemical reactor stability and sensitivity, II. Effect of parameters on sensitivity of empty tubular reactors. *A.I.Ch.E. J.* **2**, 117.

Emig, G., Hofmann, H., Hoffmann, U., and Fiand, U. 1980. Experimental studies on runaway of catalytic fixed-bed reactors (vinyl-acetate-synthesis). *Chem. Eng. Sci.* **35**, 249.

Haldar, R., and Rao, D. P. 1992. Experimental studies on parametric sensitivity of a batch reactor. *Chem. Eng. Technol.* **15**, 34.

Hidy, G. M., and Mueller, P. K. 1976. The design of the sulfate regional experiment. *Report EC-125*, Vol. 1. Palo Alto, CA: Electric Power Research Institute.

Lewis, B., and von Elbe, G. 1961. *Combustion, Flames and Explosions of Gases.* New York: Academic.

Semenov, N. N. 1928. Zur theorie des verbrennungsprozesses. *Z. Phys.* **48**, 571.

2

Introduction to Sensitivity Analysis

MATHEMATICAL THEORIES OF SENSITIVITY ANALYSIS are well developed and described in monographs available in the literature (Tomovic and Vukobratovic, 1972; Frank, 1978). In this chapter we first introduce concepts that are relevant to sensitivity studies of various chemical systems treated in this book, and then illustrate techniques that are most commonly used for evaluating the corresponding sensitivity indices.

In developing the various aspects of sensitivity analysis, we will refer to a variety of chemical systems that exhibit different characteristics. In particular, we are interested in their description in mathematical terms, which is typically provided by a model that gives an explicit or implicit relationship between the system behavior and the input parameters. This behavior is described by the state or output variables, which we indicate in general as *dependent variables* changing in time and/or in space. The *input parameters* include the physicochemical parameters of the model (such as those related to reaction kinetics, thermodynamic equilibria, and transport properties) as well as initial conditions, operating conditions, and geometric parameters of the systems. The physicochemical parameters are measured experimentally or estimated theoretically and therefore are always subject to uncertainties. On the other hand, the initial and operating conditions may change in time for a variety of reasons. Both of these obviously affect the system behavior. In particular, by *parametric sensitivity* we mean the effect of variations of the input parameters on the system behavior. The *sensitivity analysis* provides effective tools to study the parametric sensitivity of chemical systems.

2.1   Sensitivity Indices

2.1.1   Local Sensitivity

Let us consider a chemical system described by a single variable y, which changes in

9

time according to the following general differential equation,

$$\frac{dy}{dt} = f(y, \phi, t) \tag{2.1}$$

with initial condition (IC),

$$y(0) = y^i \tag{2.2}$$

where y is the dependent variable, t is the time, and ϕ represents the vector containing the m system input parameters. The function f is assumed to be continuous and continuously differentiable in all its arguments. Note that the above conditions on f are satisfied for virtually all chemical systems. They ensure that the above equation has a unique solution (see Chapter 2 in Varma and Morbidelli, 1997), called the *nominal solution*, which is continuous in t and ϕ, represented by

$$y = y(t, \phi) \tag{2.3}$$

Let us now change the jth parameter in the parameter vector ϕ, from ϕ_j to $\phi_j + \Delta\phi_j$. Then, the corresponding solution for y, called the *current solution*, becomes

$$y = y(t, \phi_j + \Delta\phi_j) \tag{2.4}$$

where for brevity, only ϕ_j, the parameter changed, is mentioned explicitly. Since y is a continuous function of ϕ_j, the current solution (2.4) can be expanded into a Taylor series as follows:

$$y(t, \phi_j + \Delta\phi_j) = y(t, \phi_j) + \frac{\partial y(t, \phi_j)}{\partial \phi_j} \cdot \Delta\phi_j + \frac{\partial^2 y(t, \phi_j + \theta \cdot \Delta\phi_j)}{\partial \phi_j^2} \cdot \frac{\Delta\phi_j^2}{2} \tag{2.5}$$

where $0 < \theta < 1$. If $\Delta\phi_j$ is sufficiently small, *i.e.*, $\Delta\phi_j \ll \phi_j$, the Taylor series can be truncated after the linear term, leading to

$$\Delta y = y(t, \phi_j + \Delta\phi_j) - y(t, \phi_j) \approx \frac{\partial y(t, \phi_j)}{\partial \phi_j} \cdot \Delta\phi_j \tag{2.6}$$

where Δy represents the variation of y due to the change of the input parameter ϕ_j, given by $\Delta\phi_j$. If we consider an infinitesimal variation ($\Delta\phi_j \to 0$), it follows from Eq. (2.6) that

$$s(y; \phi_j) = \frac{\partial y(t, \phi_j)}{\partial \phi_j} = \lim_{\Delta\phi_j \to 0} \frac{y(t, \phi_j + \Delta\phi_j) - y(t, \phi_j)}{\Delta\phi_j} \tag{2.7}$$

This is defined as the first-order local sensitivity, or simply *local sensitivity* of the dependent variable, y, with respect to the input parameter, ϕ_j. Although higher-order local sensitivities can be defined in a similar fashion, we will limit the treatment to first-order local sensitivities, since most applications are based on linear sensitivity analysis. The local sensitivity, $s(y; \phi_j)$, is also called the *absolute sensitivity*.

Another quantity related to local sensitivity, commonly used in sensitivity analysis, is the *normalized sensitivity*. The normalized sensitivity of y with respect to ϕ_j, $S(y; \phi_j)$ is defined as

$$S(y; \phi_j) = \frac{\phi_j}{y} \cdot \frac{\partial y}{\partial \phi_j} = \frac{\partial \ln y}{\partial \ln \phi_j} = \frac{\phi_j}{y} \cdot s(y; \phi_j) \tag{2.8}$$

which serves to normalize the magnitudes of the input parameter ϕ_j and the variable y. In the literature, the normalized sensitivity is sometimes also referred to as the *relative sensitivity*. Once the local sensitivity $s(y; \phi_j)$ is known, the calculation of $S(y; \phi_j)$ is straightforward.

When the sensitivity of y with respect to each one of the parameters in the m vector ϕ is considered, we obtain m sensitivity indices. These are defined as the *row sensitivity vector*,

$$s^T(y; \phi) = \frac{\partial y}{\partial \phi} = \left[\frac{\partial y}{\partial \phi_1} \frac{\partial y}{\partial \phi_2} \cdots \frac{\partial y}{\partial \phi_m} \right] = [s(y; \phi_1)s(y; \phi_2) \cdots s(y; \phi_m)] \tag{2.9}$$

Example 2.1 Conversion sensitivity in a batch reactor. As an example to illustrate the above definitions, we consider a batch reactor, where the following elementary bimolecular reaction

$$A + B \xrightarrow{k} C + D$$

with rate given by $r = kC_A C_B$ occurs. The initial concentrations of reactants A and B are indicated by C_A^i and C_B^i, respectively. Let us compute the local sensitivities of the conversion of reactant A, defined by $x_A = (C_A^i - C_A)/C_A^i$, with respect to various input parameters.

From the mass balance of reactant A, the following differential equation to describe the conversion of A can be derived immediately:

$$\frac{dx_A}{dt} = f(x_A, t, \phi) = k \cdot C_A^i \cdot (1 - x_A) \cdot (R^i - x_A) \tag{E2.1}$$

with IC

$$x_A(0) = 0 \tag{E2.2}$$

where R^i represents the inlet reactant concentration ratio, $R^i = C_B^i/C_A^i$. These equations can be solved analytically, leading to

$$\ln \frac{R^i - x_A}{R^i \cdot (1 - x_A)} = k \cdot C_A^i \cdot (R^i - 1) \cdot t \quad \text{for } R^i \neq 1 \tag{E2.3a}$$

$$x_A = k \cdot C_A^i \cdot t / \left(1 + k \cdot C_A^i \cdot t\right) \quad \text{for } R^i = 1 \tag{E2.3b}$$

Since the input parameters in this model are the reaction rate constant, k, the inlet concentration of A, C_A^i, and the inlet reactant concentration ratio, R^i, the input parameter vector may be defined as $\phi = (k \ \ C_A^i \ \ R^i)^T$. Accordingly, the row sensitivity vector of reactant A conversion is given by

$$s^T(x_A; \phi) = \left[s(x_A; k) \quad s\left(x_A; C_A^i\right) \quad s(x_A; R^i) \right] = \left[\frac{\partial x_A}{\partial k} \quad \frac{\partial x_A}{\partial C_A^i} \quad \frac{\partial x_A}{\partial R^i} \right] \tag{E2.4}$$

Differentiating both sides of Eq. (E2.3) with respect to each input parameter in ϕ, the following expressions for each local sensitivity are obtained, say for $R^i \neq 1$:

$$s(x_A; k) = C_A^i \cdot (1 - x_A) \cdot (R^i - x_A) \cdot t \tag{E2.5}$$

$$s\left(x_A; C_A^i\right) = k \cdot (1 - x_A) \cdot (R^i - x_A) \cdot t \tag{E2.6}$$

$$s(x_A; R^i) = \frac{\left[k \cdot C_A^i \cdot R^i \cdot (R^i - x_A) \cdot t - x_A\right] \cdot (1 - x_A)}{R^i \cdot (R^i - 1)} \tag{E2.7}$$

These, together with the model solution (E2.3a), allow us to compute the sensitivity of the conversion x_A with respect to each input parameter, at any time or conversion value.

In the case of a system described by n dependent variables, the dynamics are in general given by a set of differential equations,

$$\frac{dy}{dt} = f(y, \phi, t), \qquad y(0) = y^i \tag{2.10}$$

For a chemical system, the n vector of the dependent variables, y, may include the concentrations of the involved chemical species, temperature, pressure, and other state variables, while ϕ is the m vector containing the system input parameters. Now, for a chosen input parameter, ϕ_j, the local sensitivity of each variable, y_i, can be computed, based on Eq. (2.7). Thus, we have n sensitivity indices with respect to the same input parameter, which constitute the *column sensitivity vector*,

$$s(y; \phi_j) = \frac{\partial y}{\partial \phi_j} = \left[\frac{\partial y_1}{\partial \phi_j} \quad \frac{\partial y_2}{\partial \phi_j} \quad \cdots \quad \frac{\partial y_n}{\partial \phi_j} \right]^T$$

$$= [s(y_1; \phi_j) \quad s(y_2; \phi_j) \quad \cdots \quad s(y_n; \phi_j)]^T \tag{2.11}$$

Now combining all the row and column sensitivity vectors, we obtain an $n \times m$ matrix of the sensitivity indices, which is usually referred to as the *sensitivity matrix*,

$$S(y; \phi) = \frac{\partial y}{\partial \phi}$$

$$= \begin{bmatrix} \frac{\partial y_1}{\partial \phi_1} & \frac{\partial y_1}{\partial \phi_2} & \cdots & \frac{\partial y_1}{\partial \phi_m} \\ \frac{\partial y_2}{\partial \phi_1} & \frac{\partial y_2}{\partial \phi_2} & \cdots & \frac{\partial y_2}{\partial \phi_m} \\ \vdots & \vdots & \ddots & \vdots \\ \frac{\partial y_n}{\partial \phi_1} & \frac{\partial y_n}{\partial \phi_2} & \cdots & \frac{\partial y_n}{\partial \phi_m} \end{bmatrix} = \begin{bmatrix} s(y_1; \phi_1) & s(y_1; \phi_2) & \cdots & s(y_1; \phi_m) \\ s(y_2; \phi_1) & s(y_2; \phi_2) & \cdots & s(y_2; \phi_m) \\ \vdots & \vdots & \ddots & \vdots \\ s(y_n; \phi_1) & s(y_n; \phi_2) & \cdots & s(y_n; \phi_m) \end{bmatrix}$$

$$(2.12)$$

2.1.2 Objective Sensitivity

In many practical applications, when performing the sensitivity analysis, we are interested in a specific characteristic of the system, referred to as the *objective* or *objective function*. This can be represented either by one of the system-independent variables or by a performance index that can be evaluated from the system-independent variables, such as

- conversion of a reactant at a specific time or position;
- magnitude of the temperature maximum in time or in space;
- time needed by a reactant to reach a certain conversion value;
- concentration maximum of an intermediate product in a complex reaction network;
- decay rate of the catalyst activity in time;
- selectivity of a desired product at the reactor outlet;
- ignition limit for an explosive system.

The first four performance indices above are obtained from a direct solution of the relevant model equations, while the latter three are computed, through a proper definition, from the model solution. Assuming that the objective function, I, is a continuous function of a chosen jth parameter, ϕ_j in the parameter vector, ϕ, the corresponding sensitivity with respect to ϕ_j, $s(I; \phi_j)$, is defined as

$$s(I; \phi_j) = \frac{\partial I}{\partial \phi_j} = \lim_{\Delta \phi_j \to 0} \frac{I(\phi_j + \Delta \phi_j) - I(\phi_j)}{\Delta \phi_j} \tag{2.13}$$

which will be referred to as the *objective sensitivity*, although it is also called *performance-index sensitivity* (Frank, 1978) or *feature sensitivity* (Yetter *et al.*, 1985).

Similar to Eq. (2.8), the *normalized objective sensitivity*, $S(I; \phi_j)$, is defined as

$$S(I; \phi_j) = \frac{\phi_j}{I} \cdot \frac{\partial I}{\partial \phi_j} = \frac{\partial \ln I}{\partial \ln \phi_j} = \frac{\phi_j}{I} \cdot s(I; \phi_j) \tag{2.14}$$

C_A, C_B and C_c

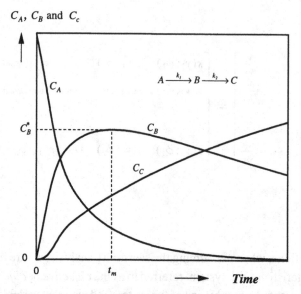

Figure 2.1. Typical time evolution of the species concentrations in a batch reactor where two consecutive reactions occur.

Example 2.2 Sensitivity of the maximum yield in an isothermal batch reactor with consecutive reactions. For two consecutive reactions

$$A \xrightarrow{k_1} B \xrightarrow{k_2} C$$

occurring in a batch reactor with zero initial concentrations of B and C, i.e., $C_B^i = C_C^i = 0$, the values of the species concentrations as functions of time are shown qualitatively in Fig. 2.1. The intermediate product B exhibits a maximum, C_B^*, at a certain time value, t_m, where the reactor yield toward B ($Y_B = C_B/C_A^i$) is maximized. Let us now perform the sensitivity analysis by taking the maximum yield, Y_B^*, as the objective, i.e., we compute the objective sensitivity of Y_B^* with respect to each input parameter.

Assuming that the consecutive reactions are first order, the dynamics of the process are described by the following two differential equations:

$$\frac{dC_A}{dt} = -k_1 \cdot C_A \tag{E2.8}$$

$$\frac{dC_B}{dt} = k_1 \cdot C_A - k_2 \cdot C_B \tag{E2.9}$$

with the ICs

$$C_A = C_A^i, \qquad C_B = 0, \quad \text{at } t = 0 \tag{E2.10}$$

The solution of Eq. (E2.8) can be readily obtained as

$$\frac{C_A}{C_A^i} = e^{-k_1 t} \tag{E2.11}$$

Dividing Eq. (E2.9) by Eq. (E2.8) gives

$$\frac{dC_B}{dC_A} = -1 + \frac{k_2 \cdot C_B}{k_1 \cdot C_A} \tag{E2.12}$$

which is a linear first-order differential equation with C_A as the independent variable. This can be solved analytically, giving the following expression for the yield of B:

$$Y_B = \frac{C_B}{C_A^i} = \frac{k_1}{k_1 - k_2} \left[\left(\frac{C_A}{C_A^i} \right)^{k_2/k_1} - \frac{C_A}{C_A^i} \right] \tag{E2.13}$$

Substituting Eq. (E2.11), we obtain

$$Y_B = \frac{k_1}{k_1 - k_2} \left(e^{-k_2 t} - e^{-k_1 t} \right) \tag{E2.14}$$

The time for Y_B to reach its maximum value, t_m, is obtained by differentiating Eq. (E2.14) with respect to t and setting $dY_B/dt = 0$, i.e.,

$$\frac{dY_B}{dt} = 0 = \frac{k_1}{k_1 - k_2} \left(-k_2 \cdot e^{-k_2 t_m} + k_1 \cdot e^{-k_1 t_m} \right) \tag{E2.15}$$

which leads to

$$t_m = \frac{\ln k_1 - \ln k_2}{k_1 - k_2} \tag{E2.16}$$

The corresponding maximum yield, Y_B^*, is obtained by substituting Eq. (E2.16) into Eq. (E2.14),

$$Y_B^* = \left(\frac{k_1}{k_2} \right)^{k_2/(k_2-k_1)} \tag{E2.17}$$

For the present system, besides the initial concentration of species B, which here is fixed to be zero, there are three input parameters: the initial reactant concentration, C_A^i, and the rate constants of the two reactions, k_1 and k_2. Accordingly, the input parameter vector is given by $\phi = (C_A^i \ k_1 \ k_2)^T$. The objective row sensitivity vector is then $s^T(Y_B^*; \phi) = [s(Y_B^*; C_A^i) \ s(Y_B^*; k_1) \ s(Y_B^*; k_2)]$, where each element has the following form obtained by differentiating Eq. (E2.17) with respect to the specific input parameter:

$$s\left(Y_B^*; C_A^i\right) = \frac{dY_B^*}{dC_A^i} = 0 \tag{E2.18a}$$

$$s\left(Y_B^*; k_1\right) = \frac{dY_B^*}{dk_1} = \frac{Y_B^* \cdot k_2}{k_1 - k_2} \cdot \left[\frac{\ln(k_1/k_2)}{k_1 - k_2} - \frac{1}{k_1} \right] \tag{E2.18b}$$

$$s\left(Y_B^*; k_2\right) = \frac{dY_B^*}{dk_2} = -\frac{Y_B^* \cdot k_1}{k_1 - k_2} \cdot \left[\frac{\ln(k_1/k_2)}{k_1 - k_2} - \frac{1}{k_1} \right] \tag{E2.18c}$$

Since the objective Y_B^*, given by Eq. (E2.17), is independent of C_A^i, $s(Y_B^*; C_A^i)$ is zero, as expected for first-order reactions. Moreover, it can be seen that the sensitivity to the rate constant of the first reaction, $s(Y_B^*; k_1)$, is always positive, if $k_1 > k_2$, which is the typical case in practice. This means that, as the rate of the first reaction increases, the maximum yield of B increases. Comparing Eq. (E2.18c) with Eq. (E.2.18b), it is found that the sensitivity to the rate constant of the second reaction, $s(Y_B^*; k_2)$, always has the opposite sign of that relative to the rate constant of the first reaction, $s(Y_B^*; k_1)$. Thus for $k_1 > k_2$, as the rate of the second reaction increases, the maximum yield of B decreases.

2.1.3   Global Sensitivity

As discussed above, local sensitivities, $s(y_i; \phi_j)$, provide information on the effect of a *small* change in *each* input parameter, ϕ_j, around a fixed nominal value, on each dependent variable, y_i. Global sensitivities instead describe the effect of simultaneous *large* variations of *all* parameters, ϕ, on the dependent variables. As shown in Eq. (2.7), when the imposed variation, $\Delta\phi_j$ of parameter ϕ_j, is small, the Taylor expansion can be truncated after the linear term, and then local sensitivities coincide with partial derivatives. Thus, for a given ϕ_j value, $s(y; \phi_j)$ can be considered eventually as a function of only the independent variable, t. On the other hand, global sensitivities involve the simultaneous variation of all the input parameters over a wide range of values, and the consequent changes of the dependent variables of the system, which determine the global sensitivities, are then functions of the width of such a range as well. As an example, let us consider Fig. 2.2, where the response of a system in terms of the dependent variable y_i is investigated as a function of two input parameters ϕ_1

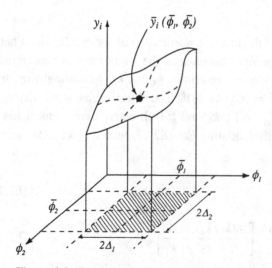

Figure 2.2. Response surface over the domain of change of two input parameters.

and ϕ_2. In the figure, they are shown to change by $\pm\Delta_1$ and $\pm\Delta_2$, thus causing y_i to vary over the illustrated surface. The local sensitivities of y_i with respect to ϕ_1 and ϕ_2, computed at the point $(\overline{\phi}_1, \overline{\phi}_2)$, correspond to the slopes of the surface at this point in the ϕ_1 and ϕ_2 directions, respectively. Thus, they reflect only the local behavior of the system around the considered point. On the other hand, the global sensitivities describe the behavior of the entire solution surface over the domain of change of the two parameters. Thus global sensitivities cannot be defined by a simple mathematical formula like the derivative for the local sensitivity and can be evaluated only through detailed numerical calculations, as will be shown later in this chapter.

2.2 Computation of Sensitivity Indices

2.2.1 Local Sensitivity

Direct differential method

This is perhaps the most natural method for computing sensitivities, and in fact it has already been applied in the examples described in the previous section. Let us now review this method in more general terms. For the single-variable system (2.1), in order to compute the local sensitivity of y with respect to the jth input parameter, ϕ_j, we first differentiate both sides of the system equation (2.1) with respect to ϕ_j. Then, considering the definition (2.7) for the local sensitivity leads to

$$\frac{d(\partial y/\partial \phi_j)}{dt} = \frac{ds(y; \phi_j)}{dt} = \frac{\partial f}{\partial y} \cdot \frac{\partial y}{\partial \phi_j} + \frac{\partial f}{\partial \phi_j} = \frac{\partial f}{\partial y} \cdot s(y; \phi_j) + \frac{\partial f}{\partial \phi_j} \qquad (2.15)$$

which represents the local *sensitivity equation*. Its initial condition can be obtained similarly by differentiating the initial condition (2.2). Depending on which input parameter in the vector ϕ is chosen, we have

$$s(y; \phi_j)|_{t=0} = \begin{cases} 0, & \phi_j \neq y^i \\ 1, & \phi_j = y^i \end{cases} \qquad (2.16a)$$

or in more concise form:

$$s(y; \phi_j)|_{t=0} = \delta(\phi_j - y^i) \qquad (2.16b)$$

where δ is the Kronecker delta function. By simultaneously solving the model equation (2.1) and the sensitivity equation (2.15), along with ICs (2.2) and (2.16), we obtain the dependent variable y and the corresponding local sensitivity $s(y; \phi_j)$ both as functions of time. This method is referred to as the *direct differential method* (DDM) for computing local sensitivities.

For the model involving n dependent variables given by Eq. (2.10), in order to obtain the sensitivity of the ith variable, y_i, to the jth input parameter, ϕ_j, we have to in principle compute the sensitivities of *all* the n variables to ϕ_j, since they may interact with each other. Thus, in this case, we have to solve n sensitivity equations together with the n model equations. The n sensitivity equations can be written in the form

$$\frac{ds(y;\phi_j)}{dt} = J(t) \cdot s(y;\phi_j) + \frac{\partial f(t)}{\partial \phi_j}$$

(2.17)

where $s(y;\phi_j)$ is the column sensitivity vector defined by Eq. (2.11), and

$$J(t) = \frac{\partial f}{\partial y} = \begin{bmatrix} \frac{\partial f_1}{\partial y_1} & \frac{\partial f_1}{\partial y_2} & \cdots & \frac{\partial f_1}{\partial y_n} \\ \frac{\partial f_2}{\partial y_1} & \frac{\partial f_2}{\partial y_2} & \cdots & \frac{\partial f_2}{\partial y_n} \\ \vdots & \vdots & \ddots & \vdots \\ \frac{\partial f_n}{\partial y_1} & \frac{\partial f_n}{\partial y_2} & \cdots & \frac{\partial f_n}{\partial y_n} \end{bmatrix}, \qquad \frac{\partial f(t)}{\partial \phi_j} = \begin{bmatrix} \frac{\partial f_1}{\partial \phi_j} \\ \frac{\partial f_2}{\partial \phi_j} \\ \vdots \\ \frac{\partial f_n}{\partial \phi_j} \end{bmatrix}$$

(2.18)

are usually referred to as the $n \times n$ *Jacobian matrix* and the $n \times 1$ *nonhomogeneous term*, respectively.

Example 2.3 Sensitivity analysis of an isothermal batch reactor with consecutive reactions of arbitrary order. Let us consider two consecutive reactions

$$A \xrightarrow{k_1} B \xrightarrow{k_2} C$$

occurring in an isothermal batch reactor with arbitrary orders n_1 and n_2, respectively. The model equations for this system are given by the mass balances of the chemical species A and B:

$$\frac{dC_A}{dt} = f_1 = -k_1 \cdot C_A^{n_1}$$

(E2.19a)

$$\frac{dC_B}{dt} = f_2 = k_1 \cdot C_A^{n_1} - k_2 \cdot C_B^{n_2}$$

(E2.19b)

or in vector form:

$$\frac{dy}{dt} = f(y, \phi, t)$$

(E2.20)

with the ICs

$$C_A = C_A^i, \qquad C_B = C_B^i, \qquad \text{at } t = 0$$

(E2.21)

where $y = [C_A \ C_B]^T$ and $\phi = [k_1 \ k_2 \ n_1 \ n_2 \ C_A^i \ C_B^i]^T$. Let us now derive the sensitivity equations with respect to the input parameters, k_1, n_2, and C_A^i.

Consider first the input parameter k_1. Differentiating both sides of Eq. (E2.19a) with respect to k_1 gives

$$\frac{ds(C_A;k_1)}{dt} = \frac{\partial f_1}{\partial C_A} \cdot s(C_A;k_1) + \frac{\partial f_1}{\partial C_B} \cdot s(C_B;k_1) + \frac{\partial f_1}{\partial k_1}$$

$$= -k_1 \cdot n_1 \cdot C_A^{n_1-1} \cdot s(C_A;k_1) - C_A^{n_1}$$

(E2.22a)

Similarly, differentiating both sides of Eqs. (E2.19b) and (E2.21), we obtain

$$\frac{ds(C_B; k_1)}{dt} = k_1 \cdot n_1 \cdot C_A^{n_1-1} \cdot s(C_A; k_1) - k_2 \cdot n_2 \cdot C_B^{n_2-1} \cdot s(C_B; k_1) + C_A^{n_1} \quad \text{(E2.22b)}$$

and the ICs

$$s(C_A; k_1) = s(C_B; k_1) = 0 \quad \text{at } t = 0 \tag{E2.23}$$

The above sensitivity equations in vector form become

$$\frac{d}{dt}\begin{bmatrix} s(C_A; k_1) \\ s(C_B; k_1) \end{bmatrix} = J(t) \cdot \begin{bmatrix} s(C_A; k_1) \\ s(C_B; k_1) \end{bmatrix} + \frac{\partial f}{\partial k_1} \tag{E2.24}$$

where the Jacobian matrix is

$$J(t) = \frac{\partial f}{\partial y} = \begin{bmatrix} \frac{\partial f_1}{\partial C_A} & \frac{\partial f_1}{\partial C_B} \\ \frac{\partial f_2}{\partial C_A} & \frac{\partial f_2}{\partial C_B} \end{bmatrix} = \begin{bmatrix} -k_1 \cdot n_1 \cdot C_A^{n_1-1} & 0 \\ k_1 \cdot n_1 \cdot C_A^{n_1-1} & -k_2 \cdot n_2 \cdot C_B^{n_2-1} \end{bmatrix} \tag{E2.25}$$

and

$$\frac{\partial f}{\partial k_1} = \begin{bmatrix} \frac{\partial f_1}{\partial k_1} \\ \frac{\partial f_2}{\partial k_1} \end{bmatrix} = \begin{bmatrix} -C_A^{n_1} \\ C_A^{n_1} \end{bmatrix} \tag{E2.26}$$

For the input parameters n_2 and C_A^i, the sensitivity equations in vector form can be written similarly as

$$\frac{d}{dt}\begin{bmatrix} s(C_A; n_2) \\ s(C_B; n_2) \end{bmatrix} = J(t) \cdot \begin{bmatrix} s(C_A; n_2) \\ s(C_B; n_2) \end{bmatrix} + \frac{\partial f}{\partial n_2} \tag{E2.27}$$

with

$$s(C_A; n_2) = s(C_B; n_2) = 0 \quad \text{at } t = 0 \tag{E2.28}$$

and

$$\frac{d}{dt}\begin{bmatrix} s(C_A; C_A^i) \\ s(C_B; C_A^i) \end{bmatrix} = J(t) \cdot \begin{bmatrix} s(C_A; C_A^i) \\ s(C_B; C_A^i) \end{bmatrix} + \frac{\partial f}{\partial C_A^i} \tag{E2.29}$$

with

$$s(C_A; C_A^i) = 1; \quad s(C_B; C_A^i) = 0 \quad \text{at } t = 0 \tag{E2.30}$$

The Jacobian matrix in both Eqs. (E2.27) and (E2.29) is given by Eq. (E2.25), while $\partial f / \partial n_2$ and $\partial f / \partial C_A^i$ have the following form:

$$\frac{\partial f}{\partial n_2} = \begin{bmatrix} \frac{\partial f_1}{\partial n_2} \\ \frac{\partial f_2}{\partial n_2} \end{bmatrix} = \begin{bmatrix} 0 \\ -k_2 \cdot n_2 \cdot C_B^{n_2-1} \end{bmatrix} \quad \text{and} \quad \frac{\partial f}{\partial C_A^i} = \begin{bmatrix} \frac{\partial f_1}{\partial C_A^i} \\ \frac{\partial f_2}{\partial C_A^i} \end{bmatrix} = \begin{bmatrix} 0 \\ 0 \end{bmatrix} \tag{E2.31}$$

It is worth noting from this example that for different input parameters, the sensitivity equations differ only in the last term of Eq. (2.17), *i.e.*, the nonhomogeneous term.

The DDM is used widely for computing local sensitivity values, since it provides complete information about each sensitivity index as a function of the independent variable. However, when the sensitivities with respect to all the m input parameters in ϕ are needed, it may become difficult to solve the n system equations (2.10) together with the $n \times m$ sensitivity equations. Even when using numerical methods, the sensitivity equations may exhibit stiffness leading to computational difficulties.

For models involving a large number of dependent variables, one may use the so-called *decoupled* approach, which divides the direct differentiation method in two steps: (1) solving first the model equation (2.10) to get the values of y, and then (2) solving the sensitivity equation (2.17). The y values obtained from the first step are interpolated numerically to estimate the Jacobian matrix $J(t)$ and the nonhomogeneous term $\partial f(t)/\partial \phi_j$ for the second step. This decoupled approach requires generally a lower numerical effort than solving the $(m+1) \times n$ differential equations simultaneously. However, the interpolation method must be properly designed in order to have an accurate and stable solution of the sensitivity equations (Gelinas and Shewes-Cox, 1977; Rabitz *et al.*, 1983).

Finite difference method

The simplest way to avoid solving simultaneously model and sensitivity equations for computing local sensitivities is to utilize the finite difference approximation. From the definition (2.7), the local sensitivity of y_i with respect to the input parameter, ϕ_j, is given by the first-order partial derivative of y_i with respect to ϕ_j. Then, for a small variation, $\Delta\phi_j$, the local sensitivity may be approximated by the corresponding difference ratio

$$s(y_i; \phi_j) = \frac{\partial y_i}{\partial \phi_j} \approx \frac{\Delta y_i}{\Delta \phi_j} = \frac{y_i(t, \phi_j + \Delta\phi_j) - y_i(t, \phi_j)}{\Delta\phi_j} \tag{2.19}$$

Accordingly, in order to determine the local sensitivities of n variables with respect to one input parameter, instead of solving simultaneously the $2 \times n$ model and sensitivity equations, we need to solve twice the n model equations for $\phi_j = \phi_j$ and $\phi_j = \phi_j + \Delta\phi_j$, to obtain $y(t, \phi_j)$ and $y(t, \phi_j + \Delta\phi_j)$, respectively. This is called the *finite difference method* (FDM).

In practice, however, since the sensitivities of different variables can exhibit rather different magnitudes, the same value of $\Delta\phi_j$ does not lead to the same accuracy in their evaluation. It is readily understood that the minimum variation $\Delta\phi_{j,\min}$ that allows for a desired accuracy in the evaluation of the sensitivity value for a given variable y_i must satisfy

$$\frac{\varepsilon_y \cdot y_i}{\Delta\phi_{j,\min}} = \varepsilon_s \cdot s(y_i; \phi_j) \tag{2.20}$$

where ε_y and ε_s are the allowable fractional errors in y_i and $s(y_i; \phi_j)$, respectively. This condition indicates that, for lower sensitivities, one can choose larger $\Delta\phi_j$. However, for high sensitivities, a large $\Delta\phi_j$ may lead to excessive errors in the sensitivity evaluation through the finite difference approximation. Thus, in general, a different variation $\Delta\phi_j$ has to be used for different variables. Moreover, since each variable and its sensitivity may change with time, even for the same variable we cannot utilize a fixed value of $\Delta\phi_j$ to estimate the corresponding sensitivity values as a function of time. Therefore, in practice it is often not true that through the FDM one needs to solve the n model equations only twice to determine the local sensitivities of all n variables with respect to one input parameter. This leads to the conclusion reached by Kramer *et al.* (1984) that in many instances the computational cost using the FDM can be quite high.

The FDM is particularly useful when we need to compute only the sensitivity of *one* among n variables with respect to *one* among m input parameters at a given point. In this case, there is no difficulty in finding a different variation $\Delta\phi_j$ for the different variables.

In addition, in practical applications, there are instances in which the DDM cannot be applied, because the sensitivity equations cannot be obtained by directly differentiating the model equations. This is the case where the sensitivity of a chosen objective is implicitly given by a complex functional form or it may not even be represented by a mathematical equation. In these cases, which we will analyze later (Chapters 7 and 8), the FDM is the only possible choice.

The Green's function method

As already observed in Example 2.3, for different input parameters, ϕ_j ($j = 1, 2, \ldots m$), the m linear sensitivity equations (2.17) differ only in the nonhomogeneous terms, $\partial f(t)/\partial\phi_j$. For this reason, several researchers (Director and Rohrer, 1969; de Jongh, 1978; Hwang *et al.*, 1978) proposed that, instead of using the DDM method that involves the solution of $(m + 1) \times n$ differential equations to obtain the sensitivities of n dependent variables with respect to m input variables, we solve first only the homogeneous part and then determine the particular solutions corresponding to each input parameter. The homogeneous part corresponds to the following Green's function problem:

$$\frac{dG(t, \tau)}{dt} = J(t)G(t, \tau), \qquad t > \tau \tag{2.21a}$$

and

$$G(\tau, \tau) = 1 \tag{2.21b}$$

where $J(t)$ is the Jacobian matrix defined by Eq. (2.18), and $G(t, \tau)$ is the *Green's function*. Then, the local sensitivities $s(y; \phi_j) = \partial y/\partial\phi_j$ can be written in terms of a linear integral transform of the nonhomogeneous terms, $\partial f(\tau)/\partial \phi_j$, with respect to $G(t, \tau)$ (Hwang et al., 1978):

$$s(y; \phi_j) = \frac{\partial y(t)}{\partial \phi_j} = G(t, 0) \cdot \delta + \int_0^t G(t, \tau) \frac{\partial f(\tau)}{\partial \phi_j} d\tau \tag{2.22}$$

where each element δ_k in vector δ is a Kronecker delta function, defined as

$$\delta_k = \delta\left(\phi_j - y_k^i\right) \tag{2.23}$$

This is called the *Green's function method* (GFM). In practical applications, one may notice that in the integral of Eq. (2.22), the first argument of $G(t, \tau)$, t, is fixed, whereas the second one, τ, varies. Then, if we need to compute the sensitivity values in a time interval $0 \le t \le t_T$, for each $t < t_T$ one repeatedly needs all the $G(t, \tau)$ values in $0 \le t \le t_T$. Thus, in order to evaluate the integral efficiently, Hwang et al. (1978) proposed to express $G(t, \tau)$ in terms of an adjoint Green's function, $\hat{G}(\tau, t)$, which is defined by

$$\frac{d}{d\tau} \hat{G}(\tau, t) = -\hat{G}(\tau, t)J(\tau), \qquad 0 \le \tau < t \tag{2.24a}$$

and

$$\hat{G}(t, t) = 1 \tag{2.24b}$$

In particular, one can show that $G(t, \tau) = \hat{G}(\tau, t)$, for use in Eq. (2.22). In this way, instead of solving Eq. (2.21), one can obtain $\hat{G}(\tau, t)$ by solving Eq. (2.24). In addition, it should be noted that $G(t, \tau)$ relies on the group property, e.g., $G(t_3, t_1) = G(t_3, t_2)G(t_2, t_1)$, where $t_1 < t_2 < t_3$.

Rabitz et al. (1983) have compared the relative advantages of the GFM and the DDM. In general, to obtain the sensitivities of n dependent variables with respect to m input parameters, the GFM requires solving $n \times n$ differential equations (2.21) plus n integrals (2.22), while the DDM involves $(m + 1) \times n$ differential equations. Since for most chemical systems, $m \gg n$, the GFM requires less numerical effort relative to the DDM. In addition, one may encounter fewer stiffness problems when solving Eqs. (2.24) together with the integral (2.22) than when solving directly the $m \times n$ sensitivity equations (2.17). Obviously, if the sensitivities of the n dependent variables with respect to only a few of the m input parameters are needed, the DDM may be preferred over the GFM.

Several applications of the GFM are discussed in Chapters 8 and 9.

Summary

As discussed above, each of the three methods (DDM, FDM, and GFM) for evaluating the local sensitivities has its specific characteristics, which define its field of application, as summarized in Table 2.1.

Table 2.1. Methods for computing local sensitivities

Method	Algorithm	Suggested Application Fields	Remarks
Direct differential method (DDM)	Solve the model and sensitivity equations simultaneously.	1. Number of dependent variables is smaller than the number of input parameters. 2. Sensitivities of dependent variables with respect to only a few of the input parameters are required.	One may encounter stiffness problems.
Finite difference method (FDM)	Solve model equations, and evaluate local sensitivities using the finite difference approximation.	1. Number of dependent variables is large, while only sensitivities of some dependent variables with respect to some of the input parameters are required. 2. Solving the model and sensitivity equations is a stiff problem. 3. The objective for sensitivity analysis is implicit or cannot be given by mathematical formulas.	One may need to find proper variations of each input parameter for sensitivity of each variable.
Green's function method (GFM)	Solve first the homogeneous part of the sensitivity equations (the Green's function problem), and then compute local sensitivities using linear integral transforms.	1. Number of dependent variables is much larger than the number of input parameters, and full sensitivity analysis of all dependent variables with respect to all input parameters is required.	One may encounter stiffness problems.

The objective sensitivity, as defined by Eq. (2.13), can in principle be calculated by any one of the methods described above. However, an objective is often a particular performance index of the system, which may be given implicitly by a complex functional form or sometimes not even represented by a mathematical formula. In these cases, since it may be difficult to obtain the sensitivity equations through direct differentiation of the model equations, the FDM becomes the most commonly used method.

In the literature, several other methods have also been reported for computing local sensitivities. Kramer *et al.* (1981) developed a modified GFM through the application of the piecewise Magnus technique (Chan *et al.*, 1968) to obtain the Green's function (2.21) and to evaluate the integral (2.22). This is referred to as the *analytically integrated Magnus (AIM) method* and is particularly useful for large systems. A computer code for the AIM algorithm has also been developed (Kramer *et al.*, 1982). A detailed comparison of required computational time between the different methods performed by Kramer *et al.* (1984) indicates that for a system of 11 dependent variables and 52 input parameters, the AIM method is the fastest, by factors of 2, 2.6, and 5.5 over the GFM, DDM, and FDM, respectively.

There are two other methods in the literature, *i.e.*, the *variational method* (Seigneur *et al.*, 1982) and the *functional analysis method* (Cacuci, 1981), that are based on the adjoint sensitivity analysis and thus can be considered as variants of the GFM. In particular, these methods are proposed for direct determination of the objective sensitivities. Note that in this case the objective has to be represented explicitly by a mathematical formula. The application of the functional analysis method to analyze the sensitivity of the temperature maximum in batch or tubular reactors has been described by Morbidelli and Varma (1988); see Chapters 3 and 4.

2.2.2 Global Sensitivity

Stochastic sensitivity analysis

According to its definition illustrated in Section 2.1.3, global sensitivity provides information about the behavior of each dependent variable y_i in response to simultaneous, large variations in all the input parameters. Typically, this information is important when we know that the input parameters are affected by uncertainties and we want to predict how much these may affect a given system dependent variable. The *stochastic sensitivity analysis*, originally developed by Costanza and Seinfeld (1981), is based on the assumption that all the input parameters are random variables whose probability density functions (PDFs) are known. In this case, the global sensitivity of the dependent variable y_i can be determined by evaluating its corresponding PDF. Note that the above assumption is required only for computing the sensitivities, since in reality the input parameters ϕ_j are not random variables but their values are uncertain.

Let us define an $n + m$ vector x composed of the n dependent variables, y, and the m input parameters, ϕ, and rewrite the system (2.10) as

$$\frac{dx}{dt} = g(x, t) \tag{2.25}$$

with initial conditions

$$x(0) = x^i$$

where

$$x = [y_1 \quad \cdots \quad y_n \quad \phi_1 \quad \cdots \quad \phi_m]^T; \quad g(x, t) = [f_1 \quad \cdots \quad f_n \quad 0 \quad \cdots \quad 0]^T;$$
$$x^i = [y_1^i \quad \cdots \quad y_n^i \quad \phi_1 \quad \cdots \quad \phi_m]^T \tag{2.26}$$

Thus, the system (2.25) contains the first n equations, which coincide with the original system equations (2.10), and the last m equations, which simply state that x_i, for $i = n+1, \ldots, n+m$, is constant in time. Accordingly, the original stochastic problem in the m input parameters becomes a stochastic problem in the $n + m$ initial values, x^i.

We consider now the initial conditions x^i in Eq. (2.25) to be random variables with known PDFs, $p^i(x)$. Then, the corresponding dependent variables $x(t)$ are also random, described by the PDF $p(x, t)$, which satisfy the following partial differential equation (Costanza and Seinfeld, 1981):

$$\frac{\partial p(x, t)}{\partial t} + \sum_{i=1}^{n} \frac{\partial}{\partial x_i}[p(x, t)f_i] = 0, \qquad p(x, 0) = p^i(x) \tag{2.27}$$

Once $p(x, t)$ is obtained, a measure of the global sensitivity of the dependent variable $x_i(t)$ can be computed as the expected value of $x_i(t)$, given by

$$E\{x_i(t)\} = \int x_i p(x, t) \, dx \tag{2.28}$$

An example illustrating the use of stochastic sensitivity analysis to analyze the sensitivity of photolysis kinetics for a complex reaction mixture has been presented by Costanza and Seinfeld (1981).

The FAST method

The *Fourier amplitude sensitivity test (FAST) method*, originally developed by Cukier, Shuler, and coworkers (Cukier *et al.*, 1973, 1975, 1978; Schaibly and Shuler, 1973) has been used to perform global sensitivity analysis for a variety of chemical systems (Boni and Penner, 1977; Koda *et al.*, 1979; Falls *et al.*, 1979; Pierce and Cukier, 1981; Smith *et al.*, 1993).

Similarly to the case of stochastic sensitivity analysis, let us assume that the m input parameters in the vector ϕ are random variables with known PDFs, $p(\phi)$. Then, a measure of the global sensitivity of each independent variable y_i with respect to the simultaneous variations of all the m parameters can be obtained by considering the ensemble mean value for y_i, given by

$$\langle y_i \rangle = \int \cdots \int y_i(\phi) p(\phi) \, d\phi_1 d\phi_2 \cdots d\phi_m \tag{2.29}$$

The key concept of the FAST method is to convert the m-dimensional integral above into an equivalent one-dimensional integral through the transformation

$$\phi_j(\xi) = G_j[\sin(\omega_j \xi)], \quad j = 1, 2, \ldots, m \tag{2.30}$$

where G_j ($j = 1, 2, \ldots, m$) is a set of known functions, corresponding to a set of parametric curves, called the search curves; ω_j is a set of incommensurate frequencies; and ξ is a scalar variable referred to as the search variable. If the G_j are appropriately chosen, it can be shown that (Weyl, 1938)

$$\langle y_i \rangle = \overline{y}_i = \lim_{T \to \infty} \frac{1}{2T} \int_{-T}^{T} y_i[\phi(\xi)] \, d\xi \tag{2.31}$$

This expression allows one to compute the relevant properties of the stochastic variable y_i, such as the mean value, $\overline{y}_i$, and variance.

The use of the incommensurate frequency set ω_j ensures that the search curve can pass arbitrarily close to every point in the parameter space. In practice, however, because of computational ease, instead of using truly incommensurate ω_j, one generally chooses integer frequencies. This involves two types of error: (1) the search curve does not pass arbitrarily close to every point in the parameter space; (2) the frequencies used to describe ϕ_j have harmonics that interfere with one another. However, if the integer frequencies are chosen properly, as described by Cukier et $al.$ (1975), the difference between $\langle y_i \rangle$ and $\overline{y}_i$ can be made arbitrarily small.

The advantage with integer frequencies is that the parameters ϕ_j become periodic functions of ξ, with period 2π. In this case, Eq. (2.31) becomes an integral in the finite interval $(-\pi, \pi)$:

$$\overline{y}_i = \frac{1}{2\pi} \int_{-\pi}^{\pi} y_i[\phi(\xi)] \, d\xi \tag{2.32}$$

which is then set equal to $\langle y_i \rangle$. For computing the total variance,

$$\sigma_i^2 = \langle y_i^2 \rangle - \langle y_i \rangle^2 \tag{2.33}$$

one can utilize the Fourier coefficients $A_{i,k}$ and $B_{i,k}$, defined by

$$A_{i,k}(t) = \frac{1}{2\pi} \int_{-\pi}^{\pi} y_i(t, \phi) \cos(k\xi) \, d\xi \tag{2.34}$$

$$B_{i,k}(t) = \frac{1}{2\pi} \int_{-\pi}^{\pi} y_i(t, \phi) \sin(k\xi) \, d\xi \tag{2.35}$$

Using Parseval's theorem, we obtain

$$\langle y_i^2(t) \rangle = \frac{1}{2\pi} \int_{-\pi}^{\pi} y_i^2(t,\xi) \, d\xi = \sum_{k=-\infty}^{\infty} \left[A_{i,k}^2(t) + B_{i,k}^2(t) \right] \tag{2.36}$$

and

$$\langle y_i(t) \rangle^2 = A_{i,0}^2 + B_{i,0}^2 = A_{i,0}^2 \tag{2.37}$$

from which the variance σ_i^2 becomes

$$\sigma_i^2(t) = 2 \sum_{k=1}^{\infty} \left[A_{i,k}^2(t) + B_{i,k}^2(t) \right] \tag{2.38}$$

If the Fourier coefficients are evaluated for the fundamental frequencies of the transformation (2.30) or its harmonics, i.e., replacing k in Eqs. (2.34) and (2.35) with $r\omega_j, r = 1, 2, \ldots$, the partial variance of y_i, $\sigma_{i,\omega_j}^2(t)$ arising from the uncertainty in the jth parameter can be computed as

$$\sigma_{i,\omega_j}^2(t) = 2 \sum_{r=1}^{\infty} \left[A_{i,r\omega_j}^2(t) + B_{i,r\omega_j}^2(t) \right] \tag{2.39}$$

Accordingly, the global sensitivity of y_i with respect to the variation of the jth parameter can be defined as

$$S_g(y_i; \phi_j) = \sigma_{i,w_j}^2(t)/\sigma_i^2(t) \tag{2.40}$$

It is worth noting that in the FAST method, the major sensitivity measures are the Fourier coefficients. For example, if both $A_{i,r\omega_j}$ and $B_{i,r\omega_j}$ are zero, the ith dependent variable y_i is insensitive to the variations of the jth input parameter ϕ_j at the rth harmonic ω_j. Moreover, if $A_{i,r\omega_j}$ and $B_{i,r\omega_j}$ are zero for all $r = 1, 2, \ldots$, it readily follows from Eq. (2.40) that $S_g(y_i; \phi_j)$ is zero, i.e., y_i is insensitive to the variations of ϕ_j.

It should be pointed out that sensitivity analysis through the FAST method requires significant computer time, due to the need for repeated solutions of the system equations at a large number of points in order to evaluate the oscillating integrals in Eqs. (2.34) and (2.35). In the following section, we describe a computational implementation of this technique.

Application of the FAST method (McRae et al., 1982)

In order to evaluate the global sensitivity defined by Eq. (2.40), we first need to compute the Fourier coefficients, $A_{i,r\omega_j}$ and $B_{i,r\omega_j}$. This implies that y_i has to be evaluated in $\xi \in [-\pi, \pi]$. If we restrict the frequency set to odd integers, then the range of ξ reduces to $[-\pi/2, \pi/2]$. In this case, we have

$$
\begin{aligned}
y_i(\pi - \xi) &= y_i(\xi) \\
y_i(-\pi + \xi) &= y_i(-\xi) \\
y_i\left(\frac{\pi}{2} + \xi\right) &= y_i\left(\frac{\pi}{2} - \xi\right) \\
y_i\left(-\frac{\pi}{2} + \xi\right) &= y_i\left(-\frac{\pi}{2} - \xi\right)
\end{aligned}
\tag{2.41}
$$

The Fourier coefficients can be expressed as

$$
A_{i,j} =
\begin{cases}
0; & \text{odd } j \\
\dfrac{1}{\pi} \displaystyle\int_0^{\pi/2} [y_i(\xi) + y_i(-\xi)] \cos(j\xi)\, d\xi; & \text{even } j
\end{cases}
\tag{2.42}
$$

and

$$
B_{i,j} =
\begin{cases}
0; & \text{odd } j \\
\dfrac{1}{\pi} \displaystyle\int_0^{\pi/2} [y_i(\xi) - y_i(-\xi)] \sin(j\xi)\, d\xi; & \text{even } j
\end{cases}
\tag{2.43}
$$

and the actual number of points R where the system must be evaluated may be derived based on the Nyquist criterion (Beauchamp and Yuen, 1979) and is given by

$$
R \geq N\omega_{\max} + 1
\tag{2.44}
$$

where N is an even integer. Moreover, for convenience in calculating the Fourier coefficients, the additional condition

$$
2R = 4q + 2
\tag{2.45}
$$

is imposed, where q is an integer. Considering equally spaced values of ξ throughout the range $[-\pi/2, \pi/2]$, the discrete points at which y_i is computed in the Fourier space are given by

$$
\xi_p = \frac{\pi}{2} \cdot \left[\frac{2p - R - 1}{R}\right], \qquad p = 1, 2, \ldots, R
\tag{2.46}
$$

Simple quadrature formulas (Cukier *et al.*, 1978) can be used to evaluate the Fourier coefficients, leading to the following expressions:

$$
A_{i,j} =
\begin{cases}
0; & \text{odd } j \\
\dfrac{1}{2q+1}\left[y_i^{(0)} + \displaystyle\sum_{k=1}^{q} \left(y_i^{(k)} + y_i^{(-k)}\right) \cos\left(\dfrac{\pi j k}{2q+1}\right)\right]; & \text{even } j
\end{cases}
\tag{2.47}
$$

and

$$B_{i,j} = \begin{cases} 0; & \text{odd } j \\ \dfrac{1}{2q+1}\left[\displaystyle\sum_{k=1}^{q}\left(y_i^{(k)} - y_i^{(-k)}\right)\sin\left(\dfrac{\pi jk}{2q+1}\right)\right]; & \text{even } j \end{cases} \qquad (2.48)$$

where the superscript k indicates the discrete point number.

It should be noted that interference between the frequencies may occur as a result of this numerical evaluation. Let us select two arbitrary parameters, ϕ_1 and ϕ_2, and their associated frequencies, ω_1 and ω_2. The interference occurs when

$$r\omega_1 = s\omega_2[\text{Mod}(N\omega_{\max} + 1)] \qquad (2.49)$$

i.e., in this case we have

$$\left|A_{i,s\omega_2}\right|^2 + \left|B_{i,s\omega_2}\right|^2 = \left|A_{i,r\omega_1}\right|^2 + \left|B_{i,r\omega_1}\right|^2 \qquad (2.50)$$

since

$$\sin\left(\frac{\pi r\omega_1}{N\omega_{\max} + 1}\right) = \pm\sin\left(\frac{\pi s\omega_1}{N\omega_{\max} + 1}\right) \qquad (2.51)$$

and

$$\cos\left(\frac{\pi r\omega_1}{N\omega_{\max} + 1}\right) = \pm\cos\left(\frac{\pi s\omega_2}{N\omega_{\max} + 1}\right) \qquad (2.52)$$

This interference, called aliasing, is eliminated when

$$R\omega_j \leq N\omega_{\max} + 1 \qquad (2.53)$$

Thus, N is the maximum number of Fourier coefficients that may be retained in calculating the partial variances without interference between the assigned frequencies. It follows that the expression of the global sensitivity, Eq. (2.49), becomes

$$S_g(y_i;\ \phi_j) = \frac{2}{\sigma_i^2}\sum_{r=1}^{N}\left(\left|A_{i,r\omega_j}\right|^2 + \left|B_{i,r\omega_j}\right|^2\right) \qquad (2.54)$$

The interference problem may also occur due to the use of an integer frequency set if the number of Fourier coefficients N in the summation (2.54) is greater than or equal to the smallest frequency. To illustrate this point, let us consider again two arbitrary parameters, ϕ_1 and ϕ_2, and their associated frequencies, ω_1 and ω_2. The two corresponding global sensitivities given by Eq. (2.54) are written as

$$S_g(y_i;\ \phi_1) \propto A_{i,\omega_1}^2 + B_{i,\omega_1}^2 + A_{i,2\omega_1}^2 + B_{i,2\omega_1}^2 + \cdots + A_{i,N\omega_1}^2 + B_{i,N\omega_1}^2$$

$$S_g(y_i;\ \phi_2) \propto A_{i,\omega_2}^2 + B_{i,\omega_2}^2 + A_{i,2\omega_2}^2 + B_{i,2\omega_2}^2 + \cdots + A_{i,N\omega_2}^2 + B_{i,N\omega_2}^2$$

If $N \geq \omega_1$, terms in the series for $S_g(y_i; \phi_1)$ and $S_g(y_i; \phi_2)$ become identical. For example, if $N = \omega_1$ and $\omega_2 > \omega_1$, there will be a term in $S_g(y_i; \phi_2)$, say the kth one, which satisfies

$$A_{i,\omega_1\omega_1} = A_{i,k\omega_2}$$

In this case, the effect of the variation of parameter ϕ_1 enters spuriously into the partial variance for the variation of parameter ϕ_2.

In general, the interference between the higher harmonics is eliminated when

$$N < \omega_{\min} - 1 \tag{2.55}$$

Since N is also related to the number of function evaluations required by Eq. (2.44), it should be set equal to the minimum possible value, i.e., $N = 2$. In this case, a minimum frequency of at least three is sufficient to remove all harmonic interference effects from the partial variances. Accordingly, the final expression for the global sensitivity becomes

$$S_g(y_i; \phi_j) = \frac{2}{\sigma_i^2} \left(\left| B_{i,1\omega_j} \right|^2 + \left| A_{i,2\omega_j} \right|^2 \right) \tag{2.56}$$

The choice of $N = 2$ restricts the number of terms in the series to two. However, this is generally sufficient because the magnitude of the higher-order terms in the Fourier series tends to decrease rapidly.

Implementation of the FAST method also requires the proper selection of the frequency set. A recursive set, as described by Cukier *et al.* (1978), may be used:

$$\omega_1 = \Omega_i \tag{2.57}$$

$$\omega_j = \omega_{j-1} + d_{i+1-j} \qquad j = 2, \ldots, m \tag{2.58}$$

where Ω_i and d_i are given in Table 2.2.

Finally, the transformation functions, G_j ($j = 1, 2, 3 \ldots m$) in Eq. (2.30), have to be given, which determine the actual search curve traversed in the ξ space. If the probabilities of occurrence for the parameters, ϕ_j ($j = 1, 2, 3 \ldots m$), are independent, the probability density describing their effects has the form

$$p(\phi) = p_1(\phi_1) p_2(\phi_2) \cdots p_m(\phi_m) \tag{2.59}$$

It then can be shown (Weyl, 1938) that the transformation functions must satisfy the following relation:

$$\pi \cdot (1 - \xi^2) \cdot p_j(G_j) \cdot \frac{dG_j}{d\xi} = 1 \tag{2.60}$$

with IC

$$G_j(0) = 0 \tag{2.61}$$

Table 2.2. Parameters used for calculating frequency sets without interference to fourth order [*i.e.*, for $N \leq 4$ in Eq. (2.54)]

i	Ω_i	d_i	i	Ω_i	d_i
1	0	4	26	385	416
2	0	8	27	157	106
3	1	6	28	215	208
4	5	10	29	449	328
5	11	20	30	163	198
6	1	22	31	337	382
7	17	32	32	253	88
8	23	40	33	375	348
9	19	38	34	441	186
10	25	26	35	673	140
11	41	56	36	773	170
12	31	62	37	875	284
13	23	46	38	873	568
14	87	76	39	587	302
15	67	96	40	849	438
16	73	60	41	623	410
17	85	86	42	637	248
18	143	126	43	891	448
19	149	134	44	943	388
20	99	112	45	1171	596
21	119	92	46	1225	216
22	237	128	47	1335	100
23	267	154	48	1725	488
24	283	196	49	1663	166
25	151	34	50	2019	0

Source: From McRae *et al.* (1982).

Table 2.3. Search curves for computational implementation of the FAST method

Application	$\phi_j(\xi)$	Mean value[a], $\bar{\phi}_j$	Nominal value, $\bar{v}_j$
Additive variation	$\phi_j(\xi) = \bar{\phi}_j \cdot [1 + \bar{v}_j \sin(\omega_j \xi)]$	$\bar{\phi}_j = \frac{\phi_{j,u} + \phi_{j,l}}{2}$	$\bar{v}_j = \frac{\phi_j - \phi_{j,l}}{\phi_j + \phi_{j,l}}$
Exponential variation	$\phi_j(\xi) = \bar{\phi}_j \cdot \exp[\bar{v}_j \sin(\omega_j \xi)]$	$\bar{\phi}_j = \sqrt{\phi_{j,u} \cdot \phi_{j,l}}$	$\bar{v}_j = \frac{1}{2} \cdot \ln\left(\frac{\phi_{j,u}}{\phi_{j,l}}\right)$
Proportional variation	$\phi_j(\xi) = \bar{\phi}_j \cdot \exp[\bar{v}_j \sin(\omega_j \xi)]$	$\bar{\phi}_j = \sqrt{\phi_{j,u} \cdot \phi_{j,l}}$	$\bar{v}_j = \ln\left(\frac{\phi_{j,u}}{\phi_j}\right)$
Skewed variation $\phi_j > (\phi_{j,u} + \phi_{j,l})/2$	$\phi_j(\xi) = v_j \left[\frac{\alpha_j + \beta_j \sin(\omega_j \xi) - 1}{\alpha_j + \beta_j \sin(\omega_j \xi)}\right]$	$\bar{\phi}_j = v_j \cdot \left(\frac{\alpha_j - 1}{\alpha_j}\right)$	$\alpha_j = \frac{1}{2}\left(\frac{\gamma_u}{\gamma_u - 1} + \frac{\gamma_l}{\gamma_l - 1}\right)^b$ $\beta_j = -\frac{\alpha_j(\gamma_u + \gamma_l - 2)}{\gamma_u - \gamma_l}$ $v_j = -\bar{\phi}_j \cdot \left(\frac{\gamma_u + \gamma_l - 2\gamma_u \gamma_l}{\gamma_u + \gamma_l - 2}\right)$

Source: From McRae *et al.* (1982).
[a] $\phi_{j,u}$ and $\phi_{j,l}$ are the upper and lower limits for the parameter, respectively.
[b] $\gamma_u = \phi_{j,u}/\bar{\phi}_j$, and $\gamma_l = \phi_{j,l}/\bar{\phi}_j$.

Table 2.3 gives four different search curve formulations and their transformation functions. The parameter probability distributions used to derive these curves are described by Cukier *et al.* (1975). The first search curve is suitable for cases with small variations in the uncertain parameters, while the others can be applied to cases with large variations.

A detailed application of the FAST method to global sensitivity analysis of photochemical air pollution is given in Chapter 9 (Section 9.3).

Nomenclature

A_i	Pre-exponential factor in Arrhenius equation of the ith reaction
$A_{i,k}$	kth Fourier coefficient with respect to the ith dependent variable, defined by Eq. (2.34)
$B_{i,k}$	kth Fourier coefficient with respect to the ith dependent variable, defined by Eq. (2.35)
$E\{x_i(t)\}$	Expected value of $x_i(t)$, defined by Eq. (2.28)
$f(y, \phi, t)$	Function vector, defined by Eq. (2.10)
$f_i(y, \phi, t)$	ith function in the function vector, $f(y, \phi, t)$
$g(x, t)$	Function vector, defined by Eq. (2.26)
G_j	Function describing the jth input parameter in the FAST method, defined by Eq. (2.30)
$G(t, \tau)$	Green's function matrix, defined by Eq. (2.21)
$\hat{G}(t, \tau)$	Adjoint Green's function matrix, defined by Eq. (2.24)
I	Objective function in the definition of the objective sensitivity $s(I; \phi_j)$
$J(t)$	$n \times n$ Jacobian matrix, defined by Eq. (2.12)
m	Number of input model parameters
n	Number of system dependent variables
$p(x, t)$	Probability density function vector for $x(t)$
r	Reaction rate
$s(I; \phi_j)$	Sensitivity of the objective I with respect to the jth input parameter ϕ_j, defined by Eq. (2.13)
$s(y; \phi_j)$	Local sensitivity of the dependent variable y with respect to the jth input parameter ϕ_j, defined by Eq. (2.7)
$s(y; \phi_j)$	Column sensitivity vector, defined by Eq. (2.11)
$s^T(y; \phi)$	Row sensitivity vector, defined by Eq. (2.9)
$S(I; \phi_j)$	$s(I; \phi_j) \cdot \phi_j / I$, normalized objective sensitivity
$S(y; \phi_j)$	$s(y; \phi_j) \cdot \phi_j / y$, normalized sensitivity of the dependent variable y with respect to the jth input parameter ϕ_j
$S(y; \phi)$	$n \times m$ sensitivity matrix, defined by Eq. (2.12)
$S_g(y_i; \phi_j)$	Global sensitivity of y_i with respect to ϕ_j from the FAST method, defined by Eq. (2.40)

t	Time
x	$n + m$ vector, defined by Eq. (2.26)
x_i	ith component in vector x
y	Dependent variable n vector
y_i	ith component in vector y

Greek Symbols

δ	Kronecker function
ϕ	Input model parameter m vector
ϕ_j	jth input parameter in vector ϕ
σ_i^2	Total variance of the ith variable, defined by Eq. (2.33) or by Eq. (2.38)
τ	Time
ω	Frequency

Superscripts

i	Initial condition

Acronyms

DDM	direct differential method
FAST	Fourier amplitude sensitivity test
FDM	finite difference method
GFM	Green's function method
PDF	probability density function

References

Beauchamp, K. G., and Yuen, C. K. 1979. *Digital Methods for Signal Analysis.* London: George Allen and Unwin.

Boni, A. A., and Penner, R. C. 1977. Sensitivity analysis of a mechanism for methane oxidation kinetics. *Comb. Sci. Tech.* **15**, 99.

Cacuci, D. G. 1981. Sensitivity theory for nonlinear systems. 1. Nonlinear functional analysis approach. *J. Math. Phys.* **22**, 2794.

Chan, S., Light, J. C., and Lin, J. L. 1968. Inelastic molecular collisions: exponential solution of coupled equations for vibration-translation energy transfer. *J. Chem. Phys.* **49**, 86.

Costanza, V., and Seinfeld, J. H. 1981. Stochastic sensitivity analysis in chemical kinetics. *J. Chem. Phys.* **74**, 3852.

Cukier, R. I., Fortuin, C. M., Shuler, K. E., Petschek, A. G., and Schaibly, J. H. 1973. Study of the sensitivity of coupled reaction systems to uncertainties in rate coefficients. I. Theory. *J. Chem. Phys.* **59**, 3873.

Cukier, R. I., Schaibly, J. H., and Shuler, K. E. 1975. Study of the sensitivity of coupled reaction systems to uncertainties in rate coefficients. III. Analysis of the application. *J. Chem. Phys.* **63**, 1140.

Cukier, R. I., Levine, H. B., and Shuler, K. E. 1978. Nonlinear sensitivity analysis of multiparameter model systems. *J. Comput. Phys.* **26**, 1.

de Jongh, D. C. J. 1978. Structural parameter sensitivity of the "limits to growth" world model. *Appl. Math. Model.* **2**, 77.

Director, S. W., and Rohrer, R. A. 1969. The generalized adjoint network and network sensitivities. *IEEE Trans. Circuit Theory* **CT-16**, 318.

Falls, A. H., McRae, G. J., and Seinfeld, J. H. 1979. Sensitivity and uncertainty of reaction mechanisms for photochemical air pollution. *Int. J. Chem. Kinet.* **11**, 1137.

Frank, P. M. 1978. *Introduction to System Sensitivity Theory*. New York: Academic.

Gelinas, R. J., and Shewes-Cox, P. D. 1977. Tropospheric photochemical mechanisms. *J. Phys. Chem.* **81**, 2468.

Hwang, J.-T., Dougherty, E. P., Rabitz, S., and Rabitz, H. 1978. The Green's function method of sensitivity analysis in chemical kinetics. *J. Chem. Phys.* **69**, 5180.

Koda, M., McRae, G., and Seinfeld, J. H. 1979. Automatic sensitivity analysis of kinetic mechanisms. *Int. J. Chem. Kinet.* **11**, 427.

Kramer, M. A., Calo, J. M., and Rabitz, H. 1981. An improved computational method for sensitivity analysis: Green's function method with "AIM." *Appl. Math. Model.* **5**, 432.

Kramer, M. A., Calo, J. M., Rabitz, H., and Kee, R. J. 1982. Sandia Tech. Rep. **82**, 8231. Livermore, CA: Sandia National Laboratory.

Kramer, M. A., Rabitz, H., Calo, J. M., and Kee, R. J. 1984. Sensitivity analysis in chemical kinetics: recent developments and computational comparisons. *Int. J. Chem. Kinet.* **16**, 559.

McRae, G. J., Tilden, J. W., and Seinfeld, J. H. 1982. Global sensitivity analysis: a computational implementation of the Fourier amplitude sensitivity test (FAST). *Comp. Chem. Eng.* **6**, 15.

Morbidelli, M., and Varma, A. 1988. A generalized criterion for parametric sensitivity: application to thermal explosion theory. *Chem. Eng. Sci.* **43**, 91.

Pierce, T. H., and Cukier, R. I. 1981. Global nonlinear sensitivity analysis using Walsh functions. *J. Comp. Phys.* **41**, 427.

Rabitz, H., Kramer, M., and Dacol, D. 1983. Sensitivity analysis in chemical kinetics. *Ann. Rev. Phys. Chem.* **34**, 419.

Schaibly, J. H., and Shuler, K. E. 1973. Study of the sensitivity of coupled reaction systems to uncertainties in rate coefficients. II. Applications. *J. Chem. Phys.* **59**, 3879.

Seigneur, C., Stephanopoulos, G., and Carr, R. W., Jr. 1982. Dynamic sensitivity analysis of chemical reaction systems: a variational method. *Chem. Eng. Sci.* **37**, 845.

Smith, J. D., Smith, P. J., and Hill, S. C. 1993. Parametric sensitivity study of a CFD-based coal combustion model. *A.I.Ch.E. J.* **39**, 1668.

Tomovic, R., and Vukobratovic, M. 1972. *General Sensitivity Theory*. New York: Elsevier.

Varma, A., and Morbidelli, M. 1997. *Mathematical Methods in Chemical Engineering*. New York: Oxford University Press.

Weyl, H. 1938. Mean motion. *Am. J. Math.* **60**, 889.

Yetter, R. A., Dryer, F. L., and Rabitz, H. 1985. Some interpretive aspects of elementary sensitivity gradients in combustion kinetics modeling. *Combust. Flame* **59**, 107.

3

Thermal Explosion in Batch Reactors

A WELL-STIRRED BATCH REACTOR is characterized by the system variables (temperature and concentrations) varying with time and an (approximately) constant volume. *Runaway* or *explosion* may occur in a batch reactor, following two main mechanisms. In the first one, the rate of heat generation is faster than the rate of heat removal by the cooling system, thus leading to a continuous rise in the reactor temperature with a consequent acceleration of the reactions, leading eventually to explosion. The second mechanism involves chain branching processes, *i.e.*, reactions that produce two or more active species from a single one, which again under certain conditions may accelerate the chemical reactions leading to an explosion. The former is generally referred to as *thermal explosion* or *thermal runaway*, and the latter as *chain branching-induced explosion*. Obviously, chain branching-induced explosion can occur only in a complex reacting system, where several elementary reaction steps are involved, as is typically the case in combustion processes.

The explosion phenomena encountered in the decomposition of azomethane (Allen and Rice, 1935) and methyl nitrate vapor (Apin *et al.*, 1936; Gray *et al.*, 1981), and the catalytic hydrolysis of acetic anhydride (Haldar and Rao, 1992), which will be examined in detail in later sections, are mainly characterized by thermal processes. A good example for the pure chain branching-induced explosion is the reaction of stoichiometric oxygen–hydrogen mixtures at very low pressure (a few Torr). In most cases, however, it is difficult to distinguish whether the explosion involves only thermal or chain branching processes, since they are frequently involved as in most oxidation systems, *e.g.*, $H_2 + O_2$ (Kopp *et al.*, 1930; Lewis and von Elbe, 1961), $PH_3 + O_2$ (Dalton, 1930), $SiH_4 + O_2$ (Emeleus and Stewart, 1935, 1936), $CS_2 + O_2$ (Leicester, 1933), $CO + O_2$ (Kopp *et al.*, 1930; Hadman *et al.*, 1932; von Elbe *et al.*, 1955), or the combustion of various hydrocarbons. These are characterized by complex reaction mechanisms and involve *both* thermal and chain branching-induced explosion phenomena. In this chapter, we deal only with thermal explosion in systems with only one reaction, while chain branching-induced explosion is discussed in detail in Chapter 7.

The development of theories for describing thermal explosion phenomena was pioneered by Semenov (1928). Several approaches have been proposed in the literature, and they can generally be classified as *geometry based* and *sensitivity based*. The former are based on some geometric feature of the profile of a system variable such as temperature or heat-release rate, while the latter relate to the parametric sensitivity concept introduced first by Bilous and Amundson (1956) in the context of chemical reactor theory. Sections 3.2 and 3.3 discuss the most significant geometry-based and sensitivity-based theories that have been developed to describe thermal explosion phenomena. Examples are given, and the advantages and limitations of these theories in practical applications are also discussed. Finally, in Section 3.4, *explicit* criteria for thermal runaway are discussed since they are particularly useful for practical applications.

3.1 Basic Equations

Let us consider an exothermic nth-order reaction occurring in a closed vessel where concentrations and temperature are assumed to be uniform. The dynamics of this system are described by a one-dimensional model including the mass and energy balance equations

$$\frac{dC}{dt} = -k(T) \cdot C^n \tag{3.1}$$

$$\rho \cdot c_v \cdot \frac{dT}{dt} = (-\Delta H) \cdot k(T) \cdot C^n - S_v \cdot U \cdot (T - T_a) \tag{3.2}$$

with initial conditions (ICs)

$$C = C^i, \qquad T = T^i \quad \text{at} \quad t = 0 \tag{3.3}$$

where t is time, C is the reactant concentration, and T is the system temperature. U is the overall heat-transfer coefficient, and T_a is the ambient temperature.

The equations can be rewritten in dimensionless form,

$$\frac{dx}{d\tau} = \exp\left(\frac{\theta}{1 + \theta/\gamma}\right) \cdot (1 - x)^n = F_1(x, \theta) \tag{3.4}$$

$$\frac{d\theta}{d\tau} = B \cdot \exp\left(\frac{\theta}{1 + \theta/\gamma}\right) \cdot (1 - x)^n - \frac{B}{\psi} \cdot (\theta - \theta_a) = F_2(x, \theta) \tag{3.5}$$

and the ICs

$$x = 0, \qquad \theta = 0 \quad \text{at} \quad \tau = 0 \tag{3.6}$$

where we have introduced the following dimensionless variables:

$$x = \frac{C^i - C}{C^i}; \qquad \theta = \frac{T - T^i}{T^i} \cdot \gamma; \qquad \tau = t \cdot k(T^i) \cdot (C^i)^{n-1} \tag{3.7}$$

and dimensionless parameters:

$$B = \frac{(-\Delta H) \cdot C^i}{\rho \cdot c_v \cdot T^i} \cdot \gamma; \qquad \gamma = \frac{E}{R_g \cdot T^i}; \qquad \psi = \frac{(-\Delta H) \cdot k(T^i) \cdot (C^i)^n}{S_v \cdot U \cdot T^i} \cdot \gamma$$

(3.8)

The dependent dimensionless variables are the reactant conversion, x, and the temperature, θ, whose definition is typical in the context of thermal explosion theory. The time variable has been made dimensionless through the characteristic time of the chemical reaction given by the reciprocal of $k(T^i) \cdot (C^i)^{n-1}$. By using these definitions, it may be seen that the system behavior is fully characterized by five parameters: n, γ, B, ψ, and θ_a. The first two characterize the reaction kinetics: n is the reaction order and γ is the dimensionless activation energy, often called the Arrhenius number. They determine the effects of reactant concentration and temperature, respectively, on the reaction rate. The remaining three parameters include all the physicochemical quantities involved in this system. In particular, B is the dimensionless heat of reaction (often also referred to as the dimensionless adiabatic temperature rise) and ψ is the Semenov number, which represents the ratio between the heat-release potential of the reaction and the heat-removal potential through cooling. Thus, for given reaction kinetics and ambient temperature, the system behavior is determined by the values of B and ψ. Accordingly in later sections, critical conditions are illustrated in the $B-\psi$ parameter plane.

3.2 Geometry-Based Criteria for Thermal Runaway

3.2.1 The Case of Negligible Reactant Consumption: Semenov Theory

The pioneering theory of thermal explosions by Semenov (1928, 1959) was developed originally based on the assumption of negligible reactant consumption. This assumption is obviously violated in most real systems; however, its simplicity and explicitness allow one to have a fundamentally correct and synthetic view of the mechanism of thermal explosion phenomena. In addition, Semenov theory is a good approximation for a number of real systems. For example, in the case of highly reactive systems, explosion occurs immediately at the very early stages of conversion, where reactant consumption is essentially negligible.

When reactant consumption is neglected, the mass balance equation (3.4) can be eliminated, and the energy balance reduces to

$$\frac{1}{B} \cdot \frac{d\theta}{d\tau} = \theta_+ - \theta_-,$$

(3.9)

where

$$\theta_+ = \exp\left(\frac{\theta}{1 + \theta/\gamma}\right)$$

(3.10)

and

$$\theta_- = \frac{1}{\psi}(\theta - \theta_a) \tag{3.11}$$

represent the rate of temperature increase due to the heat released by the reaction and the rate of temperature decrease due to heat removed by cooling, respectively. It is seen that the temperature-increase rate θ_+ increases with θ, and approaches an asymptotic value $[\theta_+ \rightarrow \exp(\gamma)]$ as $\theta \rightarrow \infty$, while the temperature-decrease rate θ_- increases linearly as θ increases with slope equal to the reciprocal of the Semenov number.

In order to have a clear illustration of the system dynamics, let us consider the temperature increase and decrease rates, θ_+ and θ_-, as functions of the system temperature θ, shown in Fig. 3.1 by curves 1 and 2, respectively. Assuming that the temperature in the reactor is initially the same as that of the surroundings, then $\theta_a = \theta^i = 0$ and, as is apparent from Fig. 3.1, $\theta_+ > \theta_-$, i.e., the temperature-increase rate is greater than the temperature-decrease rate. Thus, the system temperature first increases with time. As the temperature reaches $\theta = \theta_{ss}$, where curve 1 intersects curve 2 (point A), then the temperature-increase rate equals the temperature-decrease rate ($\theta_+ = \theta_-$). At this point, the system temperature remains constant and the reaction proceeds at steady-state conditions.

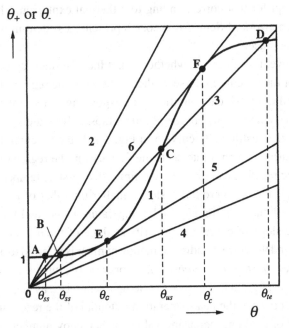

Figure 3.1. Temperature-increase rate θ_+ (curve 1) and temperature-decrease rate θ_- (curves 2 to 6) as functions of the system temperature θ: an illustration of the Semenov theory of thermal explosion.

If we keep the initial temperature unchanged and decrease the slope of the θ_- curve (curve 2), *i.e.*, increase the Semenov number, the θ_- curve (as given by curve 3) may intersect the θ_+ curve (curve 1) three times where the three intersection points from low to high θ values are indicated by B, C, and D, respectively. Point B is analogous to A, corresponding to a stable steady state, while point C at a higher temperature value, $\theta = \theta_{us}$, identifies an unstable steady state. Let us develop the argument in more detail. If the system operates at $\theta = \theta_{us}$, and the system temperature θ somehow decreases slightly to reach a value lower than θ_{us}, since in this case the temperature-decrease rate is higher than the temperature-increase rate, θ decreases even further and eventually reaches the stable steady state at θ_{ss} (point B). On the other hand, if the system temperature θ is increased slightly to reach a value higher than θ_{us}, since now $\theta_+ > \theta_-$, θ increases further and reaches the third intersection point, D at $\theta = \theta_{te}$, where the reaction proceeds again at steady-state conditions. Note that although point D is a stable steady state, it is located at an extremely high value of the system temperature, thus essentially corresponding to a thermal explosion. Therefore, *the three intersection points, B, C and D between the θ_- and θ_+ curves, indicate a stable steady state, an unstable steady state, and a thermal explosion, respectively.*

Now, let us decrease further the slope of the θ_- curve. We may encounter the situation as indicated by curve 4 in Fig. 3.1, where the θ_- curve intersects the θ_+ curve only once at an extremely high temperature value (note that this intersection point is not shown in Fig. 3.1, because it is of little interest). Now, there is no stable steady state available, except for that corresponding to a thermal explosion. This is the situation that in the framework of Semenov theory represents a system in which a thermal runaway is taking place.

From the discussion above, it is clear that whether or not thermal runaway occurs depends on whether or not the θ_- curve intersects the θ_+ curve in the region of low temperature values. Accordingly, critical conditions corresponding to the transition from nonrunaway to runaway operations are identified as those where the θ_- curve is tangent to the θ_+ curve, as illustrated by curve 5 in Fig. 3.1. If the θ_- curve has a slope greater than that of curve 5, then it intersects the θ_+ curve in the region of low temperature values, and the reaction can proceed to a finite steady-state temperature value. On the other hand, if the θ_- curve has a slope smaller than that of curve 5, it intersects the'θ_+ curve only at an extremely high temperature value, and thermal runaway occurs. Thus, the slope of curve 5 is the critical slope of the θ_- curve for thermal runaway to occur, and the corresponding Semenov number is referred to as the *critical Semenov number, ψ_c, for thermal runaway*, *i.e.*, for $\psi > \psi_c$, runaway occurs. This result is sound on physical grounds because the Semenov number, as mentioned above, represents the ratio between the heat-generation potential of the reaction and the heat-removal potential of cooling. Increasing values of Semenov number imply increasing heat-generation rates or decreasing heat-removal rates, conditions that are more favorable for thermal explosion.

There exists another situation where the θ_- curve is tangent to the θ_+ curve, as given by curve 6 in Fig. 3.1. It is seen that if the θ_- curve has a slope greater than that of curve 6 (*e.g.*, curve 2), there is only one intersection point between the θ_- and θ_+ curves, identifying a stable steady state. This is a globally intrinsic stable steady state, since no matter how large the perturbation in temperature, the system eventually returns to this steady state. On the other hand, if the θ_- curve has a slope smaller than that of curve 6 but greater than that of curve 5 (*e.g.*, curve 3), although the system can operate at a low-temperature steady state, thermal runaway is still possible for a sufficiently large perturbation in temperature leading to $\theta > \theta_{us}$. Thus, the slope defined by curve 6 can be considered as another critical slope of the θ_- curve; for a slope larger than it, the system becomes intrinsically stable. Thus, the corresponding Semenov number is referred to as the *critical Semenov number, ψ'_c, for intrinsically stable operation.*

In conclusion, the dynamic behavior is fully determined by the value of the Semenov number; specifically, *for $\psi > \psi_c$, the system undergoes thermal runaway; for $\psi'_c < \psi < \psi_c$, the reaction can operate at a low-temperature steady state but runaway is possible for large perturbations in temperature; while for $\psi < \psi'_c$ the system becomes intrinsically stable in operation temperature.*

The critical Semenov numbers, ψ_c and ψ'_c, are found readily by imposing the θ_- and θ_+ curves in Fig. 3.1 to be tangent (points E and F), *i.e.*,

$$\theta_+ = \theta_-, \qquad \frac{d\theta_+}{d\theta} = \frac{d\theta_-}{d\theta} \tag{3.12}$$

These, along with Eqs. (3.10) and (3.11), lead to

$$\frac{1}{\psi} \cdot (\theta_c - \theta_a) = \exp\left(\frac{\theta_c}{1 + \theta_c/\gamma}\right) \tag{3.13}$$

$$\frac{1}{\psi} \cdot (1 + \theta_c/\gamma)^2 = \exp\left(\frac{\theta_c}{1 + \theta_c/\gamma}\right) \tag{3.14}$$

which yield the following equation for the critical temperature θ_c:

$$(\theta_c - \theta_a) = (1 + \theta_c/\gamma)^2 \tag{3.15}$$

The corresponding critical Semenov number ψ_c is given by

$$\psi_c = (\theta_c - \theta_a) \cdot \exp\left(\frac{-\theta_c}{1 + \theta_c/\gamma}\right) \tag{3.16}$$

where the critical system temperature θ_c is computed by Eq. (3.15). Note that Eq. (3.15) gives two possible values for the critical system temperature:

$$\theta'_c = \frac{\gamma}{2} \cdot [(\gamma - 2) + \sqrt{\gamma(\gamma - 4) - 4\theta_a}] \tag{3.17}$$

and

$$\theta_c = \frac{\gamma}{2} \cdot [(\gamma - 2) - \sqrt{\gamma(\gamma - 4) - 4\theta_a}] \tag{3.18}$$

The critical temperature θ_c given by Eq. (3.18) determines the critical Semenov number ψ_c for thermal runaway, while the higher critical temperature θ'_c given by Eq. (3.17) leads to the critical Semenov number ψ'_c for intrinsically stable operation.

Note that for real values of θ_c and θ'_c, we must have

$$\gamma \cdot (\gamma - 4) > 4 \cdot \theta_a$$

a condition that is readily satisfied for systems of interest in thermal explosions.

Further simplifications of Eqs. (3.15) and (3.16) can be obtained for very large values of activation energy, *i.e.*, $\gamma \to \infty$. In this case, we have

$$\frac{\theta}{\gamma} = \frac{T - T^i}{T^i} \ll 1 \tag{3.19}$$

which leads to

$$\exp\left(\frac{\theta}{1 + \theta/\gamma}\right) \approx \exp(\theta), \tag{3.20}$$

This is commonly referred to as the *Frank-Kamenentskii approximation*, first introduced in 1939. In general, this is found to be rather satisfactory for many combustion reactions, where activation energy, γ, is often very large. Using this approximation in Eqs. (3.15) and (3.16), and considering the case $\theta_a = 0$, we have

$$\theta_c = 1, \qquad \psi_c = 1/e \tag{3.21}$$

Accordingly, the condition for thermal runaway not to occur can be expressed explicitly as follows:

$$\psi = \frac{(-\Delta H) \cdot k(T^i) \cdot (C^i)^n \cdot E}{S_v \cdot U \cdot (T^i)^2 \cdot R_g} < \frac{1}{e} = 0.368 \tag{3.22}$$

which is the criterion derived originally by Semenov (1928).

Thus, when the Frank-Kamenetskii approximation is used, *i.e.*, for large values of γ, the critical Semenov number ψ_c is a constant. The ψ_c values given by Eq. (3.16) are shown in Fig. 3.2 as a function of the Arrhenius number, γ. It may be seen that $\psi_c \to 1/e$ as $\gamma \to \infty$, but that ψ_c deviates significantly from $1/e$ for lower γ values.

Finally, it should be noted that when the assumption of $\gamma \to \infty$ is used, the system dynamics are different from those illustrated in Fig. 3.1. The temperature-increase rate θ_+ then increases exponentially as θ increases, without approaching an asymptotic value. In this case, two intersections of the θ_+ and θ_- curves occur always, and one cannot define a critical Semenov number for intrinsic stable operations.

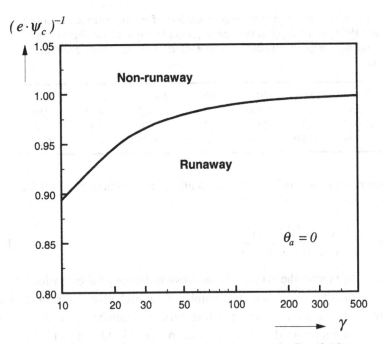

Figure 3.2. The critical Semenov number, ψ_c, given by Eq. (3.16), as a function of the Arrhenius number, γ.

Example 3.1 Application of Semenov criterion to thermal explosion of methyl nitrate. The decomposition of methyl nitrate (CH_3ON_2) in the vapor phase is a highly exothermic reaction, and may be treated as an irreversible first-order reaction. The reaction kinetics can be expressed as follows (Gray *et al.*, 1981)

$$r = 3.30 \times 10^{13} \cdot \exp\left(-\frac{1.51 \times 10^5}{R_g \cdot T}\right) \cdot C, \quad \text{mol/m}^3\text{/s}$$

The explosion phenomena related to this reaction in a batch reactor were first examined by Apin *et al.* (1936) and subsequently by Gray *et al.* (1981) at different conditions. The critical initial pressure values for explosion (*i.e.*, the *explosion limit*) measured by Gray *et al.* (1981) in a spherical reactor ($R_v = 0.064$ m) for various values of the initial temperature are given in Table 3.1. Let us now use the Semenov criterion described above to predict the explosion limits of this reaction. The heat of reaction at $T = 298$ K is given by $-\Delta H_{298} = 1.505 \times 10^5$ J/mol, and may be assumed to be independent of temperature. The overall heat-transfer coefficient at the wall is $U = 3.0$ J/m$^2 \cdot$ s $\cdot$ K, and the ambient temperature equals the initial temperature in the reactor (*i.e.*, $\theta_a = 0$).

The Semenov number for a first-order reaction is given by Eq. (3.8) as

$$\psi = \frac{(-\Delta H) \cdot k(T^i) \cdot C^i \cdot E}{S_v \cdot U \cdot (T^i)^2 \cdot R_g} \tag{E3.1}$$

43

Table 3.1. Explosion limits for the decomposition of methyl nitrate measured by Gray et al. (1981) in a spherical reactor (R_v = 0.064 m) and corresponding values of γ and ψ_c computed through the criterion (3.16) for various initial temperatures

T^i (K)	510	520	530	540	550	560	570
P^i (k Pa)	2.26	1.09	0.66	0.36	0.22	0.11	0.0625
γ	35.6	34.9	34.3	33.6	33.0	32.4	31.9
ψ_c	0.379	0.379	0.379	0.379	0.380	0.380	0.380

which, when reorganized and at critical conditions, for a spherical vessel (i.e., $S_v = 3/R_v$) gives

$$P^i = \psi_c \cdot \frac{3 \cdot U \cdot (T^i)^3 \cdot (R_g)^2}{R_v \cdot (-\Delta H) \cdot k(T^i) \cdot E} \tag{E3.2}$$

Thus, in order to predict the critical initial pressure for thermal explosion, the first step is to compute the critical Semenov number ψ_c at each given value of the initial temperature T^i and hence the corresponding Arrhenius number, $\gamma = E/(R_g \cdot T^i)$. These results are summarized in Table 3.1, where Eqs. (3.16) and (3.18) have been used to calculate ψ_c. It can be seen that, due to the relatively large values of γ, the computed values of ψ_c do not differ significantly from the classical Semenov value (3.21), $\psi_c = 1/e = 0.368$.

The critical initial pressure calculated by using Eq. (E3.2), where ψ_c is predicted by the Semenov criteria [i.e., Eqs. (3.16) and (3.18), or Eq. (3.21)] is compared in Fig. 3.3 with the experimental data shown in Table 3.1, as a function of the initial temperature. It is seen that the predictions are in good agreement with the experimental results, indicating that the assumption of negligible reactant consumption is appropriate for predicting explosion limits in the case of methyl nitrate decomposition. This is so, because this reaction involves relatively large values for the heat of reaction (i.e., $B > 80$) and the activation energy (i.e., $\gamma > 30$), which imply that explosion occurs at the very early stages of reactant conversion, so that reactant consumption can in fact be neglected.

3.2.2 Criteria Accounting for Reactant Consumption

When reactant consumption is accounted for, the system dynamics become more complex. In particular, the temperature-increase rate θ_+ in Eq. (3.10) now becomes, according to Eq. (3.5),

$$\theta_+ = \exp\left(\frac{\theta}{1 + \theta/\gamma}\right) \cdot (1 - x)^n \tag{3.23}$$

that is, the temperature-increase rate θ_+ decreases as the reactant conversion x increases with time, so that thermal runaway is less likely. Thus, the Semenov criterion for thermal runaway in this case becomes too conservative. For this reason, alternative

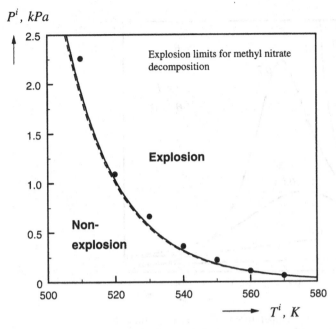

P^i, kPa

Figure 3.3. Explosion limits for methyl nitrate decomposition as measured experimentally by Gray *et al.* (1981) (•) and predicted by Semenov criteria in the form of Eqs. (3.16) and (3.18) (solid curve) or Eq. (3.21) (broken curve).

definitions for a system operating under thermal runaway conditions have been proposed in the literature.

Reactant consumption was first considered by Todes and coworkers (Todes, 1933, 1939; Kontorova and Todes, 1933; Melent'ev and Todes, 1939; Todes and Melent'ev, 1940) in defining critical conditions for thermal runaway. However, since numerical calculations were difficult then, from the limited integrations carried out manually, they failed to reach a conclusion about the critical conditions. In the following, we introduce three different runaway criteria developed by Thomas and Bowes (1961), Adler and Enig (1964), and van Welsenaere and Froment (1970), which are all based on some *geometric* feature of the temperature profile in time or in conversion.

Thomas and Bowes (TB) criterion

Based on physical intuition, Thomas and Bowes (1961) proposed to identify thermal runaway as the situation in which a positive second-order derivative occurs *before* the temperature maximum in the *temperature-time plane*. In order to visualize this definition, let us integrate the system of Eqs. (3.4) and (3.5) to compute the dimensionless temperature, θ, as a function of dimensionless time, τ, for various values of Semenov number, ψ. The results for the case of an irreversible first-order reaction are shown in Fig. 3.4.

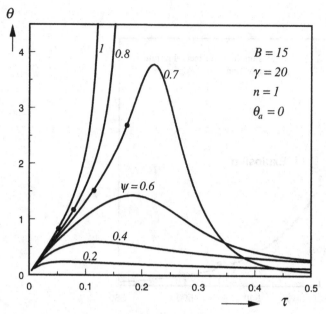

Figure 3.4. Dimensionless temperature profiles as a function of dimensionless time for various values of the Semenov number. The symbol (•) denotes inflection point.

For the three curves with lower values of the Semenov number (*i.e.*, $\psi = 0.2$, 0.4, and 0.6), the temperature maximum is relatively low and the curves before the temperature maximum are convex, *i.e.*, the second-order derivative of the temperature with respect to time is negative, indicating that the temperature increase with time is *decelerated*. Thus, these cases correspond to nonrunaway conditions.

For the other three cases with higher values of the Semenov number, the temperature maximum is much higher. In particular, in the temperature profile before the maximum, two inflection points occur, between which the curve becomes concave, *i.e.*, the second-order derivative is positive. This implies that the temperature increase with time is *accelerated*. Accordingly, these cases correspond to runaway operations.

Therefore, it appears reasonable to expect that an inflection point in the temperature profile before the maximum is necessary for runaway to occur. This observation was utilized by Thomas and Bowes (1961) to define the *critical* conditions for runaway. For this, note that under runaway conditions the temperature profile exhibits two inflection points before the maximum. At the first, the curve changes from convex to concave; at the second, it goes back to convex. Based on this, the temperature profile is concave only in the region between the two inflection points, and this region enlarges as the Semenov number increases. Therefore, Thomas and Bowes defined the critical condition for runaway to occur as the situation where the concave region first appears but its size is zero, *i.e.*, the two inflection points are coincident. This corresponds to the conditions

$$\frac{d^2\theta}{d\tau^2} = 0, \qquad \frac{d^3\theta}{d\tau^3} = 0 \tag{3.24}$$

Note that Eq. (3.24) defines only the critical inflection point but it does not say whether it is before or after the temperature maximum. Thus, with the above criterion, it is generally required to use some specific techniques to identify the inflection point that is before the temperature maximum. There exist several numerical strategies in the literature using Eq. (3.24) to find the critical conditions for runaway, which will be discussed later along with the conditions for the Adler and Enig criterion.

Adler and Enig (AE) criterion

The criterion proposed by Thomas and Bowes was examined further by Adler and Enig (1964), who found that it is more convenient to work in the *temperature-conversion plane* than in the temperature-time plane; since in this case, we need to consider only *one* equation:

$$\frac{1}{B} \cdot \frac{d\theta}{dx} = 1 - \frac{1}{\psi} \cdot \frac{\theta - \theta_a}{(1-x)^n} \cdot \exp\left(-\frac{\theta}{1+\theta/\gamma}\right) \tag{3.25}$$

with the IC

$$\theta = 0 \qquad \text{at} \quad x = 0. \tag{3.26}$$

Equation (3.25) is obtained by simply dividing Eq. (3.5) by Eq. (3.4), thereby eliminating time as the independent variable. The critical conditions for runaway to occur are then defined, similarly to the previous case, as the situation where a region with positive second-order derivative first occurs before the maximum in the temperature-conversion plane, *i.e.*,

$$\frac{d^2\theta}{dx^2} = 0, \qquad \frac{d^3\theta}{dx^3} = 0 \tag{3.27}$$

It should be noted that Lacey (1983) has shown that the criterion (3.24) based on the temperature-time plane and the criterion (3.27) based on the temperature-conversion plane may predict different critical values for the system temperature θ_c, while they both predict substantially the same critical value for the Semenov number. An example is the limiting case of $\gamma \to \infty$, in which the expressions for the critical system temperature derived by Lacey are $\theta_c = 1 + \sqrt{n}$ and $\theta_c = (\sqrt{n^2 + 4n} - n + 2)/2$, respectively, for the TB and AE criteria, while the corresponding critical values of the Semenov number are given equally by $\psi_c = 1/e$. Since the interest in practical applications is often in determining the critical conditions for thermal explosion to occur, *i.e.*, the value of the critical Semenov number, the two criteria are substantially equivalent.

Similar to the TB criterion (3.24), the AE criterion (3.27) also requires some numerical work to evaluate the critical conditions (*i.e.*, for given values of n, γ, and θ_a, to find the critical value of the Semenov number as a function of B), because Eq. (3.27) defines only the critical inflection point but does not indicate whether it is before or after the temperature maximum. To solve this problem, different approaches have been proposed in the literature (*e.g.*, Hlavacek *et al.*, 1969; van Welsenaere and

Froment, 1970; Morbidelli and Varma, 1982), which are often erroneously referred to as different criteria for runaway. A detailed comparison of these approaches has been reported by Morbidelli and Varma (1985). Among them, the approach developed by Morbidelli and Varma (1982), based on the method of isoclines, is rigorous and can be applied for all positive-order exothermic reactions with finite activation energy. In this case, the procedure to evaluate the critical conditions requires only a single numerical integration.

The critical values of the Semenov number, computed through the Adler and Enig criterion, are shown in Fig. 3.5 as a function of the heat-of-reaction parameter B,

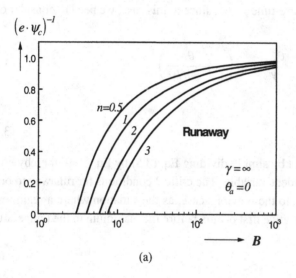

(a)

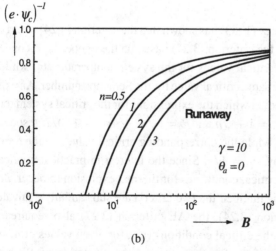

(b)

Figure 3.5. Critical values of the Semenov number, ψ_c, as a function of the heat-of-reaction parameter B, computed through the Adler and Enig criterion, for various values of the reaction order n. (a) $\gamma = \infty$; (b) $\gamma = 10$.

for various values of the reaction order, n, for the case $\theta_a = 0$, and $\gamma = \infty$ (a) or $\gamma = 10$ (b). It appears that as $B \to \infty$, for all values of the reaction order, the critical values of the Semenov number approach $1/e$ and 0.4115, i.e., the critical values predicted by the Semenov criterion, Eqs. (3.21) and (3.18), respectively. This is because, at very large values of B, the heat of reaction is extremely high, and then the explosion occurs when only a very small amount of reactant is consumed. In this case, we can safely neglect reactant consumption, and thus the AE criterion approaches the Semenov criterion – a result that supports the reliability of the AE criterion.

However, when the value of B is finite, the critical values of the Semenov number ψ_c for all cases in Fig. 3.5 become greater than those predicted by the Semenov criterion; moreover, the runaway boundary moves toward higher values of the Semenov number as the reaction order increases. In this case, the influence of reactant consumption on thermal runaway is significant, and the Semenov criterion becomes too conservative. These results indicate that *runaway becomes less likely as the reaction order increases*. This is consistent with the observation that the influence of reactant consumption on the temperature history increases with increase of the reaction order, n. Figures 3.6a and b show the runaway regions for various values of γ and θ_a for fixed values of the other parameters. As expected, *runaway becomes more likely as the activation energy and the ambient temperature increase.*

van Welsenaere and Froment (VF) criterion

The runaway criterion derived by van Welsenaere and Froment (1970), originally for runaway in a homogeneous tubular reactor, defines criticality using the locus of temperature maxima in the temperature-conversion plane.

The temperature maximum in the temperature-conversion plane is obtained by setting $d\theta/dx = 0$ in Eq. (3.25), which yields

$$(1 - x_m)^n = \frac{\theta^* - \theta_a}{\psi} \cdot \exp\left(-\frac{\theta^*}{1 + \theta^*/\gamma}\right) \tag{3.28}$$

where x_m is the conversion value where the temperature maximum θ^* occurs. For each set of values of the involved parameters, one can find by integrating Eq. (3.25) a pair of values for x_m and θ^*, which satisfy Eq. (3.28). Thus, the locus of the temperature maximum, referred to as the *maximum curve*, is constructed in the temperature-conversion plane, as shown in Fig. 3.7 for the case where parameter B has been varied. It is seen that the maximum curve exhibits a minimum with respect to θ, which can be found by differentiating Eq. (3.28) with respect to θ^* and setting the result equal to zero:

$$\theta_c = \frac{\gamma}{2} \cdot [(\gamma - 2) - \sqrt{\gamma \cdot (\gamma - 4) - 4 \cdot \theta_a}] \tag{3.29}$$

van Welsenaere and Froment defined criticality as the situation where the $\theta - x$ trajectory goes through the minimum of the maximum curve, and θ_c is the critical temperature.

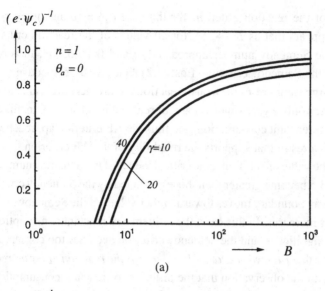

(a)

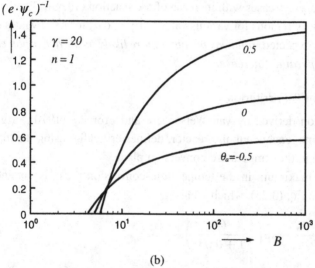

(b)

Figure 3.6. Critical values of the Semenov number, ψ_c, as a function of the heat-of-reaction parameter B, computed through the Adler and Enig criterion, for various values of (a) activation energy, γ; and (b) ambient temperature, θ_a.

Equation (3.29) gives only the critical value for the system temperature. However, we need a procedure to identify which set of values for the system parameters leads to this critical temperature. In general, a numerical technique with a trial-and-error procedure is required. In the case of a first-order reaction, van Welsenaere and Froment used an extrapolation procedure to derive an explicit expression for the critical value of a system parameter. For example, let us consider the variation of the heat-of-reaction parameter B while maintaining all other parameters fixed. Using the extrapolation

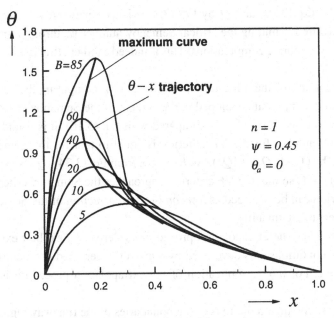

Figure 3.7. The $\theta - x$ trajectories and the maximum curve.

procedure, one can find the lower and upper bounds of the critical value of B as follows:

$$B_l = (\theta_c - \theta_a) \cdot (1 + Q^2) \tag{3.30}$$

$$B_u = (\theta_c - \theta_a) \cdot (1 + Q)^2 \tag{3.31}$$

where

$$Q = \sqrt{\frac{B}{\psi} \cdot \exp\left(-\frac{\theta_c}{1 + \theta_c/\gamma}\right)} \tag{3.32}$$

Then, the critical value of B is defined as the mean value of the lower and upper bounds:

$$B = (B_l + B_u)/2 = (\theta_c - \theta_a) \cdot (1 + Q + Q^2) \tag{3.33}$$

Since it is usually more convenient to compute the critical Semenov number ψ as a function of B, rather than *vice versa*, we can compute the quantity Q from Eq. (3.33) as follows:

$$Q = \frac{\sqrt{1 + 4 \cdot [B/(\theta_c - \theta_a) - 1]} - 1}{2} \tag{3.34}$$

and then using Eq. (3.32), the critical Semenov number ψ_c is obtained as

$$\psi_c = \left(1 + \frac{1}{Q} + \frac{1}{Q^2}\right) \cdot (\theta_c - \theta_a) \cdot \exp\left(-\frac{\theta_c}{1 + \theta_c/\gamma}\right) \tag{3.35}$$

where θ_c is given by Eq. (3.29) and Q by Eq. (3.34). Thus, for given values of B, γ, and θ_a, the procedure for finding the critical Semenov number ψ_c through the VF criterion implies only algebraic computations. This is indeed a rather attractive feature of this criterion.

It is worth noting that, for the VF criterion, Eq. (3.29) for computing the critical temperature θ_c is identical to that developed in the context of the Semenov criterion, Eq. (3.18). Moreover, when Eq. (3.35) is compared with Eq. (3.16), it is found that the critical Semenov number for the VF criterion is equal to that for the Semenov criterion multiplied by $(1+1/Q+1/Q^2)$. As $B \to \infty$, from Eq. (3.34), $Q \to \infty$ and $(1+1/Q+1/Q^2) \to 1$, so that the VF criterion approaches the Semenov criterion. Thus, the VF criterion can be regarded as a *second-order correction* of the Semenov criterion with respect to the quantity $1/Q$.

It should be noted that the extrapolation procedure to derive the explicit expression for the critical conditions was developed by van Welsenaere and Froment only for the case of $n = 1$. For $n \neq 1$, application of the extrapolation procedure is not straightforward.

A comparison between the predictions of the boundaries of the runaway region as given by the VF and AE criteria in the case of $n = 1$ is shown in the ψ^{-1}–B parameter plane (Fig. 3.8) for two different values of γ. It is seen that the two criteria predict the same critical conditions *only* at very large B values. As the B value decreases, the predictions of the VF criterion become progressively more conservative with respect to those of the AE criterion. For example, let us consider two situations, $B = 20$ and

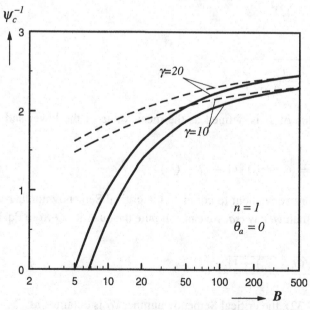

Figure 3.8. Boundaries of the runaway regions predicted by the Adler and Enig (solid curves) and van Welsenaere and Froment (broken curves) criteria for a first-order reaction; $\gamma = 10$ and 20.

Table 3.2. Explosion limits for the decomposition of methyl nitrate measured by Gray et al. (1981) in a spherical reactor ($R_v = 0.064$ m) and corresponding values of γ and B at each initial temperature

T^i (K)	510	520	530	540	550	560	570
P^i (kPa)	2.26	1.09	0.66	0.36	0.22	0.11	0.0625
γ	35.6	34.9	34.3	33.6	33.0	32.4	31.9
B	100.7	96.9	93.3	89.8	86.6	83.5	80.6

$B = 100$ in Fig. 3.8 for $\gamma = 10$, and compute the corresponding temperature profiles in the temperature-conversion plane around the critical values of the Semenov number predicted by the two criteria. The obtained results are shown in Figs. 3.9a and b for $B = 20$ and $B = 100$, respectively. In these cases, the critical values of the Semenov number, ψ_c^{-1}, predicted by the AE and VF criteria are, respectively, 1.4 and 1.89 for $B = 20$ and 2.08 and 2.17 for $B = 100$. It may be seen that around the critical value ψ_c^{-1} given by the VF criterion, the temperature maximum is rather low, and before the maximum the temperature curve is always convex, *i.e.*, the temperature increase with conversion is decelerated, thus indicating a nonrunaway behavior. On the other hand, around the critical value ψ_c^{-1} given by the AE criterion, the temperature maximum is much higher, and in particular for ψ^{-1} just smaller than ψ_c^{-1}, a concave region of the profile occurs, which indicates the temperature increase with conversion being accelerated. *Thus, the Semenov number predicted by the AE criterion is more reasonable to be considered as the critical Semenov number for reactor runaway.*

Example 3.2 Application of AE and VF criteria to thermal explosion of methyl nitrate. Let us now apply the AE and VF criteria to the case of methyl nitrate decomposition considered earlier in Example 3.1, and compare the obtained results with those given by the Semenov criterion. The values of the physicochemical parameters have been reported earlier in Example 3.1, except for the specific heat capacity of the reactant, c_v, which is not involved in the case of the Semenov criterion but is required in the AE and VF criteria for computing the value of parameter B. We use $c_v = 104.3$ J/(K · mol) (Boddington et al., 1983). The B values corresponding to each value of the initial temperature used in the experimental vessel of Gray et al. (1981) are listed in Table 3.2.

As for the Semenov criterion (Example 3.1), for both the AE and VF criteria, in order to find the critical value of the initial pressure for runaway, P_c^i, we need to first compute the critical value of the Semenov number ψ_c. For the VF criterion, ψ_c is computed through Eq. (3.35), while for the AE criterion, the method of isocline mentioned above was used (cf. Morbidelli and Varma, 1982). The obtained results are shown in Fig. 3.10, together with the experimental data and the predictions of the Semenov criterion. It may be seen that both the AE and VF criteria give somewhat better predictions than the Semenov criterion. However, it should be noted that owing to the rather large values of the parameters B and γ in this example, all three criteria

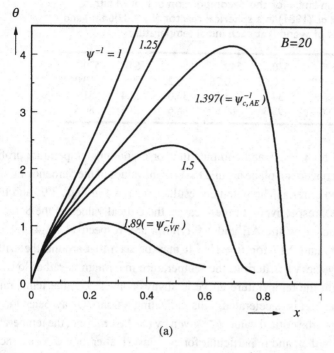

(a)

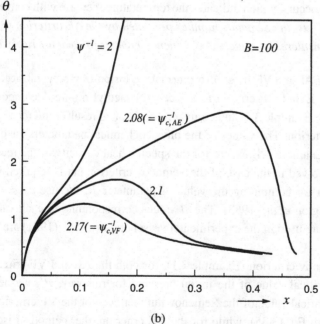

(b)

Figure 3.9. Dimensionless temperature profiles as a function of conversion for various values of the Semenov number, where $\psi_{c,AE}$ and $\psi_{c,VF}$ are the critical values of the Semenov number predicted by the Adler and Enig and the van Welsenaere and Froment criteria, respectively. $\gamma = 10$, $n = 1$, $\theta_a = 0$. (a) $B = 20$, (b) $B = 100$.

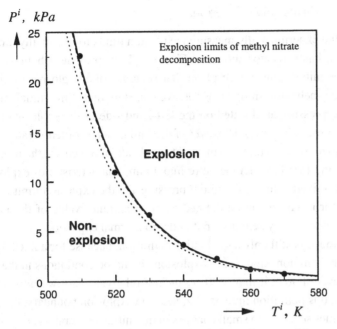

P^i, kPa

Figure 3.10. Explosion limits for methyl nitrate decomposition as measured experimentally by Gray *et al.* (1981) (•) and predicted by the Semenov (dotted curve), the van Welsenaere and Froment (broken curve) and the Adler and Enig (solid curve) criteria.

provide similar results. This would not be the case for lower values of B and γ, as indicated by the comparison discussed above (see Fig. 3.8).

3.3 Sensitivity-Based Criteria for Thermal Runaway

The criteria discussed in the previous section are all based on the idea of defining runaway operations using some *geometric* feature of the temperature profile in time or in conversion. For example, this feature is given by the absent intersection point between the temperature increase and decrease curves in the low-temperature region in Semenov criterion, and by the occurrence of a positive second-order derivative before the maximum in the AE criterion. Each of these definitions appeals to physical intuition and, at least for appropriate ranges of the operating variables, provides similar answers. However, an obvious limitation of these criteria is that they can be applied only to reacting systems where a temperature profile exists, which is not always the case in applications, as we will discuss in subsequent sections. In addition, these criteria do not give any measure of the extent or intensity of the runaway. In order to overcome these limitations, a new series of criteria have been developed, based on the concept of *parametric sensitivity*. For this, we first need to bring together two different concepts: parametric sensitivity on one side, and explosion or thermal runaway on the other.

3.3.1 The Morbidelli and Varma (MV) Criterion.

Let us reconsider the experimentally measured explosion limits for the methyl nitrate decomposition in a batch reactor shown in Fig. 3.3. They provide a boundary in the initial pressure–initial temperature plane that separates the explosive from the nonexplosive system behavior. In other words, we can state that if the initial values of temperature and pressure are located on the left-hand side of the explosion limit curve, the reaction proceeds slowly. However, if the initial values of temperature and pressure are increased, so as to move to the right-hand side of the curve, the reaction proceeds fast, leading to a large temperature maximum, characteristic of explosive systems. Thus, for a fixed value of the initial pressure P^i, the explosion limit or the critical condition for runaway can be defined as the maximum value of the initial temperature T^i at which the system does not undergo thermal explosion.

The statement above is still only a qualitative definition, because for a fixed value of P^i the transition from nonexplosion to explosion should be continuous in the initial temperature (even though it may well occur in a relatively narrow temperature interval). However, this definition implies that, near the explosion boundary, the system behavior becomes sensitive to small changes in the initial temperature; *i.e.*, small changes in the initial temperature can lead to dramatic changes in the qualitative behavior of the system (from nonexplosion to explosion, or *vice versa*). On the other hand, if the given value of the initial temperature is located far away from the explosion boundary, the system is insensitive to small changes in the initial temperature, which simply lead to correspondingly small differences in the system variables (temperature and conversion). The above observation indicates that the boundary between runaway (explosive) and nonrunaway (nonexplosive) behavior is constituted by the situations where the system behavior becomes sensitive to small changes in the operating parameters. This is usually referred to as a *parametrically sensitive region* for the system. Thus, in the following, we will develop criteria for identifying this region in order to locate the boundaries between explosive and nonexplosive behavior.

In order to investigate the sensitivity behavior of a given system, we use the normalized objective sensitivity defined in Chapter 2. Clearly, when thermal explosion in a batch reactor is considered, we are interested in the sensitivity of the temperature maximum, θ^*. In this case, the normalized objective sensitivity has the form

$$S(\theta^*; \phi) = \frac{\phi}{\theta^*}\left(\frac{\partial \theta^*}{\partial \phi}\right) = \frac{\phi}{\theta^*} \cdot s(\theta^*; \phi) \qquad (3.36)$$

where ϕ is one of the model parameters. In the MV criterion, the parametrically sensitive region of the system or criticality for thermal runaway to occur is defined as that where the normalized sensitivity of the temperature maximum, $S(\theta^*; \phi)$ reaches a maximum. For example, if we consider ϕ to be the Semenov number, ψ, and keep all the other parameters fixed, the criticality for thermal runaway to occur is located at the value of the Semenov number, ψ_c, where $S(\theta^*; \psi)$ is maximized.

It is worth noting that the value of the normalized objective sensitivity $S(\theta^*; \phi)$ can be either positive or negative. In the case of negative values of $S(\theta^*; \phi)$, the criticality for runaway arises when the normalized objective sensitivity $S(\theta^*; \phi)$ is a minimum. *Therefore, the general definition of criticality for runaway is that the absolute value of the normalized objective sensitivity reaches its maximum.*

The evaluation of the normalized objective sensitivity $S(\theta^*; \psi)$ through Eq. (3.36) requires first the computation of local sensitivity, and specifically that of the maximum temperature value, $s(\theta^*; \phi)$. The numerical methods for computing the sensitivity values have been described in Chapter 2, and any of them can be used in this simple case. In the following, we use the direct differential method, and derive the corresponding sensitivity equations by differentiating the model equations (3.4) and (3.5) with respect to the generic input parameter ϕ. By dividing both sensitivity equations thus obtained by Eq. (3.4), so as to have conversion as the independent variable, we obtain the sensitivity equations in the form

$$\frac{ds(x;\phi)}{dx} = \left(\frac{\partial F_1}{\partial x} \cdot s(x;\phi) + \frac{\partial F_1}{\partial \theta} \cdot s(\theta;\phi) + \frac{\partial F_1}{\partial \phi} \right) \Big/ F_1 \tag{3.37}$$

$$\frac{ds(\theta;\phi)}{dx} = \left(\frac{\partial F_2}{\partial x} \cdot s(x;\phi) + \frac{\partial F_2}{\partial \theta} \cdot s(\theta;\phi) + \frac{\partial F_2}{\partial \phi} \right) \Big/ F_1 \tag{3.38}$$

with the ICs

$$s(x;\phi) = 0, \qquad s(\theta;\phi) = \delta(\phi - \theta^i), \quad \text{at} \quad x = 0 \tag{3.39}$$

where δ is the Kronecker delta function. Thus, in order to evaluate $s(\theta^*; \phi)$, we need to integrate the sensitivity equations (3.37)–(3.39) together with the model equations (3.25) and (3.26), up to the conversion value where the temperature reaches its maximum, *i.e.*, $\theta = \theta^*$.

The results of the above procedure are illustrated in Fig. 3.11, where the normalized objective sensitivity of the temperature maximum with respect to the Semenov number, $S(\theta^*; \psi)$ is shown as a function of the Semenov number, ψ, for fixed values of all the other parameters. It is seen that the value of $S(\theta^*; \psi)$ exhibits a sharp maximum at $\psi = 0.615$, which clearly indicates a parametrically sensitive region of the system. Thus, according to the discussion above we can take this as the boundary separating the nonrunaway or nonexplosive ($\psi < \psi_c$) from the runaway or explosive region ($\psi > \psi_c$), with $\psi_c = 0.615$.

The above result was first obtained independently by Lacey (1983) and Boddington *et al.* (1983), who proposed to use the sensitivity maximum of the temperature maximum with respect to Semenov number, to define the critical conditions for thermal explosion. However, since only sensitivity with respect to Semenov number was used, the question naturally arises about the possibility of considering the other physico-chemical parameters of the reacting system in the definition of sensitivity. In principle, one should expect different values of ψ_c, depending on whether the sensitivity to, say,

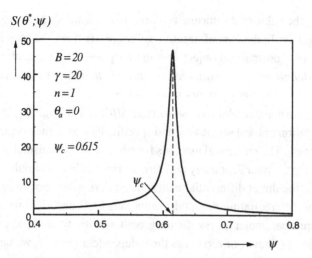

$S(\theta^*;\psi)$

Figure 3.11. Normalized sensitivity of the temperature maximum with respect to the Semenov number, $S(\theta^*;\psi)$, as a function of the Semenov number, ψ.

the heat of reaction, $S(\theta^*; B)$, or the activation energy, $S(\theta^*; \gamma)$, or any other physicochemical parameter, is considered. However, if the criterion is intrinsic, it should be *independent* of the particular choice of the parameter ϕ considered in the sensitivity definition, which inevitably involves some arbitrariness. It was in fact shown by Morbidelli and Varma (1988) that the critical Semenov number, ψ_c, defined as the value where $S(\theta^*; \phi)$ is maximum, is the same for any possible choice of ϕ.

As an example, in Fig. 3.12 the behavior of the normalized objective sensitivities, $S(\theta^*; \phi)$, defined with respect to each of the five independent parameters of the system (3.25) (*i.e.*, $\phi = B$, ψ, θ_a, γ, and n), is shown as a function of the Semenov number ψ for the same set of values for B, γ, θ_a, and n used in Fig. 3.11. In the case of $\phi = B$, ψ, θ_a, or γ, the sensitivity value exhibits a sharp maximum for a specific value of ψ, while for $\phi = n$, the sensitivity value exhibits a sharp minimum. In all cases, the specific value of ψ corresponding to the maximum or minimum is the same (*i.e.*, $\psi_c = 0.615$, up to three significant digits!). This indicates that when the system enters into the parametrically sensitive region, it becomes simultaneously sensitive to *all* the parameters that affect its behavior. This fact provides great generality to the definition of the parametrically sensitive region, which may then be taken as the boundary between explosive and nonexplosive system behavior. Hence this boundary will be referred to as the *generalized (MV) criterion for thermal runaway*.

Note that the sign of the sensitivity value has a particular meaning. A positive (negative) value of the normalized sensitivity of the temperature maximum with respect to a given system parameter indicates that the temperature maximum increases (decreases) as the magnitude of this parameter increases. Thus, if the sensitivity is positive, the transition from nonrunaway to runaway behavior occurs as this parameter

$S(\theta^*;\phi)$

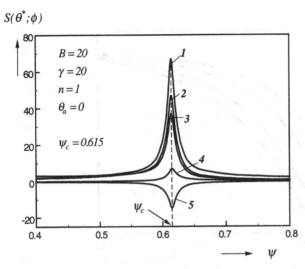

Figure 3.12. Normalized objective sensitivity, $S(\theta^*;\phi)$, as a function of the Semenov number, ψ for various model parameters, ϕ: (1) $\phi = B$, (2) $\phi = \psi$, (3) $\phi = \theta_a$, (4) $\phi = \gamma$, (5) $\phi = n$.

is increased, while if the sensitivity is negative, the same transition occurs when the corresponding parameter is decreased. The results shown in Fig. 3.12 indicate that all the parameters present in the batch reactor model (3.25) exhibit positive sensitivities, with the only exception the reaction order. This is in good agreement with the consideration that while increasing the value of the parameters B, ψ, θ_a, or γ leads to larger temperature values, for the reaction order the opposite is true.

Let us now reconsider the same cases studied in the context of Figs. 3.5a ($\gamma = \infty$) and b ($\gamma = 10$) and compare the critical values of the Semenov number computed earlier through the AE criterion with those given by the generalized criterion developed above. In particular, the normalized sensitivity of the temperature maximum to the Semenov number $S(\theta^*;\psi)$ has been used to define the criticality. The obtained results for both criteria are shown in Figs. 3.13a and b, for various values of the reaction order and $\gamma = \infty$ and 10, respectively.

It appears that, at least for sufficiently large values of the heat-of-reaction parameter B, both the AE and the generalized criteria predict the *same* critical value for the Semenov number, thus supporting the reliability of both. However, they deviate in the region of lower values of B, where the generalized criterion predicts significantly lower values of the critical Semenov number than the AE criterion. In order to explain these deviations, we consider the normalized objective sensitivity surface as a function of B and ψ for the case of $\gamma = 10$ and $n = 1$, shown in Fig. 3.14. By considering slices of this surface for constant values of the heat-of-reaction parameter B, it is seen that for large values of B the sensitivity exhibits a sharp maximum whose value is high and increases as the heat-of-reaction B increases. On the other hand, in the

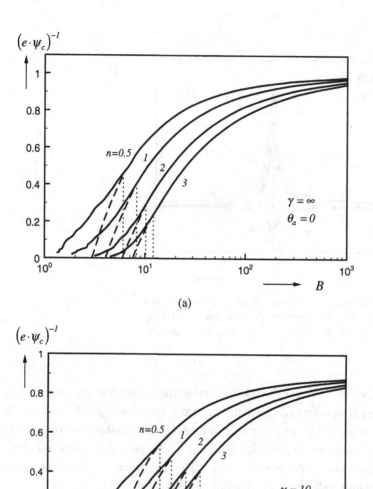

(a)

(b)

Figure 3.13. Comparison between the critical values of the Semenov number, ψ_c, computed by the Morbidelli and Varma (solid curves) and by the Adler and Enig (broken curves) criteria, as a function of the heat-of-reaction parameter B, for various values of the reaction order n. (a) $\gamma = \infty$; (b) $\gamma = 10$.

region characterized by low values of B, the sensitivity maximum exhibits a much lower value, which vanishes as B approaches zero. This indicates that the system is now relatively insensitive to changes in the ψ value. The above observations provide some information about the strength of the explosion phenomenon. For large heats of reaction, explosion is rather strong, and there is a sharp difference between explosive

Table 3.3. Critical values ψ_c as a function of B, obtained by maximizing $S(\theta^*; \phi)$ for the set of values for γ, θ_a, and n given in Fig. 3.14

B	ψ_c^a	ψ_c^b	ψ_c^c	ψ_c^d	ψ_c^e	AE*
7	1.30	3.84	1.76	1.58	1.64	10.5
10	1.08	1.57	1.21	1.19	1.21	1.48
20	0.731	0.751	0.737	0.739	0.740	0.721
30	0.614	0.616	0.615	0.616	0.616	0.607
40	0.562	0.562	0.562	0.562	0.562	0.560
50	0.533	0.533	0.533	0.533	0.533	0.533

[a,b,c,d,e]Critical values given by the generalized criterion for runaway based on $S(\theta^*; \phi)$ with $\phi = \psi$, B, θ_a, γ, and n, respectively.
*Critical value given by the Adler and Enig criterion (1964).

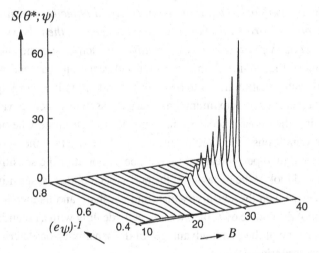

Figure 3.14. Normalized objective sensitivity $S(\theta^*; \psi)$, as a function of ψ and B; $\gamma = 10$, $\theta_a = 0$ and $n = 1$. From Morbidelli and Varma (1988).

and nonexplosive behavior. As B decreases, the explosion becomes milder and the difference between the two behaviors is also less clear. It is reasonable to expect that for sufficiently low values of the heat-of-reaction parameter (all the others being fixed), the system becomes intrinsically nonexplosive. This is indicated through the sensitivity analysis by the modest magnitude of the sensitivity maximum.

Another manifestation of the relatively insensitive behavior is that, for sufficiently small values of B, the predicted critical values, ψ_c, become dependent on the particular choice of the parameter ϕ considered in the sensitivity definition. Thus, a generalized boundary indicating the transition between the subcritical and supercritical regions ceases to exist. This may be seen in Table 3.3, where the ψ_c values, computed from the generalized criterion based on the normalized objective sensitivity, $S(\theta^*; \phi)$, with various choices of the input parameter ϕ, as well as from the AE criterion are reported. For $B > 30$, all the predicted ψ_c values are substantially identical, *i.e.*, independent of the criterion used and of the choice of ϕ in the sensitivity definition for the generalized criterion. On the other hand, for $B \leq 30$, the ψ_c values tend to differ from each other, and they do so more the lower the value of B. Again, this is due to the mild exothermicity of the system for which a really explosive behavior does not exist.

In each case of Fig. 3.13, the region where the predicted runaway boundary depends on the particular choice of the parameter ϕ is located on the left-hand side of the dotted line. It is seen that this region depends on the reaction kinetics; it enlarges as reaction order increases or the activation energy decreases. Thus, for a reaction with high reaction order and low activation energy, reactor thermal runaway occurs only for very high values of the heat of reaction.

It can then be concluded that *if for a given system the critical value of a parameter (say, the Semenov number) predicted from the normalized objective sensitivity, $S(\theta^*; \phi)$, is dependent on the particular choice of the parameter ϕ, then the system can be classified as essentially parametrically insensitive.* In such cases, one cannot define a general boundary that indicates the transition between runaway and non-runaway behavior, and each situation needs to be analyzed individually according to specific characteristics desired, *e.g.*, maximum temperature less than a specific value.

Therefore, when using the generalized criterion in order to determine whether a system is in explosive conditions, one should not only look at whether the system is located in the subcritical or supercritical region (*i.e.*, before or after the sensitivity maximum) but also should look at the behavior of the normalized objective sensitivity at critical conditions. Only when the sensitivity maximum is sharp and its location is essentially independent of the choice of ϕ, then we are really dealing with a potentially explosive system. This is one of the major advantages of the sensitivity-based criteria over the geometry-based criteria.

To complete the definition of criticality through the normalized objective sensitivity, one further observation is in order. In Fig. 3.12, the critical Semenov number, $\psi_c = 0.615$, is obtained as the ψ value where the normalized objective sensitivity $S(\theta^*; \phi)$ as a function of ψ (with fixed values for $B = 20$, $\theta_a = 0$, $\gamma = 20$, and $n = 1$) exhibits its maximum. Now, let us consider the values of $S(\theta^*; \phi)$ as a function of B, for fixed values of $\psi = \psi_c = 0.615$, $\theta_a = 0$, $\gamma = 20$, and $n = 1$, and determine the value of B where $S(\theta^*; \phi)$ exhibits its maximum. This corresponds to the critical B value according to the generalized criterion. Clearly, for consistency with the results shown in Fig. 3.12, we should find $B_c = 20$, as it is indeed indicated by the results in Fig. 3.15. Actually, this result can be generalized: *once the critical value of a given parameter is computed for fixed values of all the remaining parameters, all together they constitute a critical point in the space of the operating conditions of the system, i.e., all parameters are at their critical value.*

$S(\theta^*;\phi)$

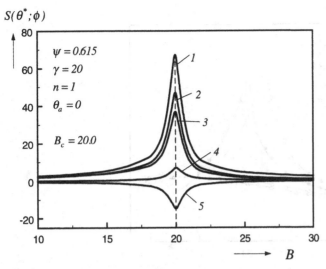

Figure 3.15. Normalized objective sensitivity, $S(\theta^*;\phi)$, as a function of the heat of reaction parameter, B, for various model parameters, ϕ: (1) $\phi = B$, (2) $\phi = \psi$, (3) $\phi = \theta_a$, (4) $\phi = \gamma$, (5) $\phi = n$.

Example 3.3 Application of the MV criterion to catalytic hydrolysis of acetic anhydride. An experimental study of thermal sensitivity in a well-stirred batch reactor has been performed by Haldar and Rao (1992) for the case of acetic anhydride hydrolysis homogeneously catalyzed by sulfuric acid. This is a relatively mild exothermic reaction ($\Delta H = -5.86 \times 10^4$ J/mol). The expression for the reaction rate, for sulfuric acid concentration of 7.86 mol/m^3, can be written as (Rao and Parey, 1988)

$$r = 1.45 \times 10^6 \cdot \exp\left(-\frac{9.35 \times 10^4}{R_g \cdot T}\right) \cdot C \quad \text{mol/m}^3\text{/s}$$

where C is the concentration of acetic anhydride and the reaction is first order with respect to this reactant. A set of experiments was carried out at fixed values of the initial concentration, $C = 5.30 \times 10^3$ mol/m^3, by changing the initial temperature and measuring the system temperature as a function of time. The experimental results are presented in Fig. 3.16. It is seen that there is a jump in the value of the temperature maximum when the initial temperature changes from 319 K to 319.5 K, indicating a transition from subcritical to supercritical conditions. These data allow one to simulate the parametric sensitivity behavior of the system, and to predict the critical value of the initial temperature for runaway through the MV criterion. The physicochemical parameters involved have been evaluated by Haldar and Rao (1992): $\rho \cdot c_v = 2.55 \times 10^6$ J/(m^3· K), $S_v = 1.57 \times 10^2$ m^2/m^3, except for the overall heat-transfer coefficient, $U = 2.29 \times 10^{-4}$ J/(m^2 · s · K), which has been estimated by least-squares fitting of the experimental data.

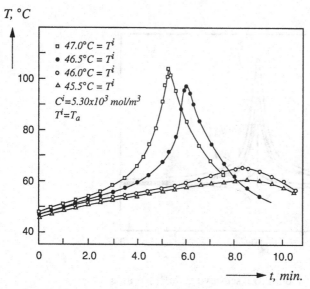

Figure 3.16. Experimental temperature profiles in a batch re-actor for acetic anhydride hydrolysis at different values of the initial temperature. From Haldar and Rao (1992).

The computed normalized sensitivity values of the temperature maximum to the initial temperature $S(T^*; T^i)$ as a function of the initial temperature are shown in Fig. 3.17. The measured and computed values of the temperature maximum are also shown in the figure, and may be observed to be close. Moreover, the sensitivity profile exhibits a sharp maximum at $T^i = 319.3$ K, which may then be defined as the critical value of the initial temperature separating subcritical from supercritical conditions. This is indeed a rather convincing experimental proof that the location of the sensitivity maximum coincides with the location of the boundary between nonrunaway and runaway system behavior. The critical values of the initial temperature predicted by the other criteria are summarized in Table 3.4. The AE and MV criteria again give substantially the same predictions. On the other hand, since in this case the value of $B(\approx 14)$ is not large, the predictions of the VF and Semenov criteria are somewhat different. In particular, from the data shown in Fig. 3.17, it may be seen that the value $T_c^i = 317.0$ K predicted by the VF criterion is actually located far away from the runaway region.

3.3.2 The Vajda and Rabitz (VR) Criterion

Along the lines of using parametric sensitivity to identify the boundary for runaway or explosive behavior, Vajda and Rabitz (1992) have considered the sensitivity of the temperature trajectory to arbitrary, unstructured perturbations applied at the temperature maximum. The linearized perturbation equation of the system (3.4) to (3.6) can

Table 3.4. Critical values of the initial temperature for runaway in the case of hydrolysis of acetic anhydride in a batch reactor predicted by various criteria

Criterion	Semenov	VF	AE	MV
$T_c^i(K)$	313.5	317.0	319.0	319.3

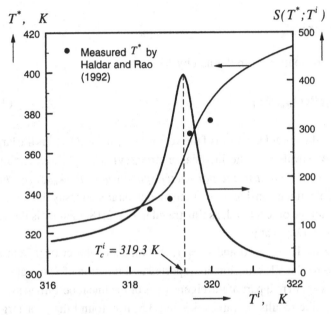

Figure 3.17. Profiles of the temperature maximum and its normalized sensitivity $S(T^*; T^i)$ as a function of the initial temperature, indicating the critical initial temperature for runaway. Data by Haldar and Rao (1992).

be written as

$$\frac{d}{d\tau}\delta y = J(y)\delta y, \qquad \delta y(0) = \delta y^i \tag{3.40}$$

where $y^T = [y_1 \ y_2] = [x \ \theta]$ and the Jacobian matrix J, with elements $a_{ij} = \partial F_i/\partial y_j$ is given from Eqs. (3.4) and (3.5) as

$$J = \begin{bmatrix} -\frac{n \cdot F_1}{1-x} & \frac{F_1}{(1+\theta/\gamma)^2} \\[2mm] \frac{n \cdot B \cdot F_1}{1-x} & \frac{B \cdot F_1}{(1+\theta/\gamma)^2} - \frac{B}{\psi} \end{bmatrix} \tag{3.41}$$

Following prior workers (Lacey, 1983; Boddington *et al.*, 1983; Morbidelli and Varma, 1988), Vajda and Rabitz (1992) also considered the sensitivity of $y^*(\tau_m)$, where τ_m denotes the time corresponding to the temperature maximum. However, they examined how a perturbation δy^* applied at time τ_m propagates for $\tau > \tau_m$. There are two possibilities. The equilibrium point $\delta y = 0$ of the perturbation equation (3.40) can be either stable where the system returns to the nominal trajectory, or unstable where the perturbation δy^* is amplified for some interval $\tau > \tau_m$. Criticality is then identified with the condition leading to the maximum of such amplification.

Let us consider a small time step $\delta \tau = \tau - \tau_m$. The solution of Eq. (3.40) is then approximated by

$$\delta y(\tau) = \exp[J(y^*)\delta\tau]\delta y^* \tag{3.42}$$

We are interested in the matrix norm defined by

$$\max \frac{\|\delta y\|}{\|\delta y^*\|} = \exp[\mathrm{Re}(\lambda_{\max})\delta\tau] \tag{3.43}$$

where $\|y(\tau)\|$ denotes the Euclidean norm of the vector $y(\tau)$, and $\mathrm{Re}(\lambda_{\max})$ is the largest real part of the two eigenvalues of the Jacobian matrix $J(y)$ at $\tau = \tau_m$. Accordingly, *criticality is defined as the point in the parameter space where $\mathrm{Re}(\lambda_{\max})$ reaches a maximum.* For example, if we consider one of the model parameters, say the Semenov number ψ, and keep all the others fixed, as illustrated in Fig. 3.18, then ψ_c is the value of ψ that maximizes $\mathrm{Re}(\lambda_{\max})$ at y^*.

For the case of $\gamma = 10$, $\theta_a = 0$ and $n = 1$, the values of the critical Semenov number ψ_c predicted by the VR criterion are reported as a function of B in Table 3.5, together with the maximum real part of the eigenvalues of the linear perturbed system, $\mathrm{Re}(\lambda_{\max})$. Comparing the results in Tables 3.3 and 3.5, it is found that, for large B

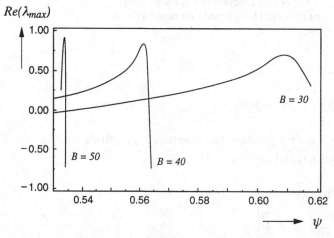

Figure 3.18. The larger real part $\mathrm{Re}(\lambda_{\max})$ of the two eigenvalues λ_1 and λ_2 of the Jacobian matrix J at the maximum temperature θ^* as a function of the Semenov number ψ; $\gamma = 10$, $\theta_a = 0$, and $n = 1$. From Vajda and Rabitz (1992).

Table 3.5. Critical values ψ_c as a function of B, predicted by the Vajda and Rabitz criterion, for the set of values for γ, θ_a, and n given in Fig. 3.16

B	$Re(\lambda_{max})$	ψ_c
7	−0.165	1.020
10	0.014	0.933
20	0.416	0.709
30	0.670	0.611
40	0.830	0.560
50	0.915	0.533

Table 3.6. Explosion limits for the decomposition of azomethane measured by Allen and Rice (1935) in a bulb (R_v = 0.0181 m) and corresponding values of γ and B at various values of the initial temperature

T^i (K)	614	620	626.5	631	636.5	643.5	645	651	659
P^i(kPa)	25.5	13.6	8.93	7.33	5.07	4.13	3.73	3.00	2.40
γ	41.7	41.3	40.9	40.6	40.2	39.8	39.7	39.4	38.9
B	113.6	111.4	109.2	107.6	105.7	103.5	103.0	101.3	98.8

values, all three criteria (AE, MV, and VR), even though apparently different in nature, provide essentially the same prediction for the critical Semenov number. Again, it is found that, for smaller values of B, the ψ_c values predicted by the three criteria deviate significantly. It is worth noting that when the B values are low, corresponding to intrinsically nonexplosive systems, the values of $Re(\lambda_{max})$ in Table 3.5 become very small or even negative. Thus, using the VR criterion, we can conclude that *when the maximum real part of the eigenvalues is very small or negative, the system is parametrically insensitive.* Therefore, similar to the MV criterion, the VR criterion also provides a measure of the strength of the explosion phenomenon.

Example 3.4 A comparison between various criteria in predicting explosion limits in azomethane decomposition. The thermal explosion involved in the decomposition of azomethane [$(CH_3)_2N_2$] was investigated by Allen and Rice (1935) for different values of initial temperature and pressure. The explosion limits measured in a bulb of 200 ml (R_v = 0.0363 m) are summarized in Table 3.6. The reaction kinetics, according to Ramsperger (1927), can be expressed as follows:

$$r = 7.41 \times 10^{15} \cdot \exp\left(-\frac{2.13 \times 10^5}{R_g \cdot T}\right) \cdot C, \quad \text{mol/m}^3\text{/s}$$

The heat of reaction is given by $\Delta H = -1.80 \times 10^5$ J/mol and the specific-heat capacity by $c_v = 107.6$ J/(K · mol) (Rice *et al.*, 1935). The overall heat-transfer coefficient has been determined by fitting the experimental data as $U = 7.31 \text{ J}/ (\text{m}^2 \cdot \text{s} \cdot \text{K})$, while the initial temperature in the reactor has always been set equal to the ambient temperature. Let us now compare predictions of the various criteria with the measured explosion limits.

Considering that the reaction under examination is first order (*i.e.*, $n = 1$) and $\theta_a = 0$ in all the experimental runs, the thermal runaway criteria have been applied by computing the critical Semenov number, ψ_c, as a function of γ and B, whose values for each experimental run are summarized in Table 3.6. From the calculated ψ_c, along with the values of the involved physicochemical parameters reported above, the initial pressure leading to explosion has been computed. The critical values of the Semenov number for the VF and the AE criteria were calculated as discussed in Examples 3.1 and 3.2. For the MV and VR criteria, appropriate numerical techniques were used. In particular, for the MV criterion, the normalized sensitivity of the temperature maximum to the initial temperature, $S(\theta^*; T^i)$, was considered and the direct differential method was used to compute the normalized sensitivities.

The critical values of the initial pressure for explosion predicted by the various criteria are shown as functions of the initial temperature in Fig. 3.19, together with the experimental data. The predictions of the AE, MV, and VR criteria are substantially

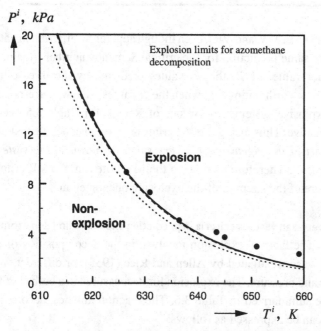

Figure 3.19. Explosion limits for azomethane decomposition as measured by Allen and Rice (1935) (•) and calculated by using various criteria: Semenov criterion (dotted curve); Van Welsenaere and Froment criterion (broken curve); Adler and Enig, Morbidelli and Varma, and Vajda and Rabitz criteria (solid curve).

identical, and close to the experimental results. Also, the predictions of the VF criterion are similar to the others, although slightly conservative. As in previous examples, the only significant difference is given by the original Semenov criterion. Again, it should be observed that since the values of the heat-of-reaction parameter B and the Arrhenius number given in Table 3.6 are large, such small differences in the predictions of all criteria are fully expected, based on our previous discussion.

3.3.3 The Strozzi and Zaldivar (SZ) Criterion

Another sensitivity-based criterion has been presented by Strozzi and Zaldivar (1994), who used the *Lyapunov exponents* to define sensitivity. It is well known that the Lyapunov exponent can monitor the behavior of two neighboring points of a system in a direction of the phase space as a function of time: If the Lyapunov exponent is positive, then the points diverge from each other; if the exponent becomes negative, they converge; when the exponent tends to zero, they remain at the same distance. Such behavior of the Lyapunov exponent is indicative of the system sensitivity.

Consider an m-dimensional dynamic system, which due to evolution changes from its original (m-sphere) state to a new (m-ellipsoid) state. The jth Lyapunov exponent, λ_j at time t, is generally defined as (Wolf *et al.*, 1985)

$$\lambda_j = \lim_{t \to \infty} \frac{1}{t} \cdot \log_2 \frac{L_j(t)}{L_j(0)} \quad \text{for } j = 1, 2, \ldots m \tag{3.44}$$

where $L_j(0)$ and $L_j(t)$ are lengths of the jth axis of the ellipsoid at $t = 0$ and $t = t$, respectively. However, for a chemical reaction occurring in a batch reactor, as $t \to \infty$, all states in the phase space always converge to a specific fixed point, *i.e.*, reactant conversion is complete and internal temperature is equal to the ambient temperature. Then, the Lyapunov exponents as defined above cannot give us any information about reactor runaway. On the other hand, before reaching the final fixed point, the state may diverge locally. Thus, for a batch reactor, it is better to define the Lyapunov exponents as functions of time, *i.e.*,

$$\lambda_j(t) = \frac{1}{t} \cdot \log_2 \frac{L_j(t)}{L_j(0)}, \quad \text{for } j = 1, 2, \ldots m \tag{3.45}$$

The exponents defined in this manner can give a measure of the volume evolution of the m-sphere:

$$V(t) = 2^{[\lambda_1(t) + \lambda_2(t) + \cdots + \lambda_m(t)] \cdot t} \tag{3.46}$$

If the volume increases, trajectories of two neighboring points in the phase space are separating, indicating that the dynamic behavior of the system undergoes a divergence. Therefore, Strozzi and Zaldivar define sensitivity using Lyapunov exponents as follows:

$$s(V;\phi) = \frac{\Delta\left\{\max_t 2^{[\lambda_1(t)+\lambda_2(t)+\cdots+\lambda_m(t)]\cdot t}\right\}}{\Delta\phi} \tag{3.47}$$

which is referred to as *Lyapunov sensitivity*, where ϕ is a chosen system input parameter of interest. The criticality for thermal runaway to occur is then defined as the ϕ value for which the Lyapunov sensitivity has an extreme (maximum or minimum).

For a given system, the complete Lyapunov exponent spectrum can be computed by using the technique developed by Wolf *et al.* (1985). For a batch reactor given by Eqs. (3.4) to (3.6), it was shown by Strozzi and Zaldivar (1994) that this criterion gives the same predictions of the critical conditions for thermal runaway as the MV and VR criteria. An experimental verification of this criterion was also reported recently by Strozzi *et al.* (1994).

3.4   Explicit Criteria for Thermal Runaway

Although the geometry-based (AE and TB) and the sensitivity-based (MV, VR, and SZ) criteria are intrinsic and give fundamentally correct descriptions of thermal runaway, they are *implicit*, and a significant amount of computation is generally required to determine the critical conditions. The only exception is the VF criterion in the case of a first-order reaction, which is explicit but provides unsatisfactory results when the values of γ and B are not large. For practical applications, it would be convenient to have explicit, even if approximate, criteria that allow a quick evaluation of the boundaries of the runaway region.

Several attempts in this direction have been reported in the literature. The first formal asymptotic analysis of the classical Semenov result to account for reactant consumption was developed by Thomas (1961), who derived the following explicit expression for the explosion boundaries:

$$\psi_c = \frac{e^{-1}}{[1 - 2.85(n/B)^{2/3}]} \tag{3.48}$$

which applies to positive-order reactions in the limit of infinite activation energy ($\gamma \to \infty$). A comparison of the critical values of the Semenov number given by Eq. (3.48) (broken curves) with those computed numerically according to the AE criterion (solid curves) is shown in Fig. 3.20 for various values of the reaction order, n. It appears that the predictions of the Thomas expression improve the Semenov result ($\psi_c = 1/e$ at $\gamma = \infty$), but are reliable only for reaction order $n \le 0.5$. When $n > 0.5$, the predicted runaway boundaries are always in the runaway region predicted by the AE criterion.

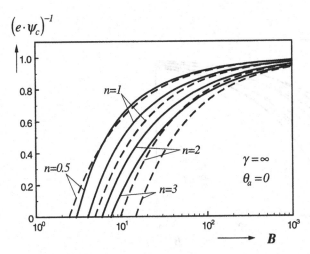

Figure 3.20. The critical values of Semenov number ψ_c, as a function of the heat-of-reaction parameter B for various values of the reaction order, n: Adler and Enig criterion (solid curves); Thomas equation (3.48) (broken curves).

The asymptotic expression

$$\psi_c = e^{-1}\left[1 + 2.946\left(\frac{n}{B}\right)^{2/3}\right] \tag{3.49}$$

derived by Kassoy and Linan (1978) through singular perturbation analysis, is more accurate, and it has been generalized further by Boddington *et al.* (1983) and Morbidelli and Varma (1988), leading to

$$\psi_c = \left(\frac{\theta_c}{h}\right)\cdot\left[1 + 2.946\left(\frac{h}{h''}\right)^{1/3}\cdot\left(\frac{n}{B}\right)^{2/3}\right] \tag{3.50}$$

where

$$h = \exp\left(\frac{\theta_c}{1 + \theta_c/\gamma}\right); \qquad h'' = \frac{d^2 h}{d\theta^2} \quad \text{at} \quad \theta = \theta_c \tag{3.51}$$

and the critical system temperature θ_c is computed through Eq. (3.18). This expression accounts for finite values of the activation energy. A comparison of the predictions from Eq. (3.50) with those computed numerically according to the AE criterion are shown in Fig. 3.21. Even though Eq. (3.50) yields better results than Eq. (3.48), it is reliable only for large values of B.

A modification in the expression (3.48) introduced by Gray and Lee (1965) deserves a special mention. They changed the constant 2.85 in Eq. (3.48) to 2.52 ($= 4^{2/3}$), in order to reproduce correctly the adiabatic condition, $\psi_c \to \infty$, of a first-order reaction ($n = 1$), for which the critical value of B can be derived analytically as $B_c^o = 4$ (Adler

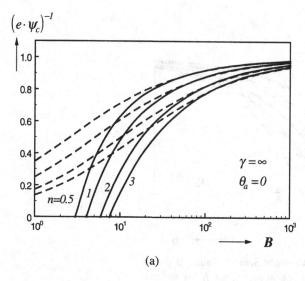

(a)

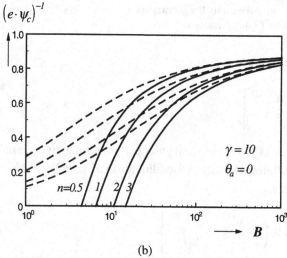

(b)

Figure 3.21. The critical values of Semenov number, ψ_c, as a function of the heat-of-reaction parameter B for various values of the reaction order, n: Adler and Enig criterion (solid curves); Eq. (3.50) (broken curves). (a) $\gamma \to \infty$; (b) $\gamma = 10$.

and Enig, 1964). Equation (3.48) is then modified to give

$$\psi_c = \frac{e^{-1}}{[1 - 2.52(n/B)^{2/3}]} \tag{3.52}$$

As shown in Fig. 3.22, this simple modification improves substantially the accuracy of the predicted ψ_c values. In particular, for $n = 1$, not only are the extreme values well reproduced but also those in the entire range of B [the maximum deviation between

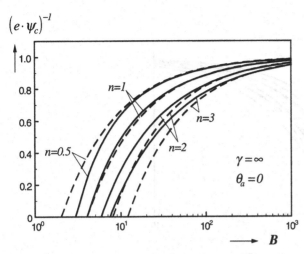

$(e \cdot \psi_c)^{-1}$

Figure 3.22. The critical values of Semenov number, ψ_c, as a function of B, for various values of the reaction order, n: Adler and Enig criterion (solid curves); Eq. (3.52) (broken curves). From Wu *et al.* (1998).

Eq. (3.52) and the AE criterion is less than 5%]. Accordingly, the approximate criterion (3.52) can be recommended for predicting the critical value ψ_c for first-order reactions and for large values of γ ($\gamma \rightarrow \infty$).

Morbidelli and Varma (1985) generalized the idea of Gray and Lee to the case of any reaction order by changing the constant 2.85 in Eq. (3.48) to $(B_c^o/n)^{2/3}$, thus leading to

$$\psi_c = \frac{e^{-1}}{\left[1 - \left(B_c^o/B\right)^{2/3}\right]} \qquad (3.53)$$

where

$$B_c^o = (1 + \sqrt{n})^2 \qquad (3.54)$$

B_c^o is again the critical B value for adiabatic reactors ($\psi_c \rightarrow \infty$). The results of Eq. (3.53) are compared in Fig. 3.23 with those given by the AE criterion (maximum error less than 5% for $n = 1, 2$ and less than 10% for $n = 0.5, 3$). This relation can then be recommended for the case of $\gamma \rightarrow \infty$ and any reaction order.

The approximate criterion (3.53) can be regarded as a correction of Semenov criterion (3.21) valid for $\gamma \rightarrow \infty$, through the introduction of the term $[1 - (B_c^o/B)^{2/3}]$ in the denominator. Then, Wu *et al.* (1998) apply the same correction to the Semenov criterion (3.16) valid for finite γ values. In the case of $\theta_a = 0$, this leads to

$$\psi_c = \frac{\theta_c}{\exp\left(\frac{\theta_c}{1 + \theta_c/\gamma}\right) \cdot \left[1 - \left(B_c^o/B\right)^{2/3}\right]} \qquad (3.55)$$

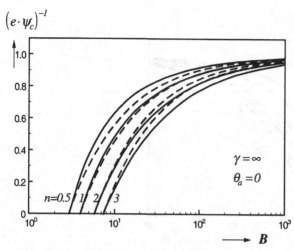

Figure 3.23. The critical values of Semenov number, ψ_c, as a function of B, for various values of the reaction order, n: Adler and Enig criterion (solid curves); Eq. (3.53) (broken curves). From Wu *et al.* (1998).

where θ_c is given by Eq. (3.18). For finite values of γ, Eq. (3.54) cannot be used to provide the critical value B_c^o for adiabatic reactors. The appropriate expression for B_c^o in this case has been derived by Morbidelli and Varma (1985):

$$B_c^o = \theta_c^o - \frac{n \cdot \theta_c^o \left(1 + \theta_c^o/\gamma\right)^2}{\left(1 + \theta_c^o/\gamma\right)^2 - \theta_c^o} \tag{3.56}$$

Note that Eq. (3.55) now involves two types of critical system temperature, θ_c and θ_c^o, introduced by Eq. (3.56). The former is the critical system temperature corresponding to the Semenov asymptotic region and is given by Eq. (3.18). The latter is the critical system temperature for adiabatic reactors, which is the solution of the following quartic equation (Morbidelli and Varma, 1985):

$$(n - 1)\left(\theta_c^o\right)^4 + 2\gamma(n - 1)(2 - \gamma)\left(\theta_c^o\right)^3 + [2(n - 1)(3 - \gamma)$$
$$- \gamma(\gamma - 2)]\gamma^2\left(\theta_c^o\right)^2 + 2[2(n - 1) + \gamma]\gamma^3\theta_c^o + (n - 1)\gamma^4 = 0 \tag{3.57}$$

in the range $\theta_c^o \in [\theta_-^o, \theta_+^o]$, where $\theta_\pm^o = [(\gamma - 2) \pm \sqrt{\gamma(\gamma - 4)}]\gamma/2$.

A comparison between the results of Eq. (3.55) and the AE criterion is shown in Figs. 3.24a and b for various values of γ and n. In all cases the agreement is satisfactory. The deviation increases as the reaction order decreases in the range $n \in (0, 3]$, but it always remains lower than 5% in the whole range of B values for $n \geq 1$. In addition, for the case $\gamma \to \infty$, Eq. (3.55) reduces to Eq. (3.53), and then Fig. 3.23 shows the accuracy of this relation for $\gamma \to \infty$. In conclusion, *Eq. (3.55) can be recommended*

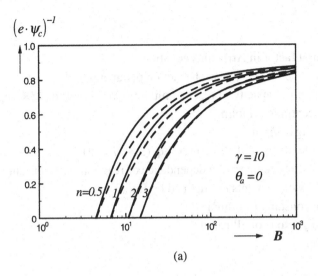

(a)

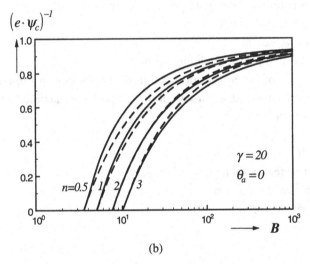

(b)

Figure 3.24. The critical values of Semenov number, ψ_c, as a function of B, for various values of n and γ: Adler and Enig criterion (solid curves); Eq. (3.55) (broken curves). (a) $\gamma = 10$; (b) $\gamma = 20$. From Wu *et al.* (1998).

as an approximate expression for the runaway boundaries in batch reactors, valid for all values of the involved physicochemical parameters.

It should be emphasized, however, that since Eq. (3.55) is only an approximate representation of the AE criterion, it retains the same limitations inherent with this criterion. Specifically, it fails to give a measure of the parametric sensitivity of the system. Thus, care should be used when it is applied to systems characterized by low B and γ values. In these cases, as discussed earlier, in order to assess the strength of the runaway phenomenon, one should use the sensitivity-based criteria, such as the MV, VR, or the SZ criterion.

Nomenclature

A	Pre-exponential factor in Arrhenius equation, $1/s$
B	$(-\Delta H) \cdot C^i \cdot \gamma / \rho \cdot c_v \cdot T^i$, heat-of-reaction parameter
c_v	Mean specific-heat capacity of reactant mixture, $J/(K \cdot mol)$ or $J/(K \cdot kg)$
C	Reactant concentration, mol/m^3
E	Activation energy, J/mol
F_i	Right-hand side of Eq. (3.4) ($i = 1$) or Eq. (3.5) ($i = 2$)
h	$\exp[\theta_c/(1 + \theta_c/\gamma)]$, temperature dependence of reaction rate constant
J	Jacobian matrix of elements a_{ij}, defined by Eq. (3.41)
k	Reaction rate constant, $(mol/m^3)^{1-n}/s$
L_j	Length of the jth axis of ellipsoid
n	Reaction order
P	Pressure, Pa
Q	Dimensionless parameter, defined by Eq. (3.32) or (3.34)
R_g	Ideal gas constant, $J/(K \cdot mol)$
R_v	Diameter of the vessel, m
S_v	External surface area per unit volume, m^2/m^3
$s(\theta^*; \phi)$	$\partial\theta^*/\partial\phi$, sensitivity of the temperature maximum θ^* to the input parameter ϕ
$S(\theta^*; \phi)$	$s(\theta^*; \phi) \cdot \phi/\theta^*$, normalized sensitivity of the temperature maximum θ^* to the input parameter ϕ
t	Time, s
T	Temperature, K
U	Overall heat transfer coefficient, $J/(m^2 \cdot s \cdot K)$
x	$1 - C/C^i$, reactant conversion
y	Dependent-variable vector

Greek Letters

δy	Perturbation of the nominal trajectory, defined by Eq. (3.40)
ΔH	Heat of reaction, J/mol
γ	$E/R_g \cdot T^i$, dimensionless activation energy
λ	Eigenvalue of the Jacobian matrix J
λ_j	jth one-dimensional Lyapunov exponent
λ_{max}	Eigenvalue of the Jacobian matrix J with the larger real part
θ	$\gamma \cdot (T - T^i)/T^i$, dimensionless temperature
θ^*	Temperature maximum
θ_+	$\exp[\theta/(1 + \theta/\gamma)]$, temperature-increase rate
θ_-	$(\theta - \theta_a)/\psi$, temperature-decrease rate
ρ	Density of the fluid mixture, kg/m^3
τ	$t \cdot k(T^i) \cdot (C^i)^{n-1}$, dimensionless time
ψ	$(-\Delta H) \cdot k(T^i) \cdot (C^i)^{n-1} \cdot \gamma / S_v \cdot U \cdot T^i$, Semenov number

Subscripts

a	Ambient condition
c	Critical condition
m	Quantity corresponding to the temperature maximum
ss	Stable state
us	Unstable state

Superscripts

o	Adiabatic condition
i	Initial condition

Acronyms

AE	Adler and Enig (1964)
MV	Morbidelli and Varma (1988)
SZ	Strozzi and Zaldivar (1994)
TB	Thomas and Bowes (1961)
VF	van Welsenaere and Froment (1970)
VR	Vajda and Rabitz (1992)

References

Adler, J., and Enig, J. W. 1964. The critical conditions in thermal explosion theory with reactant consumption. *Comb. Flame* **8**, 97.

Allen, A. O., and Rice, O. K. 1935. The explosion of azomethane. *J. Am. Chem. Soc.* **57**, 310.

Apin, A. Ya., Todes, O. M., and Kharitonov, Yu. B. 1936. Termicheskoe razlozhenie i vspyshka parov metilnitrata. *Zh. Fiz. Khim.* **8**, 866.

Bilous, O., and Amundson, N. R. 1956. Chemical reactor stability and sensitivity. II. Effect of parameters on sensitivity of empty tubular reactors. *A.I.Ch.E. J.* **2**, 117.

Boddington, T., Gray, P., Kordylewski, W., and Scott, S. K. 1983. Thermal explosions with extensive reactant consumption: a new criterion for criticality. *Proc. R. Soc. London A* **390**, 13.

Dalton, R. H. 1930. Oxidation of phosphine. *Proc. R. Soc. London. A* **128**, 263.

Emeleus, H. J., and Stewart, K. 1935. Oxidation of the silicon hydrides – I. *J. Chem. Soc.* 1182.

Emeleus, H. J., and Stewart, K. 1936. Oxidation of the silicon hydrides – II. *J. Chem. Soc.* 677.

Frank-Kamenentskii, D. A. 1939. Raspredelenie temperatur v reaktsionnom sosude i statsionarnaya teoriya teplovogo vzryva. *Zh. Fiz. Khim.* **13**, 738.

Gray, P., Griffiths, J. F., and Hasegawa, K. 1981. Nonisothermal decomposition of methyl nitrate: anomalous reaction order and activation energies and their correction. *Int. J. Chem. Kinet.* **13**, 817.

Gray, P., and Lee, P. R. 1965. Thermal explosions and the effect of reactant consumption on critical conditions. *Comb. Flame* **9**, 201.

Hadman, G., Thompson, H. W., and Hinshelwood, C. N. 1932. The explosive oxidation of carbon monoxide at lower pressures. *Proc. R. Soc. London A* **138**, 297.

Haldar, R., and Rao, D. P. 1992. Experimental studies on parametric sensitivity of a batch reactor. *Chem. Eng. Technol.* **15**, 34.

Hlavacek, V., Marek, M., and John, T. M. 1969. Modeling of chemical reactors – XII. *Coll. Czech. Chem. Commun.* **34**, 3868.

Kassoy, D. R., and Linan, A. 1978. The influence of reactant consumption on the critical conditions for homogeneous thermal explosions. *J. Mech. Appl. Math.* **31**, 99.

Kopp, D. I., Kovalskii, A. A., Zagulin, A. V., and Semenov, N. N. 1930. Ignition limits for the mixt. $2H_2 + O_2$ and $2CO + O_2$. *Z. Phys. Chem.* **B6**, 307.

Kontorova, T. A., and Todes, O. M. 1933. *Zh. Fiz. Khim.* **4**, 81.

Lacey, A. A. 1983. Critical behavior for homogeneous reacting systems with large activation energy. *Int. J. Eng. Sci.* **21**, 501.

Lewis, B., and von Elbe, G. 1961. *Combustion, Flames and Explosions of Gases*. New York: Academic.

Leicester, F. D. 1933. Ignition temperatures of mixtures of hydrogen sulfide, carbon disulfide and air. *J. Soc. Chem. Ind.* **52**, 341.

Melent'ev, P. V., and Todes, O. M. 1939. *Zh. Fiz. Khim.* **13**, 1594.

Morbidelli, M., and Varma, A. 1982. Parametric sensitivity and runaway in tubular reactors. *A.I.Ch.E. J.* **28**, 705.

Morbidelli, M., and Varma, A. 1985. On parametric sensitivity and runaway criteria of pseudo-homogeneous tubular reactors. *Chem. Eng. Sci.* **40**, 2165.

Morbidelli, M., and Varma, A. 1988. A generalized criterion for parametric sensitivity: application to thermal explosion theory. *Chem. Eng. Sci.* **43**, 91.

Ramsperger, H. C. 1927. Decomposition of azomethane. A homogeneous unimolecular reaction. *J. Am. Chem. Soc.* **49**, 912.

Rao, D. P., and Parey, S. K. B. 1988. Modeling and simulation of an exothermic reaction in a batch reactor. *Indian Chem. Eng.* **30**, 33.

Rice, O. K., Allen, A. O., and Campbell, H. C. 1935. The induction period in gaseous thermal explosion. *J. Am. Chem. Soc.* **57**, 2212.

Semenov, N. N. 1928. Zur theorie des verbrennungsprozesses. *Z. Phys.* **48**, 571.

Semenov, N. N. 1959. *Some Problems of Chemical Kinetics and Reactivity*. London: Pergamon.

Strozzi, F., Alos, M. A., and Zaldivar, J. M. 1994. A method for assessing thermal stability of batch reactors by sensitivity calculation based on Lyapunov exponents, experimental verification. *Chem. Eng. Sci.* **49(24B)**, 5549.

Strozzi, F., and Zaldivar, J. M. 1994. A general method for assessing the thermal stability of batch chemical reactors by sensitivity calculation based on Lyapunov exponents. *Chem. Eng. Sci.* **49**, 2681.

Thomas, P. H. 1961. Effect of reactant consumption on the induction period and critical condition for a thermal explosion. *Proc. R. Soc. London A* **262**, 192.

Thomas, P. H., and Bowes, P. C. 1961. Some aspects of the self-heating and ignition of solid cellulosic materials. *Br. J. Appl. Phys.* **12**, 222.

Todes, O. M. 1933. *Zh. Fiz. Khim.* **4**, 78.

Todes, O. M. 1939. *Zh. Fiz. Khim.* **13**, 868.

Todes, O. M., and Melent'ev, P. V. 1940. *Zh. Fiz. Khim.* **14**, 1026.

Vajda, S., and Rabitz, H. 1992. Parametric sensitivity and self-similarity in thermal explosion theory. *Chem. Eng. Sci.* **47**, 1063.

van Welsenaere, R. J., and Froment, G. F. 1970. Parametric sensitivity and runaway in fixed bed catalytic reactors. *Chem. Eng. Sci.* **25**, 1503.

von Elbe, G., Lewis, B., and Roth, W. 1955. *Fifth Symposium on Combustion.* New York: Reinhold.

Wolf, A., Swift, J. B., Swinney, H. R., and Vastan, J. A. 1985. Determining Lyapunov exponents from a time series. *Physica* **16D**, 285.

Wu, H., Morbidelli, M., and Varma, A. 1998. An approximate criterion for reactor thermal runaway. *Chem. Eng. Sci.* **53**, 3341.

4

Runaway in Tubular Reactors

B Y A TUBULAR REACTOR, we mean an empty tube in which a homogeneous reacting mixture flows. Typically, the reacting system may be noncatalytic, or catalytic where the catalyst is homogeneously solubilized. In some instances, fixed-bed reactors, where heterogeneous solid catalysts are used, may also behave as empty tubes. This is the case where the characteristic times of all mass and heat transports are very fast as compared to the chemical reactions, and the fixed bed is referred to as a *pseudo-homogeneous* tubular reactor.

A tubular reactor operating at steady-state conditions is a spatial system, and the system variables are functions of position within the reactor. There is a strong similarity between tubular and well-stirred batch reactors, temporal systems discussed in Chapter 3. In the latter case, we have seen that for exothermic reactions a temperature maximum may occur as the reaction proceeds in time, while in the former at steady-state conditions a temperature maximum may be exhibited at some location along the reactor, which is generally referred to as a *hot spot*. The magnitude of this hot spot should be bounded within specific limits because it may seriously affect reactor safety and performance.

The magnitude of the hot spot depends on the system parameters such as operating conditions, physicochemical properties, and reaction kinetics. For specific values of the system parameters, the hot spot may undergo large variations in response to relatively small changes in one or more of the operating conditions or parameters. In this case, the reactor is said to operate in the *runaway or parametrically sensitive* region. In practical applications, it is clearly desirable to avoid this operating region in the earlier stages of reactor design. This provides the motivation to develop *a priori* criteria that can identify runaway conditions in these reactors.

The concept of parametric sensitivity in tubular reactors was first introduced by Bilous and Amundson (1956). Since then, various criteria have been developed to distinguish, in the system parameter space, between safe and runaway operating regions. This material has been reviewed by Hlavacek (1970), Froment (1984), and Morbidelli and Varma (1985).

In this chapter we discuss the application of the *generalized criterion for runaway* introduced in Chapter 3 to tubular reactors. The predicted boundaries of the runaway regions in the system parameter space are compared with those given by earlier criteria. It is found that most of these criteria provide rather similar results when the hot spot is present inside the reactor. When the hot spot shifts to the reactor outlet, leading to the so-called pseudo-adiabatic operation, however, earlier criteria become either invalid or conservative with respect to the generalized one.

We first discuss in Section 4.2 the parametric sensitivity behavior of tubular plug-flow reactors with constant external cooling. This simple model provides an ideal example for developing runaway criteria. Moreover, in this case, it is often possible to derive explicit expressions for the boundaries of the runaway region that give a clear insight into the role of various system parameters. Results from the simple model also provide useful guidelines for understanding the parametric sensitivity problem in more complicated reactor models. Three major sources of model complexity are studied. In Section 4.3, the case in which the temperature of the cooling fluid changes along the reactor is analyzed. In Section 4.4, the influence of radial dispersion, which is particularly important in the case of nonisothermal, nonadiabatic reactors, on the critical conditions for runaway is discussed. Finally, in Section 4.5, the common case in applications in which more than one reaction occurs in the reactor is investigated.

4.1 Basic Equations for Tubular Plug-Flow Reactors

For a tubular plug-flow reactor (PFR) with cocurrent external cooling, where an exothermic, *n*th order reaction occurs, the steady-state mass and energy balances are represented by the following equations:

$$\frac{dC}{dl} = -\frac{\rho_B}{v^o} \cdot k(T) \cdot C^n \tag{4.1}$$

$$\frac{dT}{dl} = \frac{(-\Delta H)}{\rho \cdot c_p} \cdot \frac{\rho_B}{v^o} \cdot k(T) \cdot C^n - \frac{4U}{d_t \cdot v^o \cdot \rho \cdot c_p} \cdot (T - T_{co}) \tag{4.2}$$

$$\frac{dT_{co}}{dl} = \frac{\pi \cdot d_t \cdot t_n \cdot U}{c_{p,co} \cdot w_{co}} \cdot (T - T_{co}) \tag{4.3}$$

with the inlet conditions (ICs)

$$C = C^i, \qquad T = T^i \quad \text{and} \quad T_{co} = T_{co}^i \qquad \text{at } l = 0 \tag{4.4}$$

By introducing the following dimensionless variables

$$x = \frac{C^i - C}{C^i}; \qquad \theta = \frac{T - T^i}{T^i} \cdot \gamma; \qquad \theta_{co} = \frac{T_{co} - T^i}{T^i} \cdot \gamma; \qquad z = \frac{l}{L} \tag{4.5}$$

and dimensionless parameters

$$Da = \frac{\rho_B \cdot k(T^i) \cdot (C^i)^{n-1} \cdot L}{v^o}; \qquad B = \frac{(-\Delta H) \cdot C^i}{\rho \cdot c_p \cdot T^i} \cdot \gamma; \qquad \gamma = \frac{E}{R_g \cdot T^i};$$

$$St = \frac{4 \cdot U \cdot L}{d_t \cdot v^o \cdot \rho \cdot c_p}; \qquad \tau = \frac{\pi \cdot d_t^2 \cdot t_n \cdot v^o \cdot \rho \cdot c_p}{4 \cdot c_{p,co} \cdot w_{co}} \qquad (4.6)$$

we may write Eqs. (4.1) to (4.4) in dimensionless form:

$$\frac{dx}{dz} = Da \cdot \exp\left(\frac{\theta}{1 + \theta/\gamma}\right) \cdot (1 - x)^n = f_1(x, \theta, \phi) \qquad (4.7)$$

$$\frac{d\theta}{dz} = Da \cdot B \cdot \exp\left(\frac{\theta}{1 + \theta/\gamma}\right) \cdot (1 - x)^n - St \cdot (\theta - \theta_{co}) = f_2(x, \theta, \phi) \qquad (4.8)$$

$$\frac{d\theta_{co}}{dz} = \tau \cdot St \cdot (\theta - \theta_{co}) \qquad (4.9)$$

$$x = 0, \qquad \theta = \theta^i \quad \text{and} \quad \theta_{co} = \theta_{co}^i \qquad \text{at } z = 0 \qquad (4.10)$$

where ϕ indicates the vector of all the input parameters contained in the model (*i.e.*, B, Da, St, γ, n, τ, θ^i, and θ_{co}^i). From the above equations, an algebraic relation between reactor temperature, coolant temperature, and conversion can be readily established. Substituting Eqs. (4.7) and (4.9) into Eq. (4.8) leads to

$$\frac{d\theta}{dz} = B \cdot \frac{dx}{dz} - \frac{1}{\tau} \cdot \frac{d\theta_{co}}{dz} \qquad (4.11)$$

which, when integrated with the ICs (4.10), gives

$$\theta_{co} = \theta_{co}^i + [B \cdot x - (\theta - \theta^i)] \cdot \tau \qquad (4.12)$$

This indicates that the coolant temperature θ_{co} can be readily computed at any location along the reactor, once the values of conversion x and reactor temperature θ are known. Thus, a complete description of the system is obtained by using only two differential equations, Eqs. (4.7) and (4.8), where the coolant temperature is given by Eq. (4.12).

It is worth noting that the parameter τ represents the heat-capacity ratio between the reaction mixture and the external coolant. Thus, for very small values of τ, the classical case of external coolant at constant temperature is obtained, as it clearly appears from Eq. (4.12) where, as $\tau \to 0$, we have $\theta_{co} = \theta_{co}^i$. This is typically the case when the coolant undergoes a phase transition.

4.2 Plug-Flow Reactors with Constant External Cooling

4.2.1 Runaway Criteria

Plug-flow reactors with constant external cooling are the most widely studied in the literature. When taking conversion, instead of the axial coordinate, as the independent variable, *i.e.*, by dividing Eq. (4.8) by Eq. (4.7), the model reduces to a *single* equation

$$\frac{1}{B} \cdot \frac{d\theta}{dx} = 1 - \frac{St}{B \cdot Da} \cdot \frac{\theta - \theta_{co}}{(1 - x)^n} \cdot \exp\left(-\frac{\theta}{1 + \theta/\gamma}\right) \tag{4.13}$$

with the IC

$$\theta = \theta^i \quad \text{at } x = 0 \tag{4.14}$$

Thus, the tubular reactor model becomes identical to the batch reactor model, as given by Eq. (3.25) in Chapter 3 where the Semenov number ψ and θ_a are replaced, respectively, by

$$\psi = \frac{B \cdot Da}{St}, \qquad \theta_a = \theta_{co} \tag{4.15}$$

Accordingly, all the criteria for runaway in a batch reactor, discussed in Chapter 3, can be directly used also for tubular reactors with constant external cooling. Note that for the sensitivity-based, MV generalized (Morbidelli and Varma, 1988) and VR (Vajda and Rabitz, 1992) criteria, the sensitivity objective, given by the temperature maximum in the case of a batch reactor, is now obviously given by the hot-spot value along the reactor. In addition, the explicit (E) expression for the boundaries of the runaway regions developed in Chapter 3 [Eq. (3.54)] can be extended directly to the present case, using the redefinition of the Semenov number ψ given by Eq. (4.15), leading to

$$\frac{B \cdot Da}{St} = \frac{\theta_c}{\exp\left(\frac{\theta_c}{1+\theta_c/\gamma}\right) \cdot \left[1 - \left(B_c^o/B\right)^{2/3}\right]} \tag{4.16}$$

where θ_c is the critical temperature arising from the Semenov theory of thermal explosion, given by

$$\theta_c = \frac{\gamma}{2} \cdot [(\gamma - 2) - \sqrt{\gamma(\gamma - 4) - 4\theta_{co}}] \tag{4.17}$$

B_c^o is the critical B value for adiabatic reactors, given by the following expression (Morbidelli and Varma, 1985):

$$B_c^o = \theta_c^o - \frac{n \cdot \theta_c^o \left(1 + \theta_c^o/\gamma\right)^2}{\left(1 + \theta_c^o/\gamma\right)^2 - \theta_c^o} \tag{4.18}$$

Note that θ_c^o in Eq. (4.18) represents the critical temperature in the case of an adiabatic reactor. Thus, it is different from θ_c defined by Eq. (4.17) and is given by the solution of the following quartic equation (Morbidelli and Varma, 1985):

$$(n-1)(\theta_c^o)^4 + 2\gamma(n-1)(2-\gamma)(\theta_c^o)^3 + [2(n-1)(3-\gamma) - \gamma(\gamma-2)]$$
$$\times \gamma^2(\theta_c^o)^2 + 2[2(n-1)+\gamma]\gamma^3\theta_c^o + (n-1)\gamma^4 = 0 \qquad (4.19)$$

in the range $\theta_c^o \in [\theta_-^o, \theta_+^o]$, where $\theta_\pm^o = [(\gamma-2)\pm\sqrt{\gamma(\gamma-4)}]\gamma/2$. In some particular cases, Eq. (4.18) reduces to simple explicit relationships for B_c^o:

$$B_c^o = (1+\sqrt{n})^2 \quad \text{for } \gamma \to \infty \qquad (4.20a)$$

$$B_c^o = \frac{4 \cdot \gamma}{\gamma - 4} \quad \text{for } n = 1 \qquad (4.20b)$$

Some of the earlier runaway criteria were derived specifically for tubular plug-flow reactors with constant external cooling. The first of these was proposed by Barkelew (1959) in the case of a first-order irreversible reaction, based on a geometric property of the temperature trajectories under runaway conditions, deduced from an empirical analysis of a large number of numerical solutions of the model. A simplified temperature dependence of the reaction rate, based on the Frank-Kamenetskii approximation for large activation energies, was used. Dente and Collina (1964) proposed a criterion (DC) of intrinsic nature considering runaway as the situation where the temperature profile exhibits a region with positive second-order derivative somewhere before the hot spot in the temperature–axial coordinate phase plane. They define criticality as the first appearance of a region with positive second-order derivative before the hot spot, which corresponds to the conditions

$$\frac{d^2\theta}{dz^2} = \frac{d^3\theta}{dz^3} = 0 \qquad (4.21)$$

Although developed independently, this criterion is identical to that of Thomas and Bowes (1961) for batch reactors [see Eq. (3.24)] when time is simply replaced by axial coordinate. Van Welsenaere and Froment (1970) developed two intrinsic criteria (VF). The first one defines the criticality for runaway according to the locus of the temperature maxima in the temperature-conversion plane and, as discussed in detail in Chapter 3, is relatively conservative with respect to the other criteria mentioned above. The second one is similar to the Adler and Enig (1964) criterion (AE) for batch reactors, again replacing time with axial coordinate.

The intrinsic criterion (HP) proposed by Henning and Perez (1986) is based on the behavior of local sensitivity of reactor temperature to the inlet temperature:

$$s(\theta; \theta^i) = \frac{\partial\theta}{\partial\theta^i} \qquad (4.22)$$

whose value is obtained by integrating the system equations (4.7) to (4.8) over z, together with the following sensitivity equations, obtained by direct differentiation of Eqs. (4.7) and (4.8) with respect to θ^i:

$$\frac{ds(x;\theta^i)}{dz} = \frac{dx}{dz} \cdot \left[\frac{s(\theta;\theta^i)}{(1+\theta/\gamma)^2} - \frac{n \cdot s(x;\theta^i)}{1-x} \right] \tag{4.23}$$

$$\frac{ds(\theta;\theta^i)}{dz} = B \cdot \frac{ds(x;\theta^i)}{dz} - St \cdot s(\theta;\theta^i) \tag{4.24}$$

and ICs

$$s(x;\theta^i) = 0; \qquad s(\theta;\theta^i) = 1 \quad \text{at } z = 0 \tag{4.25}$$

Typical temperature profiles along the reactor are shown in Fig. 4.1a for various values of the Stanton number, St, while the corresponding sensitivity values of the system temperature to the inlet temperature are shown in Fig. 4.1b. It can be seen that a safe (nonrunaway) reactor operation, which from Fig. 4.1a appears to be obtained for St greater than 76, is characterized by sensitivity values that decrease monotonically with z (Fig. 4.1b). When runaway occurs, the $s(\theta;\theta^i)$ curve first decreases to reach a minimum and then increases with z. Thus, the HP criterion defines the critical condition as the first occurrence of a minimum on the $s(\theta;\theta^i) - z$ curve before the temperature maximum on the $\theta - z$ trajectory, which corresponds to

$$\frac{d^2 s(\theta;\theta^i)}{dz^2} = \frac{ds(\theta;\theta^i)}{dz} = 0 \tag{4.26}$$

For this criterion, similarly to the AE one, a trial-and-error procedure is generally required to compute the critical conditions.

Let us now compare the boundaries of the runaway regions predicted by the various criteria discussed above. These are shown in Fig. 4.2 for the case of a first-order reaction with $\gamma = \infty$ (Fig. 4.2a) and $\gamma = 10$ (Fig. 4.2b). The predictions of the Barkelew criterion are not shown in Fig. 4.2b because it is applicable only for very high γ values. It is seen that the criteria of HP and Barkelew are relatively conservative, although the most conservative one is the VF criterion. The predictions of the DC criterion are close to those of the AE, MV, and VR criteria. The only deviations occur in the region of low B values, which, as discussed in Chapter 3, is intrinsically insensitive thus making irrelevant the transition between safe operation and runaway. In the same figure, the values obtained using the E criterion (4.16) for the critical boundaries are also shown. It can be seen that this provides quite reasonable results, particularly close to those of the AE criterion. Thus, the use of this expression, which does not require any trial and error, can be particularly recommended in practical applications, at least as a first approximation.

Example 4.1 Runaway behavior in the naphthalene oxidation reactor. Naphthalene oxidation is a highly exothermic reaction, carried out commercially on a V_2O_5 catalyst in

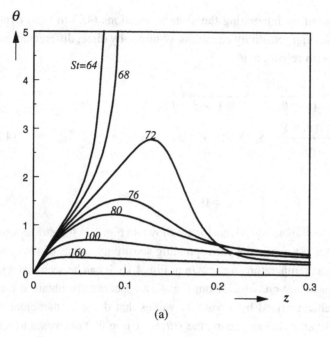

(a)

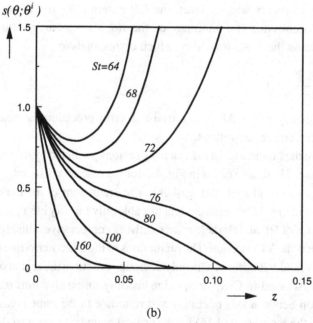

(b)

Figure 4.1. Profiles of (a) the system temperature θ and (b) its local sensitivity to the inlet temperature $s(\theta; \theta^i)$ along the reactor axis at various St values. The case with constant external cooling: $B = 40$; $\gamma = 10$; $Da = 1$; $n = 1$; $\theta_{co} = 0$.

$St/(DaB)$

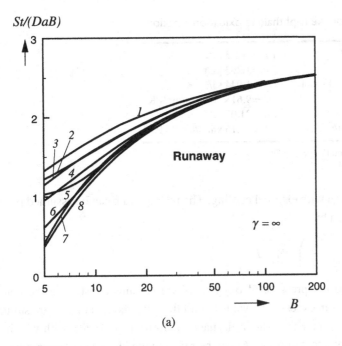

(a)

$St/(DaB)$

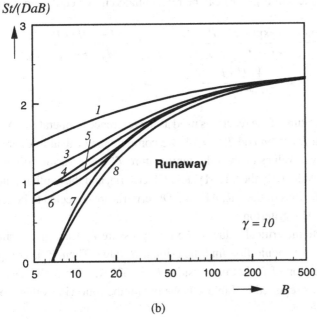

(b)

Figure 4.2. Critical conditions for runaway in the $St/(DaB) - B$ parameter plane for constant external cooling, predicted by various criteria: curve 1, van Welsenaere and Froment (1970); curve 2, Barkelew (1959); curve 3, Henning and Perez (1986); curve 4, Dente and Collina (1964); curve 5, Morbidelli and Varma (1988); curve 6, Vajda and Rabitz (1992); curve 7, Adler and Enig (1964); curve 8, the explicit [Eq. (4.16)]. $n = 1$; $Da = 1$; $\theta_{co}^{i} = 0$. (a) $\gamma = \infty$; (b) $\gamma = 10$.

Table 4.1. Data for the naphthalene oxidation reaction

$M_w = 29.48$ kg/kmol	$P_T = 101.3$ kPa
$\rho_B = 1300$ kg/m^3	$\rho = 1.293$ kg/m^3
$\Delta H = -1.284 \times 10^6$ kJ/kmol	$C_p = 1.044$ kJ/kg/K
$v^o = 1$ m/s	$U = 9.61 \times 10^{-2}$ kJ/m^2/s/K
$d_t = 0.025$ m	$P_O = 21.07$ kPa
$E = 1.134 \times 10^5$ kJ/kmol	$A = 11.16$ kmol/kg/s/(kPa)^2

From van Welsenaere and Froment (1970).

multitubular reactors with external cooling. The reaction rate can be assumed pseudo first-order and given by

$$r = A \cdot \exp\left(-\frac{E}{R_g \cdot T}\right) \cdot P_O \cdot P$$

where P_O is the partial pressure of oxygen, which remains constant in the reactor since this reactant is in excess. The values of all the involved parameters are summarized in Table 4.1, as given by van Welsenaere and Froment (1970), with which the dimensionless parameters in Eq. (4.6) can be reformulated in the equivalent form

$$Da = \frac{M_w \cdot P_T \cdot \rho_B \cdot A \cdot \exp(-\gamma) \cdot P_O \cdot L}{\rho \cdot v^o}; \qquad B = \frac{(-\Delta H) \cdot P^i \cdot \gamma}{M_w \cdot P_T \cdot T^i \cdot C_p}$$

$$\gamma = \frac{E}{R_g \cdot T^i}; \qquad St = \frac{4 \cdot U \cdot L}{d_t \cdot v^o \cdot \rho \cdot C_p}$$

where the inlet temperature of the reactor is used as the reference temperature. Assuming the reactor length $L = 2$ m and $T_{co} = T^i$, we compute the critical inlet pressure P_c^i for runaway at given values of the inlet temperature T^i using various criteria. In the following, we consider only the VF, MV, and HP criteria, and the explicit criterion (4.16), since the predictions of the VR, AE, and DC criteria are expected to be similar to those given by the MV criterion.

In order to compute the critical value of the inlet pressure P_c^i for a given value of the inlet temperature, we should first find the critical B value. Then P_c^i is computed from the above expression of B. Let us consider $T^i = 625$ K. With the values of all the other parameters as reported in Table 4.1, we obtain the following values for the dimensionless parameters:

$$\gamma = 21.8, \quad Da = 0.470, \quad St = 22.8 \ (St \text{ is independent of } T^i)$$

The application of the VF criterion has been described in Chapter 3. Upon substituting Eq. (4.15) into Eq. (3.35), the parameter Q is given by

$$Q = \sqrt{\frac{St}{Da}} \cdot \exp\left(-\frac{\theta_c}{1 + \theta_c/\gamma}\right)$$

where θ_c is the critical temperature arising from the Semenov theory of thermal explosion, as given by Eq. (4.17). Using the parameter values reported above and $\theta_{co} = 0$, we obtain $\theta_c = 1.104$ and $Q = 4.11$. With these, Eq. (3.33) gives the critical B value, $B_c = 24.3$, and then $P^i_{c,VF} = 1.69\,\text{kPa}$.

The HP criterion requires the integration of system equations (4.7) to (4.10), together with the sensitivity equations (4.23) to (4.25), coupled with a proper trial-and-error procedure to compute the critical B value. This leads to $B_c = 25.6$, which implies that $P^i_{c,HP} = 1.78\,\text{kPa}$.

In order to apply the explicit criterion (4.16), we should first evaluate the values of the parameters θ_c and B^o_c. The former is given by Eq. (4.17), while the latter for a first-order reaction is simply given by Eq. (4.20b), leading to $\theta_c = 1.104$ and $B^o_c = 4.90$. Then, the implicit algebraic equation (4.16) can be solved numerically leading to $B_c = 27.4$, and then $P^i_{c,E} = 1.90\,\text{kPa}$.

For the sensitivity-based, MV generalized criterion, we need to compute the normalized sensitivity of the temperature maximum to the inlet pressure, $S(\theta^*; P^i)$, as a function of the inlet pressure. This implies first to compute the local sensitivity $s(\theta; P^i)$ as a function of conversion by the direct differential method described in Chapter 2, i.e., by simultaneously solving the system equations (4.13) and (4.14) along with the corresponding sensitivity equations (3.40) to (3.42) given in Chapter 3. Then the value of $S(\theta^*; P^i)$ is computed from the value of $s(\theta; P^i)$ at $\theta = \theta^*$ through Eq. (3.39) in Chapter 3. The values of the normalized objective sensitivity and the temperature maximum are shown in Fig. 4.3 as functions of the inlet pressure. The critical inlet pressure is defined as that where the normalized objective sensitivity reaches its maximum, i.e., $P^i_{c,MV} = 1.85\,\text{kPa}$.

A significant, qualitative comparison among the various criteria applied above can be readily made by considering the behavior of the temperature maximum shown in Fig. 4.3. When the inlet pressure increases across the critical value $P^i_{c,MV}$, a very sharp increase of the temperature maximum occurs, thus indicating a parametrically sensitive behavior for the reactor. In Fig. 4.3 are also indicated the critical values of the inlet pressure given by the other criteria: VF, HP, and E. It is seen that both the VF and HP criteria (particularly the first one) are conservative relative to the MV generalized criterion. For the inlet pressure value indicated as critical by the VF and HP criteria, the temperature maximum is actually not very high, and it increases rather slowly with the inlet pressure. This behavior does not appear to be sensitive, leading to runaway. The explicit criterion (4.16) is closer to that given by the MV criterion. These findings are confirmed by the critical inlet pressure values computed for other inlet temperatures and summarized in Table 4.2.

4.2.2 The Region of Pseudo-Adiabatic Operation

Most of the criteria discussed above are based on the observation of reactor behavior in the temperature-conversion phase plane. In other words, in the reactor model, instead

Table 4.2. Critical values of the inlet pressure P_c^i for runaway in the naphthalene oxidation process obtained by different criteria

T^i (K)	$P_{c,VF}^i$ (kPa)	$P_{c,HP}^i$ (kPa)	$P_{c,MV}^i$ (kPa)	$P_{c,E}^i$ (kPa)
623	1.79	1.88	1.94	2.00
625	1.69	1.78	1.85	1.90
628	1.56	1.65	1.73	1.78
630	1.48	1.57	1.65	1.69

$P_{c,VF}^i$ from van Welsenaere-Froment criterion; $P_{c,HP}^i$ from Henning-Perez criterion; $P_{c,MV}^i$ from Morbidelli-Varma generalized criterion; $P_{c,E}^i$ from explicit criterion (4.16).

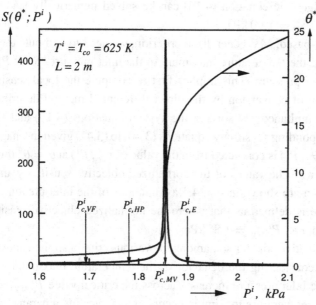

Figure 4.3. Behavior of the temperature maximum θ^* and its normalized sensitivity $S(\theta^*; P^i)$ as a function of the inlet pressure for naphthalene oxidation reaction. $P_{c,VF}^i$, $P_{c,HP}^i$, $P_{c,MV}^i$, and $P_{c,E}^i$ are values of the critical inlet pressure predicted by the VF, HP, MV, and the explicit criteria.

of the axial coordinate, the reactant conversion [Eqs. (4.13) and (4.14)] has been used as the independent variable. This implicitly assumes that the temperature maximum (hot spot) is always present in the reactor. Indeed, in the case of batch reactors, this is a safe approach, since, as time passes, conversion increases from 0 to 1; in the typical case where the coolant temperature does not exceed the initial temperature, a temperature maximum is always reached. For tubular reactors, such approach ignores the constraint imposed on the system behavior by the finite reactor length that limits conversion. For example, in the case of constant external cooling, since in practice $\theta^i > \theta_{co}$, for exothermic reactions, we see that the model using conversion as the independent variable always predicts the occurrence of a maximum in the temperature profile, which can be readily seen from Eqs. (4.13) and (4.14). This is indicated by $d\theta/dx < 0$ as $x \to 1$. On the other hand, when taking the axial coordinate as the independent variable, we may have the situation where no maximum occurs in the temperature profile, because of the specific length of the reactor considered. This operation, which is usually referred to as *pseudo-adiabatic operation (PAO)*, is indeed intrinsically safe for the reactor. However, in principle it is possible that if the reactor had been longer, the temperature maximum would have been reached and runaway would have occurred. In order to better understand the connection between these two situations, we next investigate how the pseudo-adiabatic conditions compare with those corresponding to runaway as predicted by criteria based on the behavior in the temperature-conversion phase plane.

Let us first discuss briefly the characteristics of PAO. The occurrence of PAO in the case of constant external cooling is shown clearly in Fig. 4.4, where for fixed values of B and all the other parameters, the temperature profiles as a function of the axial coordinate are shown for various values of St. The corresponding conversion values are shown in Fig. 4.5. In particular, in Fig. 4.4a it can be seen that for $St = 0.2$, the hot spot is located at about $z = 0.72$, and the corresponding temperature value is very high. In this case, conversion is practically complete right after the hot spot. Thus, the occurrence of a maximum in the temperature profile, which in the following will be referred to as *hot-spot operation (HSO)*, is due to the complete conversion of the reactant. In other words, in the portion of the reactor following the hot spot, heat generation is negligible while heat removal is continuing, thus leading to a decrease of the temperature value. Let us now increase, say, the wall heat-transfer coefficient, which corresponds to larger values of St. Since the temperature values along the reactor decrease, leading to lower reaction rates, a longer portion of the reactor is required to complete the conversion, and the hot spot moves toward the reactor outlet. When the St value increases further, a critical value is reached (1.48 in Fig. 4.4a) where the hot spot is washed out from the reactor and PAO is obtained. It is worth noting that in the PAO region, not only the temperature values in the reactor decrease as St increases, but also the shape of the temperature profile changes from concave (*e.g.*, $St = 2$ and 2.5 in Fig. 4.4a) to convex (*e.g.*, $St = 3$ and 3.5 in Fig. 4.4b), so that when St increases further, a second critical value is reached, *i.e.*, 3.66 in Fig. 4.4b, where a maximum appears again in the reactor temperature profile. However, in this region of reactor operating conditions, both the temperature value at the hot spot and the conversion at the reactor outlet are very low. Thus, this region of the HSO is characterized by a very low reaction rate in the reactor, and the occurrence of a temperature maximum is due to the heat removal rate overtaking the rate of heat generation even before significant reactant depletion.

In the example discussed above, we have seen that, by increasing the Stanton number, the system undergoes two transitions in the operation region, going from HSO to PAO and then back to HSO. By repeating this exercise for various values of the

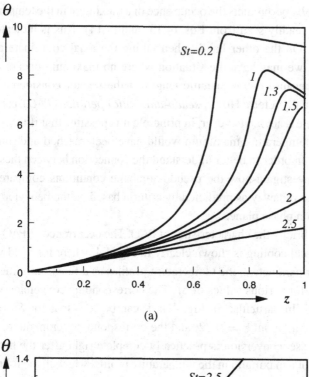

(a)

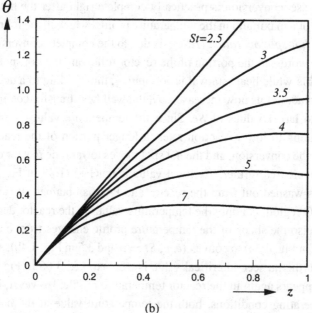

(b)

Figure 4.4. Profiles of the system temperature along the reactor axis for various values of St, indicating the occurrence of the pseudo-adiabatic operation (PAO) region. $Da = 0.2$; $B = 10$; $\gamma = 20$; $n = 1$; $\tau = 0$. From Wu *et al.* (1998).

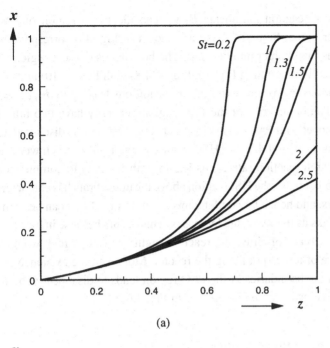

(a)

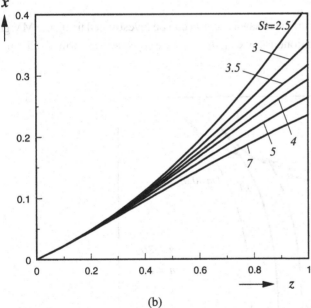

(b)

Figure 4.5. Profiles of the reactant conversion along the reactor axis for various values of St, corresponding to the conditions in Fig. 4.4. From Wu *et al.* (1998).

dimensionless heat-of-reaction parameter B, we can identify a region of PAO in the St–B parameter plane. Some of these are shown in Fig. 4.6 corresponding to various values of the Damkohler number, Da. The boundaries of these regions were obtained by numerically integrating Eqs. (4.7) and (4.8) with Eqs. (4.10) and (4.12), and using a trial-and-error procedure to find the conditions leading to $d\theta/dz = 0$ at the reactor outlet. The boundaries of the PAO region generally have two branches, one lower and one upper, and PAO occurs in the enclosed region. As discussed above in the context of Figs. 4.4 and 4.5, the HSO region to the right of the lower branch is characterized by high reaction rates, thus leading substantially to complete outlet conversion. On the other hand, the HSO region above the upper branch is characterized by low reaction rates, and hence low outlet conversion. In Fig. 4.6, it can be seen that the PAO region shrinks as the Da value increases. This occurs because increasing Da values implies the increase of either the reactor length or reaction rate, which both favor the appearance of a hot spot along the reactor. Moreover, the existence of two distinct branches of the boundaries of the PAO region tends to disappear at high Da values as, for example, is the case for $Da = 1$ in Fig. 4.6.

4.2.3 Influence of PAO on the Runaway Region

The influence of PAO on the runaway region can be investigated using the MV generalized criterion. As mentioned above, in this case we consider the normalized sensitivity

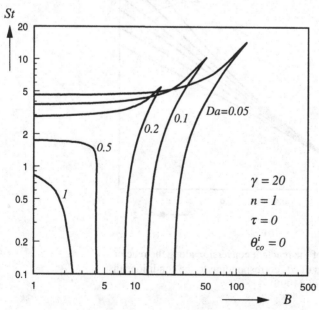

Figure 4.6. Pseudo-adiabatic operation (PAO) boundaries plotted in the St–B parameter plane for various values of Da, for the case with the external coolant at a constant temperature. From Wu *et al.* (1998).

of the temperature maximum, θ^*, defined as

$$S(\theta^*; \phi) = \frac{\phi}{\theta^*} \cdot \left(\frac{\partial \theta^*}{\partial \phi}\right) = \frac{\phi}{\theta^*} \cdot s(\theta^*; \phi) \tag{4.27}$$

where ϕ is one of the independent model parameters and $s(\theta^*; \phi)$ is the local sensitivity, $s(\theta; \phi)$, at the hot spot ($\theta = \theta^*$). The criticality for runaway is defined as the situation where the value of $S(\theta^*; \phi)$ is maximum or minimum. Note that when the reactor operates in the PAO region with no maximum in the temperature profile along the reactor, θ^* is taken as the temperature value at the reactor outlet.

We have seen earlier that when taking conversion as the independent variable, the tubular reactor model becomes identical to the batch reactor model, and then we can compute $s(\theta^*; \phi)$ by the direct differential method using the same sensitivity equations derived in Chapter 3, as shown in Example 4.1. When taking instead the axial coordinate as the independent variable, we can compute $s(\theta^*; \phi)$ using the following sensitivity equations over the axial coordinate z:

$$\frac{ds(x; \phi)}{dz} = \frac{\partial f_1}{\partial x} \cdot s(x; \phi) + \frac{\partial f_1}{\partial \theta} \cdot s(\theta; \phi) + \frac{\partial f_1}{\partial \phi} \tag{4.28}$$

$$\frac{ds(\theta; \phi)}{dz} = \frac{\partial f_2}{\partial x} \cdot s(x; \phi) + \frac{\partial f_2}{\partial \theta} \cdot s(\theta; \phi) + \frac{\partial f_2}{\partial \phi} \tag{4.29}$$

with ICs

$$s(x; \phi) = 0 \quad \text{and} \quad s(\theta; \phi) = \delta(\phi - \theta^i) \quad \text{at } z = 0 \tag{4.30}$$

together with the system equations (4.7), (4.8), (4.10), and (4.12). The above equations are obtained by differentiating Eqs. (4.7), (4.8), and (4.10) with respect to ϕ. The value of the normalized objective sensitivity $S(\theta^*; \phi)$ is computed from Eq. (4.27) with the value of $s(\theta; \phi)$ at $\theta = \theta^*$.

In the following we compare the regions corresponding to PAO with the regions corresponding to runaway predicted by the MV generalized criterion, using the axial coordinate (*i.e.*, z-MV) or the reactant conversion (*i.e.*, x-MV) as the independent variable.

Let us consider the case of a first-order reaction with $Da = 0.1$ and $\gamma = 20$. The PAO occurs in a large portion of the St–B parameter plane, as shown in Fig. 4.7, where curve 2, composed of two branches, represents the boundary of the PAO region. Moreover, in the same figure the critical conditions for runaway are reported as computed using various criteria. Curves 1 and 3 are the critical conditions predicted by the z-MV and x-MV criteria, respectively. Curves 4, 5, 6, and 7 are the results given, respectively, by the criteria of DC (Dente and Collina, 1964), HP (Henning and Perez, 1986), AE (Adler and Enig, 1964), and VF (van Welsenaere and Froment, 1970). It is seen that when the reactor operates in the HSO region (*i.e.*, to the right of •), all criteria predict substantially the same critical conditions for runaway. When the reactor operates in the PAO region, however, the predictions of all criteria deviate from one another. Moreover, the boundaries of the runaway region predicted by the

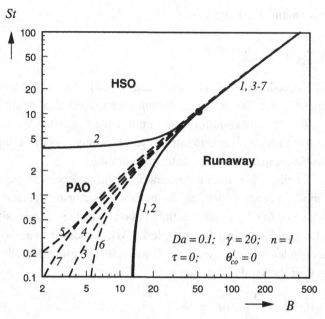

Figure 4.7. A comparison of the predicted runaway boundaries by various criteria in the case where the pseudo-adiabatic operation (PAO) occurs: 1 – z-MV; 3 – x-MV; 4 – DC; 5 – HP; 6 – AE; 7 – VF. Curve 2 corresponds to the PAO boundary. Symbol • indicates end of PAO region. From Wu *et al.* (1998).

z-MV criterion in this case are coincident with the lower branch of the PAO boundary, while those predicted by all the other criteria fall inside the PAO region.

Let us first discuss the differences in the predictions of the z-MV and x-MV criteria in the PAO region. For example, for $B = 20$ in Fig. 4.7, the critical St values predicted by the z-MV and x-MV criteria are 2.31 and 3.25, respectively. Note that under these conditions, both critical values exhibit the generalized feature, as shown in Figs. 4.8, and 4.9, where it can be seen that in both cases the normalized objective sensitivities to five different parameters (B, Da, St, γ, and n) reach a maximum or a minimum at the same value of St. Thus, the difference in the predictions of the two criteria is not due to an intrinsically insensitive behavior of the system in this region. The actual reason is the occurrence of the PAO region for the reactor. The z-MV criterion, by taking the axial coordinate as the independent variable, can properly account for this behavior. Thus it can resolve that the reactor is too short for developing a local temperature maximum, the temperature profile is monotonically increasing, and so the temperature value considered in the sensitivity analysis is that at the reactor outlet. When taking instead the reactant conversion as the independent variable, as in the x-MV criterion, we implicitly consider the possibility of complete conversion, which can only be assured in reactors of infinite length. This leads to one less parameter (*i.e.*, the reactor length) in the model. It follows that the x-MV criterion does not account

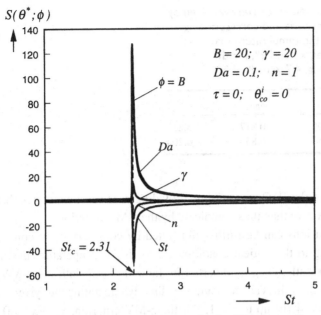

Figure 4.8. Normalized objective sensitivity $S(\theta^*;\phi)$ as a function of St for various choices of the parameter ϕ, for the case where axial coordinate is taken as the independent variable. From Wu *et al.* (1998).

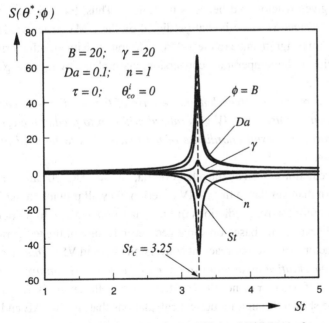

Figure 4.9. Normalized objective sensitivity $S(\theta^*;\phi)$ as a function of St for various choices of the system parameter ϕ, for the case where conversion is taken as the independent variable. From Wu *et al.* (1998).

Table 4.3. Critical St values for runaway given by the z-MV and x-MV criteria and corresponding conversion values at the temperature maximum and the reactor outlet, x_m and x^o, for the conditions shows in Fig. 4.7 with $B = 20$

Criterion	St_c	x_m	x^o
z-MV	2.306	0.817	0.817
x-MV	3.252	0.846	0.204

for the occurrence of the PAO region, thus providing runaway boundaries that are always more conservative than those predicted by the z-MV criterion.

The above conclusions can be further illustrated by considering the conversion values corresponding to the critical conditions, at both the temperature maximum (x_m) and the reactor outlet (x^o), as reported in Table 4.3. For both the z-MV and x-MV criteria, the outlet values are computed by directly integrating the system equations (4.7), (4.8), and (4.10) up to $z = 1$. For the z-MV criterion, $x_m = x^o = 0.817$, so that the temperature maximum is located at the reactor outlet, indicating that the critical condition belongs to the PAO region. For the x-MV criterion, $x_m = 0.846$, and $x^o = 0.204 < x_m$. A conversion value (x_m) at the temperature maximum higher than the conversion value (x^o) at the reactor outlet indicates that the temperature maximum occurs outside the given reactor, and hence is unrealistic. Thus, for a given reactor length, the relevant critical St value is that predicted by the z-MV criterion. However, we can retain some significance to the x-MV criterion, as with all other criteria based on the behavior in the temperature-conversion phase plane, by regarding it as a conservative estimate.

In summary, *for a reactor of given length, because of the possible existence of the PAO regime, when one uses the MV generalized criterion to predict the critical conditions for runaway, the axial coordinate of the reactor has to be used as the independent variable.*

In Fig. 4.7, the predictions given by the other criteria, *i.e.*, AE, DC, HP, and VF, are also shown. It is seen that, similar to the x-MV criterion, they all provide in the PAO regime exceedingly conservative predictions of the runaway boundary. In particular, the AE, DC, and VF criteria are based on some geometric feature of the temperature profile along the reactor axis (DC) or reactant conversion (AE and VF). *Thus, we can conclude that the conservative predictions are due to the geometric nature of these criteria*, independent of whether or not the constraint on reactor length is accounted for. In fact, it can be shown through numerical calculations that, for the AE and DC criteria (which are based on the first appearance of a region with positive second-order derivative *before* the temperature maximum), even though the temperature maximum is located outside the reactor in the case of PAO, the positive second-order derivative can indeed occur inside the reactor. A similar behavior is also exhibited by the HP criterion, which, as mentioned earlier, defines the critical condition for runaway as the first appearance of a minimum on the $s(\theta, \theta^i)$–z profile before the temperature maximum. Note that this criterion, although using the sensitivity concept, is still a geometry-based criterion, which is different from the MV generalized criterion that is instead based on the quantity (the maximum) of the normalized objective sensitivity.

Finally, note that, for large B values, as shown in Fig. 4.7, all criteria yield essentially similar results. This arises because, in such cases, runaway occurs soon after the reactor inlet, at negligible reactant conversion. Thus all criteria approach the classic criterion of Semenov [Eq. (3.16)].

Thus, summarizing, the z-MV criterion provides the best representation of the reactor runaway behavior, and this leads to the regions of reactor operation identified by curves 1 and 2 in Fig. 4.7. It appears that *three regimes are possible: two of them are safe and may either exhibit a temperature maximum along the reactor (HSO) or not (PAO), while the third regime corresponds to runaway and it always includes a temperature maximum.*

The shape of the reactor operation regions shown in Fig. 4.7 is typical for relatively low Da values. As the Da value increases, the PAO region shrinks and may even disappear, as shown in Fig. 4.10, where the PAO and runway boundaries are shown for various values of Da. This indicates that, for sufficiently large Da values, the transition from HSO to runaway occurs directly, without involving the PAO region,

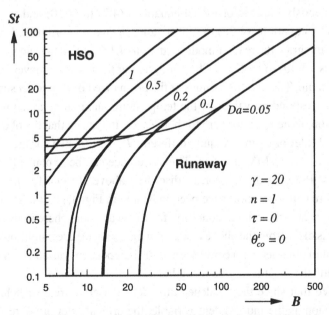

Figure 4.10. Runaway boundaries (thicker curves) predicted by the z-MV criterion are plotted in the St–B parameter plane for various Da values, together with the PAO boundaries (thinner curves) from Fig. 4.6. From Wu *et al.* (1998).

even at relatively low values of the heat-of-reaction parameter B. The reason is that for large Da values, high reactant conversion occurs, leading always to a temperature maximum inside the reactor. It should also be mentioned that, for relatively high Da, in the region of small B values, the system becomes intrinsically insensitive and the predicted critical boundary between safe and runaway operation has less physical meaning and accordingly the runaway criterion loses its generalized character. This leads to derivations between the PAO and the runaway boundaries that appear in Fig. 4.10, *e.g.*, in the case of $Da = 0.2$.

Example 4.2 Runaway behavior in a naphthalene oxidation reactor operating in the PAO region. Let us consider the naphthalene oxidation reactor discussed in Example 4.1, but with a residence time four times smaller. For this, let us take a reactor length $L = 1$ m (instead of $L = 2$ m) and a fluid velocity $v^o = 2$ m/s (instead of $v^o = 1$ m/s), without changing any other parameter. From the definition of the dimensionless parameters in Example 4.1, it follows that only the values of Da and St are affected by these changes, and in particular we now have $Da = 0.1175$ and $St = 5.7$. We now compute the critical value of the inlet pressure P_c^i for runaway at the inlet temperature $T^i = 625$ K by using both the z-MV and x-MV criteria.

As discussed in the previous example, we define the critical condition P_c^i as the inlet pressure value at which the normalized objective sensitivity $S(\theta^*; P^i)$ as a function of P^i reaches its maximum. We again use the direct differential method to compute the sensitivity values. The value of $S(\theta^*; P^i)$ for the z-MV criterion is obtained by integrating simultaneously the reactor model equations (4.7) to (4.10) and the corresponding sensitivity equations (4.28) to (4.30) along the axial coordinate. For the x-MV criterion, we integrate the reactor model equation (4.13) and (4.14) and the sensitivity given by Eqs. (3.40) to (3.42) in Chapter 3, taking the reactant conversion as the independent variable. The obtained values of the normalized objective sensitivity, together with the corresponding values of the temperature maximum, are shown in Fig. 4.11 as a function of the inlet pressure for both cases. It appears that the obtained values of the critical inlet pressure are quite different: $P_c^i = 1.93$ kPa for the z-MV and $P_c^i = 1.85$ kPa for the x-MV criterion. This difference can be explained owing to the occurrence of PAO for the reactor, as discussed above. To clarify, the three different regions of reactor operation have been indicated in Fig. 4.11. It is seen that the x-MV criterion locates the critical condition for runaway somewhere in the PAO region. However, it is clear from the above discussion that, under these conditions, the temperature maximum obtained with conversion as the independent variable (curve 4) is not achieved inside the reactor of given length.

It should be noted that since the residence time does not affect reactor behavior when using conversion as the independent variable, the critical inlet pressure value predicted by the x-MV criterion in both Figs. 4.11 and 4.3 are identical. This is readily confirmed by noting that, in the reactor model equations (4.13) and (4.14), the parameters Da and St do not appear independently, but rather always combined

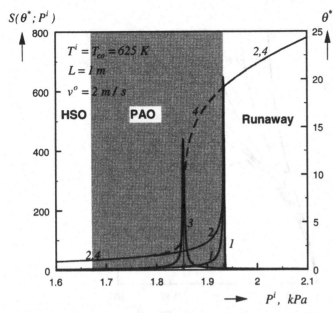

Figure 4.11. Sensitivity in naphthalene oxidation reactors. Values of the normalized objective sensitivity $S(\theta^*; P^i)$ as a function of the initial pressure obtained using either axial coordinate (curve 1) or reactant conversion (curve 3) as the independent variable. Curves 2 (z-MV) and 4 (x-MV) represent the corresponding values of the temperature maximum.

within the Semenov number ψ, Eq. (4.15), where the dependence on reactor length L and fluid velocity v^o cancel out.

4.3 Plug-Flow Reactors Varying Coolant Temperature

The case of external coolant at constant temperature considered in the previous section corresponds to $\tau = 0$ in Eq. (4.12), which requires an extremely large heat capacity $c_{p,co}$ or an extremely large coolant flow rate w_{co}. This condition is easily achieved at laboratory scale, but is difficult at industrial scale. In this section, we discuss the effects of temperature changes in the coolant, when it flows cocurrently with the reacting mixture, on both the reactor PAO and runaway regions. The basic equation for the reactor model are given by Eqs. (4.7) to (4.10), with $\tau > 0$.

4.3.1 The Regions of Pseudo-Adiabatic Operation

Let us consider the case of a first-order reaction with $\gamma = 20$ and $Da = 0.1$, whose PAO region in the St–B parameter plane was shown previously in Fig. 4.6 for constant external cooling, *i.e.*, $\tau = 0$. This has been reproduced in Fig. 4.12 and compared

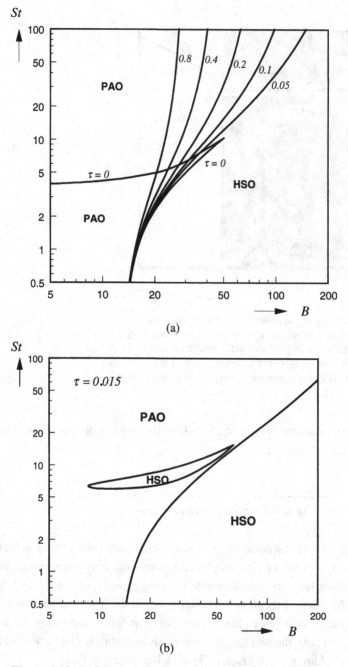

Figure 4.12. PAO boundaries in the $St–B$ parameter plane for various values of τ. $Da = 0.1$; $\gamma = 20$; $n = 1$; $\theta_{co}^i = 0$.

with the PAO regions computed for various values of $\tau > 0$. It is seen that the upper branch of the PAO boundary exists only for very small τ values, *i.e.*, $\tau < 0.04$ in this case. For $\tau \geq 0.05$, a different shape of the PAO regions is found, involving only one transition between PAO and HSO as St (B) increases for any constant value of B

(St). The transition between these two different shapes of the PAO regions is rather peculiar, such as the one shown in Fig. 4.12b for $\tau = 0.015$, where an HSO island is located inside the PAO region.

On physical grounds, this can be explained by considering that, as discussed in the previous section, the HSO region located above the upper branch of the PAO region for $\tau = 0$ is characterized by low reaction rates and relatively low outlet conversion. This is due to the strong heat removal rate (large St), which cools down the reacting mixture even before all the reactant is depleted. For $\tau > 0$, since the coolant temperature increases along the reactor axis, the driving force for heat removal is reduced, and consequently the internal temperature increases, leading to larger reaction rates. Accordingly, a PAO rather than an HSO is developed for the reactor. In other words, the decrease in the temperature driving force, arising when $\tau > 0$, requires larger St values for the same PAO to HSO transition to occur as observed at $\tau = 0$. Therefore, the upper branch of the PAO boundary moves upward as τ increases.

The occurrence of a PAO regime in the case of cocurrent external cooling ($\tau > 0$) was first identified by Soria Lopez et al. (1981) and then discussed in the context of orthoxylene catalytic oxidation by De Lasa (1983). Soria Lopez et al. (1981) also derived an explicit expression to predict the PAO boundary for a first-order reaction, which with our notation becomes

$$\frac{\tau \cdot St}{Da} = \exp\left(\frac{\theta_\infty}{1 + \theta_\infty/\gamma}\right) \tag{4.31}$$

where θ_∞ was considered as the limiting system temperature for PAO (namely, it is the temperature maximum corresponding to the PAO boundary) and was derived to have the form

$$\theta_\infty = \frac{\theta_{co}^i + (B + \theta^i) \cdot \tau}{1 + \tau} \tag{4.32}$$

The derivation of this expression is based on the consideration that, at the PAO boundary, the internal temperature at the reactor outlet approaches the coolant temperature, and the conversion is complete. This corresponds to the assumption of the reactor length $L \to \infty$. Note that this expression, as is also evident from Eq. (4.31), breaks down in the case of constant external cooling, i.e., $\tau = 0$, since it implies that at the PAO boundary the internal temperature will always be equal to the coolant temperature.

A comparison between the PAO regions calculated by Eq. (4.31) and those computed numerically in the τ–B parameter plane is shown in Fig. 4.13 for various values of Da. It is seen that, for large Da values, the explicit expression (4.31) indeed provides satisfactory results for a rather wide range of τ values. On the other hand, for lower values of Da, the predicted boundaries are accurate only at low τ values. The failure of the explicit criterion in predicting the PAO boundary at low Da and large τ values is due to the assumption that at the PAO boundary, the internal temperature at the reactor outlet approaches the coolant temperature. In fact, when a reactor is

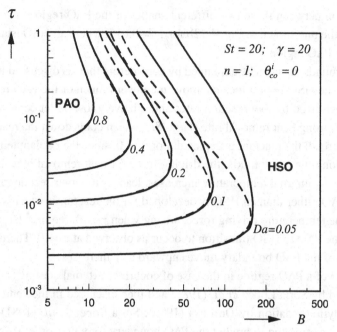

Figure 4.13. PAO boundaries in the τ–B parameter plane, calculated numerically (solid curves) or predicted by the explicit expression derived by Soria Lopez *et al.* (1981) (broken curves), for various values of *Da*. From Wu *et al.* (1998).

short (small *Da*) and heat capacity of the coolant is limited (large τ), as confirmed by our numerical calculations, the internal temperature at the reactor outlet can be substantially different from the external coolant temperature.

Note that, in Fig. 4.13, similarly to the *St*–*B* parameter plane shown in Fig. 4.7, the PAO boundaries in the τ–B parameter plane have two branches (lower and upper) with respect to *B*. However, in this case the HSO region below the lower branch is characterized by low reaction rates and relatively low outlet conversion, while the HSO region above the upper branch is characterized by high reaction rates and essentially complete outlet conversion.

4.3.2 Influence of PAO on Runaway Regions

The influence of cocurrent external cooling on reactor runaway behavior was first investigated by Soria Lopez *et al.* (1981) and later by Hosten and Froment (1986) through an extension of the VF criterion for constant external cooling. Similarly, Henning and Perez (1986) extended the HP criterion to this case, while Bauman *et al.* (1990) adopted the *x*-MV criterion and Wu *et al.* (1998) the *z*-MV criterion. According to the conclusions reached in Section 4.2.3 for reactors with constant external cooling, in the following we will focus on the application and performance of the z-MV criterion. Comparisons of the predictions of this criterion with those of the others will also be discussed.

Let us first compare the predictions given by the z-MV and the x-MV criteria. Figure 4.14 shows the computed runaway boundaries in the $St-B$ parameter plane for various τ values, where the solid and broken curves denote the results given by the z-MV and the x-MV criteria, respectively. Note that the corresponding PAO boundaries are shown in Fig. 4.12, and coincide with the runaway boundaries given by the z-MV criterion shown in Fig. 4.14.

In the case of constant external cooling shown in Fig. 4.7, we have seen that the z-MV and the x-MV criteria give the same predictions for the runaway boundaries, when the adjacent safe reactor operation is in the HSO regime. On the other hand, when the adjacent safe region is in the PAO regime, the x-MV criterion gives conservative results. In the case of cocurrent external cooling, it is seen from Fig. 4.14a that the transition from safe to runaway operation occurs always when the reactor is in the PAO regime, and hence the predictions of the two criteria are always rather different. Similarly to the case of constant external cooling, this behavior is due to the constraint imposed by the finite reactor length, which is accounted for by the z-MV criterion but ignored by the x-MV criterion. This is confirmed by the observation that, for all the conditions corresponding to criticality for runaway predicted by the x-MV criterion, the temperature maximum occurs for conversions larger than the reactor outlet values. It is worth pointing out that the differences shown in Fig. 4.14a are related to the specific operating conditions employed. For example, by increasing values of the reactor length and maintaining all the other parameters fixed, the results of the z-MV criterion approach those given by the x-MV criterion. In particular, by increasing the value of St while keeping the ratio St/Da constant, the critical value of the heat-of-reaction parameter B predicted by the z-MV criterion decreases and approaches the value given by the x-MV criterion, which itself remains unchanged.

A further confirmation of the behavior of the predictions of the two criteria is shown in Fig. 4.14b, where, although reduced to a small island, a region exists where safe HSO is possible. It is clear that the critical St values given by the z-MV and the x-MV criteria are close and in fact coincide when the runaway boundary is close to the HSO regime. In conclusion, for reactors with cocurrent external cooling, even more than for those with constant external cooling, the use of the z-MV criterion is recommended. In other words, *the z-MV criterion should always be used to predict critical conditions for runaway, for reactors of finite length.*

In general, and particularly in the case of constant external cooling, the critical conditions for runaway are investigated in the $St-B$ or in the St/DaB (*or* ψ)$-B$ parameter plane. Here the boundaries between safe and runaway behavior provide the critical wall heat transfer rate as a function of the heat generation rate. In the case of cocurrent external cooling, the representation of the runaway regions in the $\tau-B$ parameter plane with a fixed value of St may be more useful in practical applications.

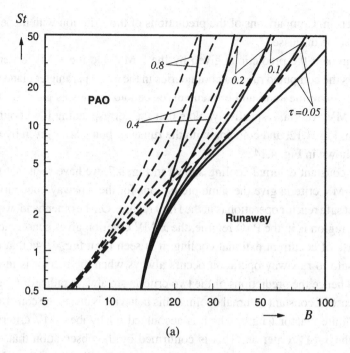

(a)

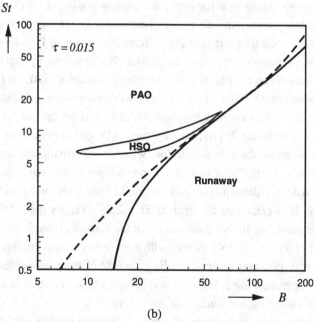

(b)

Figure 4.14. Runaway boundaries in the St–B parameter plane, predicted by the z-MV (solid curves) and the x-MV criteria (broken curves) for various values of τ. Solid curves corresponds also to the PAO boundaries. $Da = 0.1$; $\gamma = 20$; $n = 1$; $\theta_{co}^i = 0$. From Wu et al. (1998).

The critical value of the parameter τ indicates just how the heat capacity of the external coolant should be chosen in order to prevent the reactor from runaway for given heat generation and wall heat-transfer rates. Figures 4.15a, b, and c illustrate the runaway boundaries given by the various criteria in the τ–B parameter plane, together with the corresponding PAO boundaries, for various values of Da. Note that the PAO boundaries in this case have two branches (lower and upper) with respect to B, which, as mentioned above, separate the PAO regime from two different types of HSO. The one below the lower branch (low τ values) is characterized by low reaction rates and low outlet conversion, while the HSO above the upper branch is characterized by high reaction rates and essentially complete outlet conversion. In general, we see that, below the bifurcation point of the PAO boundary (curve 7), the critical conditions for runaway predicted by the various criteria are rather similar. These are given by the curve emerging from the bifurcation point and going toward low τ values (this is actually a short curve, which is best evident in Fig. 4.15c). However, in the region above the bifurcation point of the PAO boundary, the critical conditions given by all criteria are substantially different. These results are similar to those discussed for reactors with constant external cooling, with reference to the St–B parameter plane, shown in Fig. 4.7. In particular, we see again that when the transition from safe to runaway operation occurs with the reactor operating in the HSO (safe) regime, all criteria are substantially in agreement. On the other hand, if the reactor is in the PAO regime, then different criteria provide different predictions of the runaway conditions.

Moreover, in Fig. 4.15a, above the bifurcation point of the PAO boundary, the runaway boundary predicted by the z-MV criterion (curve 1) coincides with the upper branch of the PAO boundary, while those given by all the other criteria are located inside the PAO region, indicating exceedingly conservative conditions.

As the Da value increases as shown in Figs. 4.15b and c, the PAO and all the predicted runaway boundaries move toward lower B values, where the runaway phenomenon becomes intrinsically less intensive. As a result, the runaway boundaries given by the various criteria become less reliable. In Fig. 4.15c, the runaway boundary predicted by the z-MV criterion (curve 1) also tends to deviate from the PAO boundary (curve 7) and to move inside the PAO region. Moreover, for high values of τ in Fig. 4.15c, the critical B values predicted by the z-MV criterion become smaller than those predicted by the x-MV, $i.e.$, more conservative.

In order to understand better the last phenomenon observed above, let us consider the runaway boundaries predicted by both the x-MV and the z-MV criteria for a further increased Da value ($Da = 0.4$), as shown in Fig. 4.16a. It is found that, in this case, the z-MV criterion predicts two independent runaway boundaries (curves 1): one located in the PAO region and another in the HSO region that coincides with that predicted by the x-MV criterion (curve 2). The development of the double runaway boundaries for the z-MV criterion can be understood from Fig. 4.16b, where the profiles of the normalized objective sensitivity $S(\theta^*; B)$ as a function of B are shown for various values of τ. For the given set of parameters in Fig. 4.16a, it can be shown from

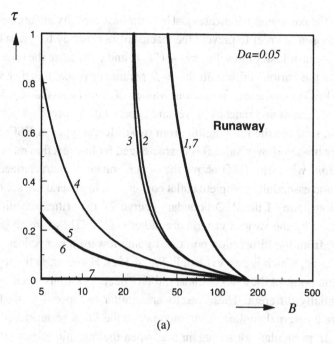

(a)

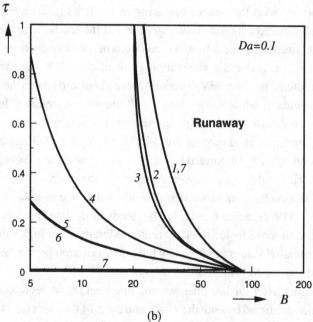

(b)

Figure 4.15. Comparison of the runaway boundaries predicted by various criteria in the τ–B parameter plane: 1 – z-MV; 2 – x-MV; 3 – VF; 4 – AE; 5 – HP; 6 – DC. Curves 7 indicate the PAO boundaries. $St = 20$; $\gamma = 20$; $n = 1$; $\theta_{co}^{i} = 0$. (a) $Da = 0.05$; (b) $Da = 0.1$; (c) $Da = 0.2$. From Wu *et al.* (1998).

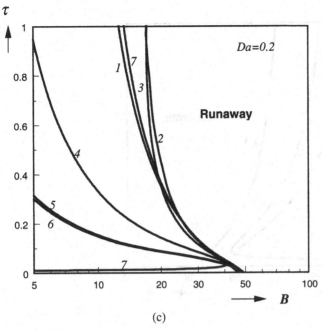

(c)

Figure 4.15. (cont.)

Fig. 4.10 that at $\tau = 0$ the reactor operates in the HSO region for any given value of B, and hence the temperature maximum θ^* for defining the normalized objective sensitivity $S(\theta^*; B)$ for both the z-MV and the x-MV criteria is the same. Thus, the $S(\theta^*; B)$ profile for $\tau = 0$ in Fig. 4.16b is for both the z-MV and the x-MV criteria, where the sharp sensitivity peak gives the critical B value for runaway. As τ increases, the PAO appears in the region of low B values. However, for each τ value if we keep computing the $S(\theta^*; B)$–B profile for B values greater than that corresponding to the PAO boundary, the sensitivity profiles for both the z-MV and the x-MV criteria are still identical, and yield the second peak in the case of $\tau = 0.1, 0.6$, or 1 in Fig. 4.16b. As a result, we obtain a merged runaway boundary (the solid curve 1, 2) that is located in the HSO region in Fig. 4.16a. On the other hand, if we compute the $S(\theta^*; B)$–B profile inside the PAO region for the z-MV criterion, we can also obtain another peak for each τ value. This peak corresponds to the first one in the case of $\tau = 0.1, 0.6$, or 1 in Fig. 4.16b, which arises from the sensitivity behavior of the reactor outlet temperature. This leads to the broken curve 1 located inside the PAO region in Fig. 4.16a. Since the temperature maximum at the reactor outlet cannot be identified by the reactor model with conversion as the independent variable, the broken curve in Fig. 4.16a cannot be produced by the x-MV criterion.

When two peaks in the $S(\theta^*; B)$–B curve exist, the first, which arises in the PAO region and occurs at low B values, is generally also of low magnitude. On the other hand, the second peak, in the HSO region, is sharper, indicating significant sensitive behavior. Thus the second peak is the true indicator of runaway behavior. It should

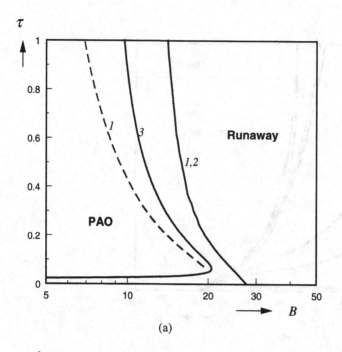

(a)

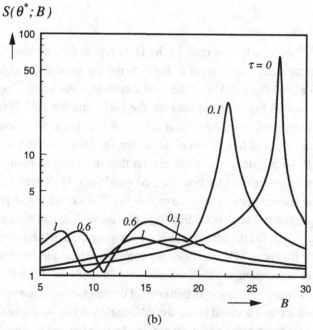

(b)

Figure 4.16. (a) Runaway boundaries predicted by z-MV (curves 1) and x-MV (curve 2) criteria and PAO boundary (curve 3) in the τ–B plane. (b) Values of the normalized objective sensitivity $S(\theta^*; B)$ as a function of B for various values of τ in (a). $Da = 0.4$. All the other parameters are as given in Fig. 4.15. From Wu *et al.* (1998).

also be mentioned that, for the runaway boundary in the case of high τ and Da values, where the critical B value is low, the reactor operates in a parametrically insensitive region, characterized by relatively low values of the sensitivity maximum, as shown in Fig. 4.16b. In such cases, one cannot define generalized boundary indicating a transition between runaway and safe operations.

Finally, we should mention the opposite mode of external cooling relative to the one considered in this section, *i.e.*, the countercurrent flow between coolant and reacting stream. Of course, the sensitivity behavior of reactors with countercurrent external cooling is quite different from the one described above for cocurrent external cooling. The situation is further complicated by the possible occurrence of multiple steady states (cf. Luss and Medellin, 1972), which are not possible for constant and cocurrent external cooling. However, the literature studies for this mode are relatively few (cf. Akella and Lee, 1983; Akella *et al.*, 1985), and do not permit one to draw a complete picture about the sensitivity behavior of these reactors. As a general comment, we note that the application of the countercurrent mode of cooling is generally aimed at a better utilization of the heat generated in the reactor, to preheat the feed. When compared with the cocurrent mode, the countercurrent mode, as observed by Degnan and Wei (1979), leads to greater possibility of reactor runaway, since near the reactor inlet the temperature difference between the internal fluid and the external coolant generally reaches a minimum, while the reaction and heat generation rates are at their maximum. Thus, when the parametric sensitivity problem is crucial, the external cooling should flow cocurrently rather than countercurrently to the reacting mixture.

4.4 Role of Radial Temperature and Concentration Gradients

In previous sections, the critical conditions for reactor runaway have been discussed using the one-dimensional pseudo-homogeneous model, where the effects of both axial and radial dispersions are neglected. Indeed, in studies of reactor simulation, the reliability of the one-dimensional pseudo-homogeneous model has been well assessed in many practical applications. For example, Carberry and Wendel (1963) have shown that, for flow velocities typically used in industrial practice, the effect of axial dispersion of heat and mass on outlet conversion can be neglected when the bed length exceeds about 50 particle diameters. Radial dispersion of mass usually has little effect on isothermal reactor performance (Froment, 1967; Carberry and White, 1969), as does the radial dispersion of heat in adiabatic reactors. In the case of externally cooled reactors, however, radial gradients of temperature and concentration may be severe and play a significant role in the reactor behavior. In this case, Finlayson (1971) has suggested that if the radial temperature profile is estimated through a one-point collocation procedure, the system behavior may be well approximated using a one-dimensional model. Yet, when investigating the reactor behavior in the vicinity of parametrically sensitive regions, where even small inaccuracies have a significant effect, it may be expected that axial and radial dispersion phenomena are important. Accordingly, in this section we discuss the effect of radial dispersion on the predicted critical conditions for runaway. The role of axial dispersion will be discussed in Section 5.3 of Chapter 5, together with the sensitivity behavior of continuous-flow stirred tank reactors (CSTRs) and PFRs, because it is well known that the axial dispersion model represents an intermediate situation between two opposite extremes, CSTRs and PFRs (Varma and Aris, 1977).

When radial dispersion is considered, the reactor model becomes two dimensional. For a single irreversible reaction with cocurrent external cooling, the steady-state mass and energy equations may be written in dimensionless form as follows:

$$\frac{\partial x}{\partial z} = \frac{\kappa}{Pe_m} \cdot \left(\frac{\partial^2 x}{\partial y^2} + \frac{1}{y} \cdot \frac{\partial x}{\partial y} \right) + Da \cdot (1-x)^n \cdot \exp\left(\frac{\theta}{1+\theta/\gamma} \right) \tag{4.33}$$

$$\frac{\partial \theta}{\partial z} = \frac{\kappa}{Pe_h} \cdot \left(\frac{\partial^2 \theta}{\partial y^2} + \frac{1}{y} \cdot \frac{\partial \theta}{\partial y} \right) + B \cdot Da \cdot (1-x)^n \cdot \exp\left(\frac{\theta}{1+\theta/\gamma} \right) \tag{4.34}$$

$$\frac{\partial \theta_{co}}{\partial z} = \tau \cdot \left(2 \cdot Bi \cdot \frac{\kappa}{Pe_h} \right) \cdot (\theta|_{y=1} - \theta_{co}) \tag{4.35}$$

with the inlet and boundary conditions

$$x = 0, \quad \theta = \theta^i, \quad \theta_{co} = \theta_{co}^i \qquad \text{at } z = 0 \tag{4.36}$$

$$\frac{\partial x}{\partial y} = \frac{\partial \theta}{\partial y} = 0 \qquad \text{at } y = 0 \tag{4.37a}$$

$$\frac{\partial x}{\partial y} = 0, \quad \frac{\partial \theta}{\partial y} = -Bi \cdot (\theta - \theta_{co}) \quad \text{at } y = 1 \tag{4.37b}$$

where

$$Pe_m = \frac{v^o \cdot d_t}{2 \cdot D_r}; \qquad Pe_h = \frac{v^o \cdot d_t \cdot \rho \cdot c_p}{2 \cdot \lambda_r}; \qquad \kappa = \frac{2 \cdot L}{d_t};$$

$$Bi = \frac{d_t \cdot h_w}{2 \cdot \lambda_r}; \qquad y = \frac{2 \cdot r}{d_t} \tag{4.38}$$

With the current capability of computers, solving the above two-dimensional model equations is no longer a problem. However, when conducting the sensitivity analysis, one needs to solve the model and sensitivity equations simultaneously, which increases computational requirements. Thus, appropriate simplifications of the above two-dimensional model are welcome for practical applications.

Hagan *et al.* (1988b) were the first to investigate the effect of radial dispersion on reactor runaway conditions. They restricted their investigation to the case of $t_{dif}/t_{reac} \ll 1$, where t_{dif} is the timescale on which heat is removed from the reactor by transfer to the cooling medium and t_{reac} is the timescale on which the reaction occurs. Based on the notation above, this implies that $t_{dif}/t_{reac} = Da \cdot Pe_h/\kappa \ll 1$. In the case of constant external cooling, with this restriction, Hagan *et al.* (1988a) derived the following one-dimensional model to approximate the above two-dimensional model:

$$\frac{dx}{dz} = Da \cdot \exp\left(\frac{\bar{\theta}}{1 + \bar{\theta}/\gamma}\right) \cdot (1 - x)^n \tag{4.39}$$

$$\frac{d\bar{\theta}}{dz} = B \cdot Da \cdot \exp\left(\frac{\bar{\theta}}{1 + \bar{\theta}/\gamma}\right) \cdot (1 - x)^n - \alpha \cdot \frac{8\kappa}{Pe_h} \cdot (1 + \bar{\theta}/\gamma)^2 \tag{4.40}$$

with the inlet conditions

$$x = 0, \qquad \bar{\theta} = \theta^i \quad \text{at } z = 0 \tag{4.41}$$

where α is a heat-loss parameter defined by the following implicit equation:

$$\frac{\bar{\theta} - \theta_{co}}{(1 + \bar{\theta}/\gamma)^2} = \frac{4 \cdot \alpha}{Bi} - \ln(1 - \alpha) - \frac{(1 + \bar{\theta}/\gamma)}{3 \cdot \gamma} \cdot \ln^2(1 - \alpha) \tag{4.42}$$

Thus, solving Eqs. (4.39) to (4.41) requires solving the algebraic equation (4.42) at each location z to obtain α, whose value is generally in the range $0 < \alpha < 0.5$. For this reason, this approximate one-dimensional model is usually referred to as the α *model*. Note that the last term in Eq. (4.42) is generally very small and can be omitted, thus leading to

$$\frac{\bar{\theta} - \theta_{co}}{(1 + \bar{\theta}/\gamma)^2} = \frac{4 \cdot \alpha}{Bi} - \ln(1 - \alpha) \tag{4.42'}$$

The obtained $\bar{\theta}(z)$ is an average value of the radial temperature at any given axial location, z. The approximate radial temperature profile at each z can be obtained from $\bar{\theta}(z)$ through the following explicit expression:

$$\theta(z, y) = \theta_{co} + [1 + \bar{\theta}(z)/\gamma]^2 \cdot \left[\frac{4 \cdot \alpha}{Bi} - 2 \cdot \ln(1 - \alpha + \alpha \cdot y^2)\right] \tag{4.43}$$

Figure 4.17 shows a comparison between the hot-spot values in the reactor predicted by the α model and the exact two-dimensional and the standard one-dimensional plug-flow ($Pe_m = Pe_h = \infty$) models. It is apparent that the predictions of the α model are excellent. Significant errors occur only at low B values. The average temperature and conversion profiles along the reactor axis given by the α model are also close to those given by the exact two-dimensional model, as shown in Fig. 4.18. The largest errors occur downstream near the hot spot. Further investigations by Hagan *et al.* indicate that the accuracy of the α model improves as B and γ increase and as the reaction order decreases. Thus, it may be concluded that the α model can be applied to substitute for the exact two-dimensional model in predicting the critical conditions for runaway, particularly when parametric sensitivity becomes most severe, *i.e.*, for large B and γ values with low reaction order.

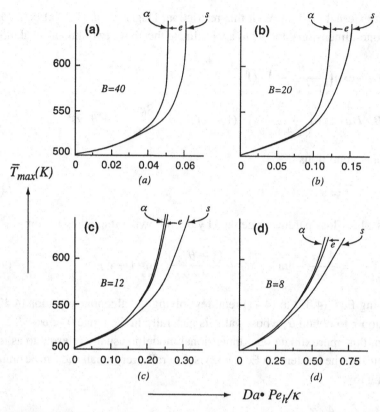

Figure 4.17. Hot-spot value in the reactor as a function of $Da \cdot Pe_h/\kappa$ for various B values predicted by different reactor models: the α model (α); the exact two-dimensional model (e); the standard one-dimensional model (s). $\gamma = 10$; $Bi = 4.8$; $Le = 1.25$; $n = 1$; $T^i = T_{co} = 500\,K$. From Hagan et al. (1988a).

The model indicated above as the standard one-dimensional model is that given by Eqs. (4.7) to (4.10) in the case of constant external cooling, where the Stanton number is computed by

$$St = \frac{\kappa}{Pe_h} \cdot \frac{6 \cdot Bi}{3 + Bi} \tag{4.44}$$

according to the one-point orthogonal collocation approximation proposed by Finlayson (1971). As can be seen in Figs. 4.17 and 4.18, the conversion, the hot-spot value and the temperature profile along the reactor axis are significantly different from those given by the exact two-dimensional model. Therefore, caution should be applied in using this procedure for reactor parametric sensitivity studies.

Although the one-dimensional α model is accurate in approximating the two-dimensional model, the same authors (Hagan et al., 1988b) applied it using an in-effective criterion to identify the conditions for reactor runaway. Their criterion is geometry based and an extension of the Semenov criterion, with some additional

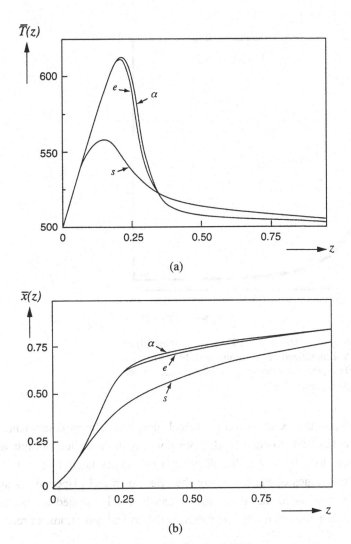

Figure 4.18. Profiles of (a) temperature and (b) conversion along the reactor axis predicted by different reactor models: the α model (α); the exact two-dimensional model (e); the standard one-dimensional model (s). Same conditions as in Fig. 4.17, with $B = 20$ and $Da \cdot Pe_h/\kappa = 0.12$. From Hagan *et al.* (1988a).

approximations. Juncu and Floarea (1995) compared the results of this criterion with those using the MV generalized criterion, and indicated that the former provides rather conservative predictions of the critical conditions for reactor runaway.

Let us now use the α model to predict the critical conditions for runaway using the MV generalized criterion. A typical runaway boundary in the Da–Pe_h/κ parameter plane is shown in Fig. 4.19, based on the normalized sensitivity of $\bar{\theta}^*$ to Pe_h/κ, i.e., $S(\bar{\theta}^*; Pe_h/\kappa)$. Figure 4.19 also shows the runaway boundaries, computed again through the MV generalized criterion but using the exact two-dimensional and the standard one-dimensional models. It appears that the runaway boundaries predicted

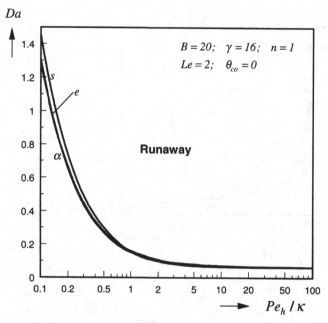

Figure 4.19. Runaway boundaries in the Da–Pe_h/κ phase plane predicted by the MV generalized criterion using different reactor models: the α model (α); the exact two-dimensional model (e); the standard one-dimensional model (s).

using the α model and the exact two-dimensional model are almost coincident. In particular, the α model leads to good predictions not only in the region of high values of Pe_h/κ, where the effect of radial dispersion is actually negligible, but also in the region of low values of Pe_h/κ, where the effect of radial dispersion is significant. Thus, it is concluded that the α model can be safely applied to approximate the exact two-dimensional model in predicting the critical conditions for reactor runaway.

On the other hand, the standard one-dimensional model derived from the one-point orthogonal collocation approximation leads to larger errors in the predicted runaway boundary, particularly in the region of low values of Pe_h/κ. Furthermore, the predicted runaway boundary is located inside the runaway region of the two-dimensional model. This can result in erroneous design of the reactor, leading to runaway operations. Thus, care should be exercised when using the standard one-dimensional model to predict the critical conditions for reactor runaway.

4.5 Complex Kinetic Schemes

In contrast to single reaction systems, a system involving competing reactions exhibits not only thermal runaway problem discussed previously, which in this case also affects the outlet *yield* and *selectivity* of desired products, but also the parametric sensitivity of the outlet yield and selectivity with respect to the reactor operating conditions. In principle, the latter may be caused by other mechanisms that may not necessarily involve the runaway of the reactor temperature. In the following we will limit the analysis of complex kinetic schemes to the cases of two *consecutive* or two *parallel* reactions. Despite this limitation, it has the merit of accounting for the most important qualitative features.

Dente *et al.* (1966a, b) were the first to investigate parametric sensitivity in a tubular reactor where two parallel or consecutive reactions occur. The sensitivity criterion used in their studies was the one discussed in Section 4.2, based on the reactor behavior in the temperature-axial coordinate plane. It was established, through extensive reactor simulations, that thermal runaway is a necessary condition for runaway of the outlet yield and selectivity. More recently, similar reacting systems were analyzed by Westerterp and coworkers (Westerterp and Ptasinski, 1984; Westerterp and Overtoom, 1985). Their focus was placed on the proper design of reactors from the yield and selectivity points of view. In particular, criteria were developed to define the region in the reactor parameter space where outlet yield and selectivity do not fall below certain *a priori* fixed values. The nonoccurrence of runaway was then checked *a posteriori* by direct solution of the reactor model. In the following, we will discuss mainly the application of the MV generalized criterion to the case of a one-dimensional pseudo-homogeneous plug-flow reactor where two consecutive or two parallel reactions occur. It has been shown (Morbidelli and Varma, 1988) that this criterion allows one to investigate reactor sensitivity not only in terms of the temperature maximum, but also relative to the outlet yield and selectivity of the desired product, which cannot be investigated directly by earlier geometry-based criteria.

In the following analysis we indicate as **A** the reactant fed to the reactor and **B** the desired product (the intermediate in the case of two consecutive reactions). The choice of the appropriate objective function for optimal design depends on the specific quantity of interest. However, in general, the ideal objective function is given by a suitable combination of yield and selectivity, defined as follows:

$$Y = u_B, \qquad S = \frac{u_B}{1 - u_A} \tag{4.45}$$

For example, for large values of ratio between the value of the reactant and its recycle cost, selectivity should be optimized, while for small values of this ratio, yield becomes the most significant reactor performance parameter. In the sequel, we analyze both of these extreme cases and investigate the sensitivity behavior of yield and selectivity at the reactor outlet.

Typical industrial processes where selectivity and/or yield are of great concern are partial oxidations. The problem is to control the oxidation processes so as to maximize the production of the desired, partially oxidized, intermediate product, while avoiding complete oxidation to CO and CO_2. Some examples of these processes are

Table 4.4. Typical values of physicochemical parameters and operating conditions for some partial oxidation processes

Process	B	St/(DaB)	H_r	R_r^i	γ_1	γ_2	n_1	n_2
Naphthalene oxidation to phthalic anhydride on V_2O_5 in excess air; two consecutive reactions (Carberry and White, 1969)	10	2.0	1.7	0.05	16	34	1	1
Ethylene oxidation to ethylene oxide on Ag in excess ethylene; two parallel reactions (Westerterp and Ptasinski, 1984)	11.2	8.0	2.2	0.50	14	21	1	1
Orthoxylene oxidation to phthalic anhydride on V_2O_5 in excess air; two parallel reactions (Froment, 1967)	15.4	2.6	3.6	0.10	22	23	1	1
Benzene oxidation to maleic anhydride on V_2O_5 in excess air; two parallel reactions	10.2	6.9	1.8	0.70	17	20	1	1
two consecutive reactions (Wohlfahrt and Emig, 1980)	10.2	6.9	0.8	0.07	17	14	1	1

listed in Table 4.4, together with typical values of the corresponding physicochemical parameters and operating conditions in terms of dimensionless parameters of obvious meaning, which will be formally defined shortly [see Eq. (4.50)]. It is worth noting that the reaction kinetics of these systems can be well approximated using two consecutive or parallel reactions, where the undesired reaction (*i.e.*, the second one) usually exhibits a larger activation energy than the desired one, *i.e.*, $\gamma_2 > \gamma_1$. Moreover, the operating conditions are usually selected so that, at least at the reactor inlet, the rate of the desired reaction is larger than that of the undesired one, *i.e.*, $R_r^i < 1$. It is apparent that, in these cases, thermal runaway involves also a dramatic decrease of the outlet selectivity. In fact, even a modest temperature increase may have detrimental effects on the process performance in terms of selectivity toward the desired product.

Other processes of industrial relevance that involve two or more reactions include hydrogenations or oligomerizations of mixtures of various olefins, isomerizations of hydrocarbons (*e.g.*, xylenes and olefins), alkylations, and others. These reacting systems can be described at least approximately through two consecutive or parallel reactions, which typically exhibit similar values for kinetic parameters (*i.e.*, $\gamma_1 \approx \gamma_2$ and $R_r^i \approx 1$). In these cases, temperature has small effect on the reactor selectivity, and the consequences of thermal runaway are then the same as in single reaction systems, *i.e.*, safety, catalyst deactivation, and so on. The range of parameter values covered in the following discussion (for both two consecutive and two parallel reaction schemes) accounts for both of the situations described above.

4.5.1 The Case of Two Consecutive Reactions ($A \xrightarrow{1} B \xrightarrow{2} C$)

Assuming that the reaction rates of the two reactions can be expressed as

$$r_1 = k_1 \cdot C_A^{n_1} \tag{R1}$$

$$r_2 = k_2 \cdot C_B^{n_2} \tag{R2}$$

we may write the steady-state mass and energy balances in a one-dimensional plug-flow reactor, with constant external cooling (*i.e.*, θ_{co} = constant) as follows:

$$\frac{du_A}{dz} = -Da \cdot R_1 \tag{4.46}$$

$$\frac{du_B}{dz} = Da \cdot \left(R_1 - R_r^i \cdot R_2\right) \tag{4.47}$$

$$\frac{d\theta}{dz} = B \cdot Da \cdot \left(R_1 + H_r \cdot R_r^i \cdot R_2\right) - St \cdot (\theta - \theta_{co}) \tag{4.48}$$

with the inlet conditions

$$u_A = 1, \quad u_B = u_B^i, \quad \theta = 0 \quad \text{at } z = 0 \tag{4.49}$$

where

$$R_1 = \exp\left(\frac{\gamma_1 \cdot \theta}{\gamma_1 + \theta}\right) \cdot u_A^{n_1}; \quad R_2 = \exp\left(\frac{\gamma_2 \cdot \theta}{\gamma_1 + \theta}\right) \cdot u_B^{n_2};$$

$$\theta = \left(\frac{T}{T^i} - 1\right) \cdot \gamma_1; \quad u_A = C_A/C_A^i; \quad u_B = C_B/C_A^i;$$

$$R_r^i = k_2(T^i) \cdot C_A^{n_2 - n_1}/k_1(T^i); \quad H_r = \Delta H_2/\Delta H_1 \tag{4.50}$$

Thermal runaway

In this subsection we analyze the sensitivity behavior of the system purely in terms of its thermal behavior. With respect to reactors involving a single reaction, discussed earlier, the only difference is the presence of a second consecutive reaction. This leads to some peculiarities in the sensitivity behavior of the reactor, which become evident when analyzing the usual plots of the sensitivity $S(\theta^*; \phi)$ as a function of the heat-of-reaction parameter B. These are shown in Fig. 4.20 for a given set of parameter values and two values of the ratio between undesired and desired reactions at the reactor inlet, *i.e.*, $R_r^i = 0.5$ and 2, respectively. It is apparent that the sensitivity curves exhibit two maxima. These arise because $\gamma_2 > \gamma_1$ and the second reaction tends to run away for B values smaller than those that would lead to runaway of the first reaction. However, such a runaway is controlled by the limited amount of the intermediate product **B** that is available, the production of which is controlled by the first reaction. Thus, the first sensitivity peak appearing in Fig. 4.20, which is due to the second

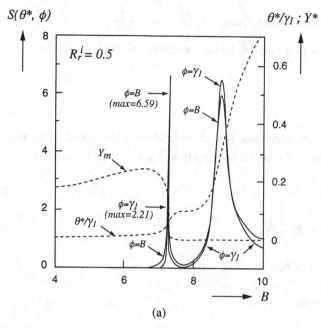

(a)

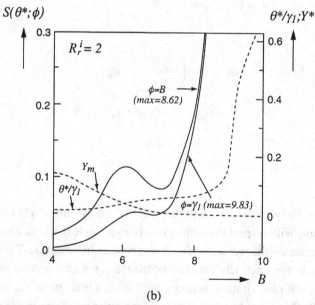

(b)

Figure 4.20. Temperature maximum θ^*, corresponding yield Y_m, and normalized objective sensitivity $S(\theta^*; \phi)$ with respect to $\phi = B$ and $\phi = \gamma_1$, as functions of the heat-of-reaction parameter B, in the case of two consecutive reactions. $n_1 = 1$; $n_2 = 1$; $\gamma_1 = 20$; $\gamma_2 = 100$; $H_r = 1$; $St/Da = 30$; $\theta^i = \theta_{co} = 0$; $u_B^i = 0$. (a) $R_r^i = 0.5$; (b) $R_r^i = 2$. From Morbidelli and Varma (1988).

reaction, may be modest, while the second peak occurs at larger B values, where the first reaction also enters the runaway region. This behavior is rather general and arises when the second reaction is intrinsically more sensitive than the first one, such as when $\gamma_2 > \gamma_1$ or $n_2 < n_1$.

To understand better the sensitivity behavior of these systems, let us first consider the limiting case in which the rate constant for the second reaction is much larger than that of the first one, *i.e.*, $R_r^i \gg 1$. Thus, the concentration of the intermediate product **B** remains small and almost constant along the reactor, so that $du_B/dz = 0$ and from Eq. (4.47) it follows that

$$R_2 \rightarrow \frac{R_1}{R_r^i} \tag{4.51}$$

and Eq. (4.48) reduces to

$$\frac{d\theta}{dz} = B \cdot Da \cdot R_1 \cdot (1 + H_r) - St \cdot (\theta - \theta_{co}) \tag{4.52}$$

By comparing this equation with Eq. (4.8), and accounting for the expression of R_1 in Eq. (4.50), it is readily seen that this corresponds to the heat balance in the case of a single reaction whose characteristics are identical to the first reaction, with the only exception being that the heat-of-reaction parameter is now multiplied by the factor $1 + H_r$, *i.e.*, it is given by $B \cdot (1 + H_r)$. From the physical point of view, this means that, when the second reaction is much faster than the first one, the system behaves as if only the first reaction is occurring, but involving a heat of reaction equal to the sum of those coming from the two consecutive reactions [*i.e.*, $B \cdot (1 + H_r)$]. Thus, we would expect that, as $R_r^i \rightarrow \infty$, the critical B value for runaway in the case of two consecutive reactions approaches the critical B value in the case of a single reaction divided by $(1 + H_r)$.

Let us now return to analyze in more detail the sensitivity values shown in Fig. 4.20a. It appears that the first peak is sufficiently strong to produce a significant temperature increase and to almost completely deplete the intermediate product **B**. This can be seen by the values of the temperature maximum, θ^*, and the corresponding yield, Y_m, which drops to zero at criticality. Moreover, in the same figure are also shown the sensitivity values with respect to two different model input parameters, *i.e.*, B and γ_1. It is apparent that both the first and the second peaks exhibit the generalized feature, *i.e.*, they occur at the same B value no matter which input parameter ϕ is used in the definition of sensitivity. This, however, is not the case for the larger R_r^i value considered in Fig. 4.20b, where the decreased availability of the reactant **B** does not allow the second reaction to fully develop its runaway behavior. By further increasing the R_r^i value, the magnitude of the first peak further decreases and eventually disappears, leading to the situation discussed above, where the sensitivity curve exhibits only one peak, similar to the case of a single reaction (*i.e.*, $u_B \approx 0$). In the case where the first peak disappears, the second one should be taken as the critical condition separating runaway from nonrunaway behavior. However, this case is usually of limited practical interest since it involves a very low concentration of the intermediate product **B**, leading to undesirable low yield and selectivity values. Thus, *in computing the critical conditions for runaway through the MV generalized criterion, the occurrence of the first sensitivity peak should be considered.*

In the previous section, we have seen that in the case of a single reaction the PAO and the runaway boundaries predicted by the z-MV criterion identify in the St–B parameter plane three distinct regions: PAO, HSO, and runaway, as illustrated in Fig. 4.7. Let us now use the same set of parameter values as in Fig. 4.7 for the first of two consecutive reactions and investigate the influence of the second one on the reactor operation diagram in the St–B parameter plane. Figure 4.21 shows the computed runaway and PAO boundaries in the cases of both consecutive reactions (solid curves) and a single reaction (broken curves, from Fig. 4.7). The runaway boundary is found by the MV generalized criterion using the reactor length as the independent variable, and taking the first peak of the sensitivity curves to identify the critical condition for runaway. The PAO boundaries, similar to those of a single reaction, are obtained by numerically integrating the model equations (4.46) to (4.50) and using a trial-and-error procedure to enforce the condition $d\theta/dz = 0$ at the reactor outlet. As expected, an additional second exothermic reaction makes the runaway boundary move toward lower B values, *i.e.*, thermal runaway becomes more likely. Similar to the case of a

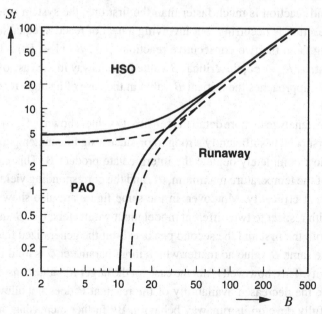

Figure 4.21. HSO, PAO, and runaway regions in the St–B parameter plane in the case of two consecutive reactions. $Da = 0.1$; $n_1 = 1$; $n_2 = 1$; $\gamma_1 = 20$; $\gamma_2 = 40$; $H_r = 1$; $R_r^i = 0.5$; $\theta^i = \theta_{co} = 0$; $u_B^i = 0$. The broken curves, from Fig. 4.7, correspond to the case of a single reaction.

single reaction, the lower branch of the PAO boundary coincides with the runaway boundary. The upper branch moves toward higher St values. This occurs because, due to the additional heat produced by the second reaction, the heat-removal rate by external cooling has to be increased in order to keep the hot spot inside the reactor. Nevertheless, the qualitative shape of the reactor operation diagram in the St–B parameter plane remains unchanged.

In order to investigate in more detail the effect of the second reaction on the runaway behavior of a reactor where two consecutive reactions occur, the critical conditions for runaway have been computed for various values of the kinetic parameters of the second reaction. In Figs. 4.22a, b, and c, the critical B values predicted by the MV generalized criterion are shown as a function of the inlet reaction-rate ratio R_r^i for various values of the activation energy γ_2, the heat-of-reaction ratio H_r, and the reaction order n_2, respectively. Again, in all cases, the first peak of the normalized sensitivity $S(\theta^*; \phi)$ is used to define criticality (including a few cases where the first peak does not fully satisfy the generalized sensitivity criterion). As expected, it is found that *the second exothermic reaction has always an enhancing effect on sensitivity, thus enlarging the region of reactor runaway.* Thus, for increasing values of the activation energy γ_2 and the heat of reaction H_r, the runaway region enlarges, while it shrinks for increasing values of the reaction order n_2. In the case where the heat generated by the second reaction is zero, *i.e.*, $H_r = 0$, the critical B value is independent of the second reaction rate (see Fig. 4.22b). For increasing values of R_r^i, the critical B value approaches asymptotically a constant value, which, as discussed above, is given by the value at $R_r^i = 0$ divided by $1 + H_r$. Moreover, in all cases, the critical B values for $R_r^i = 0$ ($B_c = 18.9$) coincide with that reported in Fig. 4.10 (at $St = 30$ and $Da = 1$) for single reaction systems.

Sensitivity of the outlet yield

Let us consider the case of a tubular reactor in which two consecutive reactions occur and we wish to maximize the reactor outlet yield of the intermediate product **B**. Thus, as before, thermal runaway is one problem that should be considered in the reactor design. However, in addition, it is important to evaluate its effect on the reactor outlet yield as well as to investigate the sensitivity of the yield itself with respect to small variations of the system parameters. In the following, we focus on these aspects, in particular, with reference to optimally designed reactors.

Through numerical integration of the model equations (4.46) to (4.49), one may evaluate the profiles of the species concentrations and temperature along the reactor length, as shown for a typical example in Fig. 4.23. It is seen that the concentration of the intermediate product **B** reaches its maximum at some location inside the reactor and then decreases with z. The situation depicted in Fig. 4.23 indicates that the selected reactor is too long, because the reactor portion after the maximum has a detrimental effect on the yield of **B**. Let us rewrite Eq. (4.48) in the equivalent form:

$$\frac{d\theta}{dz} = Da \cdot \left[B \cdot (R_1 + H_r \cdot R_r^i \cdot R_2) - (St/Da) \cdot (\theta - \theta_{co}) \right] \qquad (4.48')$$

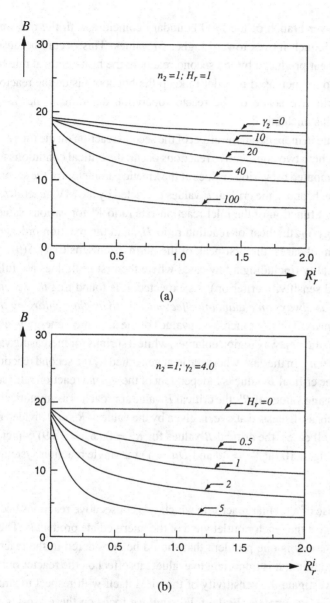

Figure 4.22. Runaway regions in the $B-R_r^i$ parameter plane in the case of two consecutive reactions: (a) effect of activation energy of the second reaction (γ_2), (b) effect of heat-of-reaction ratio (H_r), and (c) effect of the second reaction order (n_2). $n_1 = 1$; $\gamma_1 = 20$; $St/Da = 30$; $\theta^i = \theta_{co} = 0$; $u_B^i = 0$. From Morbidelli and Varma (1988).

and consider St/Da as one dimensionless parameter (which is independent of the reactor length L), so that in the model equations (4.46) to (4.49) only the Damkohler number Da contains the reactor length L. Thus, when all the other parameters are fixed, a too long reactor means a too large Da value. In general, there exists an optimal

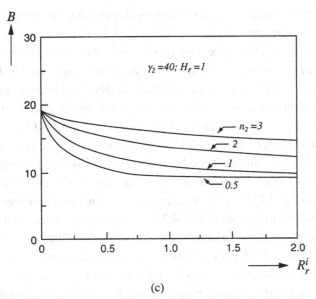

(c)

Figure 4.22. (cont.)

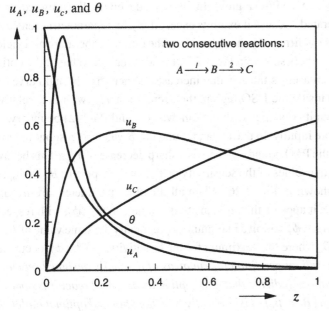

Figure 4.23. Typical profiles of species concentrations and temperature along the reactor axis in the case of two consecutive reactions.

Da value, Da_{opt}, which gives the maximum outlet yield. The necessary condition for Da to be optimal is given by

$$\left.\frac{du_B}{dz}\right|_{z=1} = 0 \quad \text{at } Da = Da_{opt} \tag{4.53}$$

125

In the sequel we perform the optimal reactor design by determining the optimal Da value (*i.e.*, the optimal reactor length L) that satisfies Eq. (4.53).

For a given set of parameter values, the numerically computed optimal Da values that satisfy Eq. (4.53) are shown as a function of the heat-of-reaction parameter B in Figs. 4.24a and b (curve 3) for $H_r = 1$ and $H_r = 5$, respectively. The obtained curves are referred to in the sequel as the *optimal curves*. The corresponding values of the outlet yield, y^o_{opt}, are shown in Figs. 4.25a and b, together with the values of the outlet conversion, x^o, and temperature, θ^o. Figure 4.24 also shows the PAO (curve 2) and run-away (curve 1) boundaries. The runaway boundaries are predicted by the MV general-ized criterion. It is seen that this reactor operation diagram, similar to that in the $St-B$ parameter plane shown in Fig. 4.21, is characterized by three distinct operation regions: PAO, HSO, and runaway. By inspection of Fig. 4.24 it appears that, for low values of the heat-of-reaction parameter B, the optimal curve is located inside the HSO region and far from the runaway boundary. Thus, the optimally designed reactor can operate safely without parametric sensitivity and runaway problems. As the B value increases, the optimal curve approaches the runaway boundary, and beyond the B value corre-sponding to the bifurcation point on the PAO boundary, it coincides with the boundary between the runaway and the PAO regions. In this case, the optimally designed reactor cannot be safely operated because it exhibits exceedingly high parametric sensitivity.

This conclusion is confirmed by the behavior of the outlet temperature and yield in optimally designed reactors shown in Fig. 4.25. It is seen that the value of the optimal outlet yield, Y^o_{opt}, decreases as the B value increases. When the optimal curve (3 in Fig. 4.24) is located inside the HSO region, the decrease of Y^o_{opt} with B is relatively small. However, when it coincides with the runaway (1) and PAO (2) boundary, Y^o_{opt} decreases with B more rapidly. In particular, around the B value corresponding to the bifurcation point on the PAO boundary, there is a sharp decrease of Y^o_{opt}. This behavior is also illustrated by the values of the sensitivity of Y^o_{opt} with respect to B, $S(Y^o_{opt}; B)$, as a function of B shown in Fig. 4.26. When all the other parameters are fixed, by changing the B value, it appears that the sensitivity values, $S(Y^o_{opt}; \phi)$, with respect to various input parameters, ϕ, exhibit a maximum at practically the same value of B, *i.e.*, B_c. Thus, the value B_c where the maximum for the sensitivity, $S(Y^o_{opt}; \phi)$, occurs may be considered as *the critical condition for runaway of optimally designed reactors*.

The above observations indicate that if the optimally designed reactor is operating inside the HSO region ($B < B_c$), it is generally safe, the obtained optimal outlet yield is relatively high, and, thus, the design is correct. Instead, if the optimally designed reactor operates at the boundary between the runaway and the PAO regions ($B > B_c$), the reactor behavior is parametrically sensitive, the outlet yield is low, and this design cannot be used in practice.

It should be noted that the reactor operation diagram in the $Da-B$ parameter plane depends on the values of the other involved parameters, as can be seen by comparing Figs. 4.24a and b, where the H_r values are 1 and 5, respectively. With respect to practical applications, it is particularly important to investigate the effect

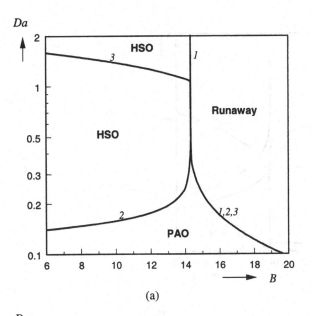

(a)

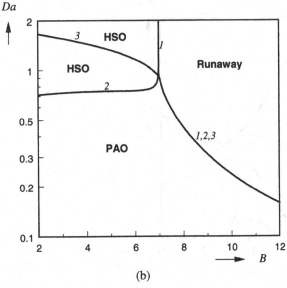

(b)

Figure 4.24. Runaway boundary (curve 1), PAO boundary (curve 2), and optimal operation curve for maximum outlet yield of **B** (curve 3) in the Da–B parameter plane for two consecutive reactions: (a) $H_r = 1$; (b) $H_r = 5$. $St/Da = 30$; $n_1 = 1$; $n_2 = 1$; $\gamma_1 = 20$; $\gamma_2 = 40$; $R_r^i = 0.25$; $\theta^i = \theta_{co} = 0$; $u_B^i = 0$.

of the parameter St/Da on the reactor operation diagram. As is apparent from its definition in Eq. (4.6), St/Da represents the ratio between the overall heat transfer coefficient to the reactor wall and the inlet reaction rate. It can be easily shown that both the runaway and the PAO regions in the Da–B parameter plane move toward higher

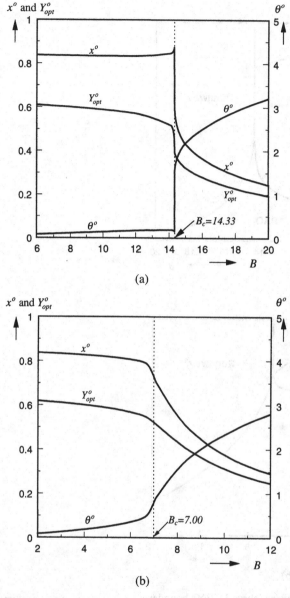

(a)

(b)

Figure 4.25. Outlet yield of $\mathbf{B}$ (Y^o_{opt}), outlet conversion (x^o), and outlet temperature (θ^o), corresponding to the optimal reactor design shown in Fig. 4.24: (a) $H_r = 1$; (b) $H_r = 5$.

B values as the St/Da value increases, *e.g.*, as the overall heat transfer coefficient to the reactor wall increases. This means that the HSO region enlarges toward higher B values as St/Da increases, and then the portion of the optimal curve located inside the HSO region also extends to higher B values. Therefore, for a given B value, if the predicted Da_{opt} value leads to a reactor operation at the boundary between the PAO and the runaway regions, it is possible, by increasing the overall heat transfer coefficient

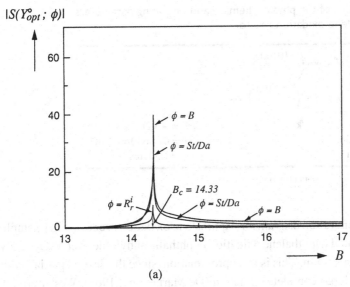

(a)

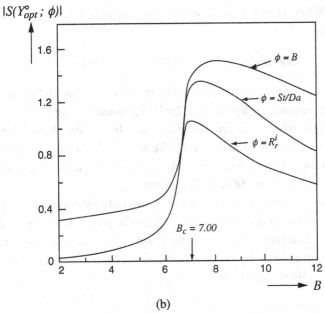

(b)

Figure 4.26. Values of the normalized sensitivity $S(Y_{opt}^o; \phi)$ as a function of the heat-of-reaction parameter B for various input parameters ϕ, where Y_{opt}^o is the outlet yield of the intermediate product **B** corresponding to the optimal reactor design shown in Fig. 4.24: (a) $H_r = 1$; (b) $H_r = 5$.

to the reactor wall, to move the operating conditions of the optimally designed reactor inside the HSO region, thus obtaining an optimal reactor that can be safely operated in practice. Thus, *for all other parameters fixed, it is always possible to make optimally designed reactors safe by increasing the wall heat-transfer coefficient.*

Table 4.5. Values of the physicochemical and operating parameters for the naphthalene oxidation reactor

$$k_1 \cdot \rho_B = 6.46 \times 10^6 \cdot \exp\left(-\frac{10100}{T}\right), 1/s$$

$$k_2 \cdot \rho_B = 37.9 \times 10^{12} \cdot \exp\left(-\frac{22100}{T}\right), 1/s$$

$\Delta H_1 = -1.881 \times 10^6$ kJ/kmol	$\Delta H_2 = -3.282 \times 10^6$ kJ/kmol
$c_p \cdot \rho = 1.352$ kJ/m^3/K	$U = 0.186$ kJ/m^2/s/K
$d_t = 0.025$ m	$v^o = 1.30$ m/s

From Westerterp and Overtoom (1984).

Example 4.3 Reactor operation diagram for naphthalene oxidation process.   In Examples 4.1 and 4.2, we treated naphthalene oxidation to phthalic anhydride over V_2O_5 catalyst as a single reaction system. This is an approximation, since the desired partial oxidation product can undergo complete oxidation (De Maria *et al.*, 1961). Westerterp (1962), and Carberry and White (1969) have proposed the following consecutive reaction scheme for this system:

$$\text{naphthalene (A)} \xrightarrow{1} \text{phthalic anhydride (B)} \xrightarrow{2} CO_2 + H_2O$$

where both reactions are assumed to be pseudo first order with respect to the hydrocarbon reactant (*i.e.*, naphthalene and phthalic anhydride, respectively) and zeroth order with respect to oxygen, which is in great excess. The values of the physicochemical parameters for this reaction system as given by Westerterp and Overtoom (1984) are summarized in Table 4.5. Let us now construct the reactor operation diagram in the $Da-B$ parameter plane at $T^i = 654$ K and assuming $T_{co} = T^i$.

When the inlet temperature is given, Da and B are only functions of L and C_A^i, respectively. Thus, the $Da-B$ parameter plane can also be regarded as a direct representation of the $L-C_A^i$ parameter plane. With the given value of the inlet temperature, the following values for the dimensionless parameters required to solve the model equations (4.46) to (4.50) are obtained:

$$\gamma_1 = 15.4; \qquad \gamma_2 = 33.8; \qquad St/Da = 17.3;$$
$$R_r^i = 0.063; \qquad H_r = 1.74$$

The corresponding reactor operation diagram in the $Da-B$ (or $L-C_A^i$) parameter plane is shown in Fig. 4.27a, where curve 1 is the runaway boundary predicted by the MV generalized criterion, curve 2 is the PAO boundary, and curve 3 is the optimal curve that gives the Da value (or reactor length L) that maximizes the outlet yield of phthalic anhydride at each given value of B (or equivalently, of the inlet value of the naphthalene concentration C_A^i). The corresponding values of the optimal outlet yield are shown in Fig. 4.27b, together with the values of the associated outlet conversion and temperature.

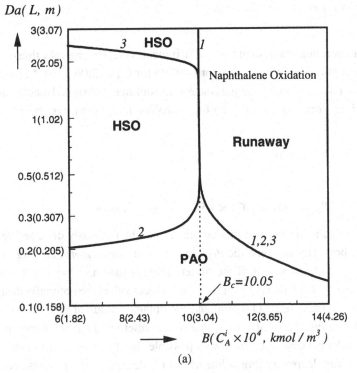

(a)

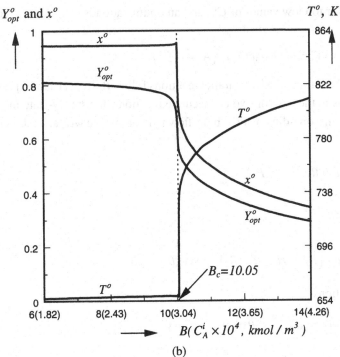

(b)

Figure 4.27. For the naphthalene oxidation reactor. (a) Reactor operation diagram in the Da–B parameter plane: curve 1, runaway boundary predicted by the MV generalized criterion; curve 2, PAO boundary; curve 3, optimal curve for outlet yield. (b) Values of the optimal outlet yield and of the corresponding outlet conversion and temperature. $T^i = T_{co} = 654$ K.

The reactor operation diagram in Fig. 4.27a indicates that the critical value of B for runaway of the optimally designed reactor is 10.05 (or $C_A^i = 3.06 \times 10^{-4}$ kmol/m^3). Thus, for the given set of the system parameters, the optimally designed reactor should operate at an inlet concentration value of naphthalene *lower* than this critical value, i.e.,

$$C_A^i < 3.06 \times 10^{-4} \text{ kmol/m}^3$$

or

$$P_A^i = C_A^i \cdot T^i \cdot R_g < 3.06 \times 10^{-4} \times 654 \times 8.314 = 1.66 \text{ kPa}$$

When the C_A^i value is higher than this critical value, the optimally designed reactor operates at the boundary between the PAO and the runaway regions, leading to both thermal runaway and low values of the outlet yield, as shown in Fig. 4.27b. On the other hand, when the C_A^i value is lower than this critical value, the optimally designed reactor operates in the safe HSO region and the obtained value of the outlet yield is much larger. From Fig. 4.27b, it is seen that if the outlet yield has to be maximized, a value of the inlet concentration as low as possible should be used. However, since reactor productivity decreases almost linearly as C_A^i decreases, the process economy does not allow for too low values of C_A^i, and an optimization is clearly required.

4.5.2 The Case of Two Parallel Reactions ($A \overset{1}{\longrightarrow} B; A \overset{2}{\longrightarrow} C$)

Assuming that the rates of the two parallel reactions follow the power-law expressions (R1) and (R2) as in the case of two consecutive reactions, the steady-state mass and energy balances in a one-dimensional plug-flow reactor may be written in dimensionless form as

$$\frac{du_A}{dz} = -Da \cdot \left(R_1 + R_r^i \cdot R_2 \right) \tag{4.54}$$

$$\frac{du_B}{dz} = Da \cdot R_1 \tag{4.55}$$

$$\frac{d\theta}{dz} = B \cdot Da \cdot \left(R_1 + H_r \cdot R_r^i \cdot R_2 \right) - St \cdot (\theta - \theta_{co}) \tag{4.56}$$

with inlet conditions

$$u_A = 1, \qquad u_B = u_B^i, \qquad \theta = 0 \quad \text{at } z = 0 \tag{4.57}$$

where

$$R_1 = \exp\left[\frac{\gamma_1 \cdot \theta}{\gamma_1 + \theta} \right] \cdot u_A^{n_1}, \qquad R_2 = \exp\left[\frac{\gamma_2 \cdot \theta}{\gamma_1 + \theta} \right] \cdot u_A^{n_2}. \tag{4.58}$$

All the other quantities are defined as in Eq. (4.50). Similarly to the case of consecutive reactions, we first analyze the thermal runaway behavior of a reactor in which two parallel reactions occur and in particular the influence of the second reaction on the reactor operation diagram, through the application of the MV generalized criterion. Next, we analyze the sensitivity of the outlet yield of the desired product B, in close connection with thermal runaway behavior.

Thermal runaway

The effect of the physicochemical parameters of the second parallel reaction, *i.e.*, activation energy, γ_2, ratio of the heat of reaction, H_r, and reaction order, n_2, on the thermal runaway region in the $B-R_r^i$ parameter plane are shown in Figs. 4.28a, b and c, respectively. As $R_r^i \to 0$, Eqs. (4.54) to (4.56) reduce to the case of a single reaction, whose characteristics are equal to those of reaction 1; thus, $B_c \to 18.9$, which coincides with the value obtained in the case of a single reaction shown in Fig. 4.10, independent of the values of γ_2, H_r, and n_2. As $R_r^i \to \infty$, Eq. (4.56) reduces to

$$\frac{d\theta}{dz} = \left(R_r^i \cdot Da\right) \cdot (H_r \cdot B) \cdot R_2 - St \cdot (\theta - \theta_{co}) \tag{4.59}$$

which is again the heat balance for the case of a single reaction, whose physicochemical parameters are now equal to those of the second reaction. In particular, the heat-of-reaction parameter now equals $H_r \cdot B$ and the Damkohler number equals $R_r^i \cdot Da$.

Thus, when the R_r^i value increases from zero to infinity, the critical parameter B_c goes from the value corresponding to the first reaction alone to that for the second one alone. In this regard, the sensitivity behavior of two parallel reactions is different from that of two consecutive reactions. In this case, both reactions have all the reactant **A** available, so that if one of them enters its own sensitivity region, it drives the entire system to thermal runaway, carrying along also the other reaction. Thus, unlike the case of two consecutive reactions shown in Fig. 4.20, *two peaks in the sensitivity-versus-B profile do not occur in the case of two parallel reactions*.

On the other hand, however, the total heat produced by two parallel reactions depends on their individual heats of reaction and their relative rates. Let us consider the effect of adding a second parallel reaction on the sensitivity behavior of a single reaction system. We see that if the second reaction has heat of reaction higher than that of the first, then the presence of the second reaction increases the total amount of heat produced by the depletion of reactant **A**, and thus thermal runaway becomes more likely. This is the case illustrated in Fig. 4.28b for $H_r > 1$, where the value of the critical parameter B_c decreases as R_r^i increases, which implies the enlarging of the runaway region. Of course, the opposite behavior is found when the second reaction is much less exothermic than the first one. In this case, the second reaction makes thermal runaway less likely, as it clearly appears in Fig. 4.28b for $H_r = 0$ or 0.1, where we see that the critical value of the heat-of-reaction parameter B increases as R_r^i increases.

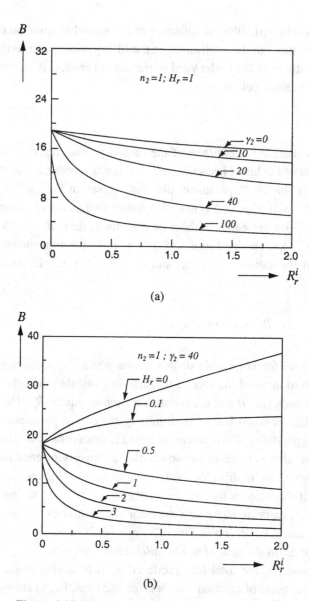

Figure 4.28. Boundaries of the runaway regions in the B–R_r^i parameter plane in the case of two parallel reactions: (a) effect of activation energy of the second reaction (γ_2), (b) effect of heat-of-reaction ratio (H_r), and (c) effect of the second reaction order (n_2). $n_1 = 1$; $\gamma_1 = 20$; $St/Da = 30$; $\theta^i = \theta_{co} = 0$; $u_B^i = 0$. From Morbidelli and Varma (1988).

In principle, it may be expected that if the heat of the second reaction is equal to that of the first, $i.e.$, $H_r = 1$, the addition of the second reaction should not affect the critical parameter B_c. Then, the case of $H_r = 1$ in Fig. 4.28b would be a horizontal line rather than a declining curve. However, if the added second reaction has a higher

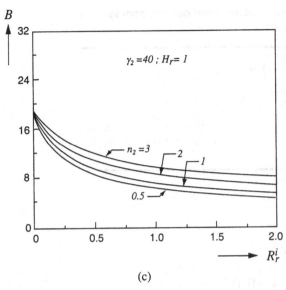

(c)

Figure 4.28. (cont.)

activation energy, *i.e.*, $\gamma_2 > \gamma_1$, it obviously enlarges the reactor runaway region. This is the situation that we have in Fig. 4.28b, where for $H_r = 1$ and $H_r = 0.5$ the critical parameter B_c decreases (*i.e.*, the runaway region enlarges) as R_r^i increases. Recall that the sensitivity behavior of systems where two consecutive reactions occur is substantially different. Here heat is produced not only by depletion of reactant **A** but also by depletion of the intermediate product **B**, and therefore the presence of the second exothermic reaction always makes thermal runaway more likely, *i.e.*, it enlarges the runaway region in the reactor operating parameter space.

Sensitivity of the outlet yield

The sensitivity behavior of the outlet yield in a tubular reactor where two parallel reactions occur simultaneously is analyzed in this section through a specific example, *i.e.*, the ethylene epoxidation reaction. It is worth noting that this system can actually be regarded as a representative of the various partial oxidation reactions that are common in practice, such as those listed in Table 4.4.

Example 4.4 Reactor operation diagram for ethylene epoxidation process. Ethylene oxide is industrially produced through gas-phase oxidation of ethylene with oxygen or air on supported silver catalyst. The main by-products are CO_2 and H_2O, which can be formed through direct oxidation of ethylene (parallel reaction) as well as through further oxidation of ethylene oxide (consecutive reaction). Voge and Adams (1967) suggested that in industrial operation conditions the consecutive path may be neglected, so that the reaction scheme can be approximated by the following two

Table 4.6. Values of the physicochemical and operating parameters for the ethylene epoxidation reactor

$k_1 \cdot \rho_B = 5.98 \times 10^4 \cdot \exp\left(-\dfrac{7200}{T}\right), 1/s$	
$k_2 \cdot \rho_B = 4.20 \times 10^7 \cdot \exp\left(-\dfrac{10800}{T}\right), 1/s$	
$\Delta H_1 = -2.10 \times 10^5$ kJ/kmol	$\Delta H_2 = -4.73 \times 10^5$ kJ/kmol
$c_p \cdot \rho = 7.03$ kJ/m^3/K	$U = 0.298$ kJ/m^2/s/K
$d_t = 0.025$ m	

From Westerterp and Ptasinski (1984).

reactions:

$$O_2 + 2C_2H_4 \xrightarrow{\ 1\ } 2C_2H_4O$$

$$O_2 + \frac{1}{3}C_2H_4 \xrightarrow{\ 2\ } \frac{2}{3}CO_2 + \frac{2}{3}H_2O$$

Since, for the oxygen-based process, ethylene in the feed is in large excess, the rate equations of both reactions are first order with respect to oxygen and zeroth order with respect to ethylene. For this system, let us construct the reactor operation diagram in the Da–B parameter plane, using values of the physicochemical parameters reported by Westerterp and Ptasinski (1984) and summarized in Table 4.6. Moreover, we assume equal inlet and coolant temperatures, *i.e.*, $T^i = T_{co} = 495$ K, and superficial velocity, $v^o = 1$ m/s.

Using these values, the dimensionless parameters in the model Eqs. (4.55) to (4.58) are given by

$$\gamma_1 = 14.5; \qquad \gamma_2 = 21.8; \qquad St/Da = 235.3; \qquad R_r^i = 0.488; \qquad H_r = 2.25$$

The corresponding reactor operation diagram in the Da–B parameter plane is shown in Fig. 4.29a, where the boundaries of the runaway (curve 1) and the PAO (curve 2) regions are reported together with the optimal curve (3) that gives the optimal outlet yield of product **B** (ethylene oxide), similar to the case of two consecutive reactions discussed earlier. The PAO boundary is computed by numerical integration of the model equations (4.54) to (4.58) through a trial-and-error procedure to enforce the condition $d\theta/dz = 0$ at $z = 1$. The runaway boundary is predicted by the MV generalized criterion, based on the sensitivity of the temperature maximum with respect to the heat of reaction parameter, *i.e.*, $S(\theta^*; B)$. The optimal curve represents for each given B the optimal Da value, Da_{opt}, that satisfies the following condition:

$$\left. \frac{du_B}{dz} \right|_{z=1} = 0 \quad \text{at } Da = Da_{opt}$$

Figure 4.29b shows the obtained values of the optimal outlet yield and the corresponding outlet temperature.

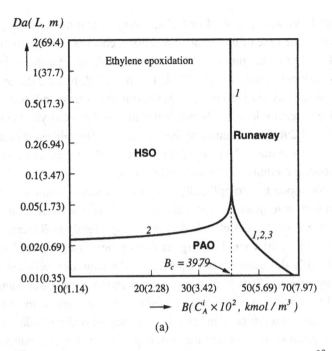

(a)

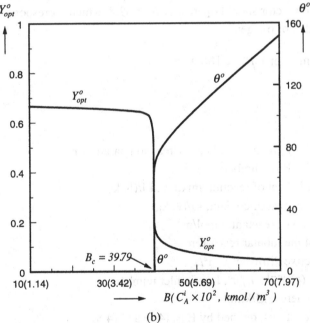

(b)

Figure 4.29. For the ethylene epoxidation reactor with ethylene-rich conditions. (a) Reactor operation diagram in the Da–B parameter plane: curve 1, runaway boundary; curve 2, PAO boundary; curve 3, optimal curve for maximum outlet yield. (b) Outlet yield and temperature for optimally designed reactors. $T^i = T_{co} = 495$ K; $v^o = 1$ m/s.

By comparing Fig. 4.29a with Fig. 4.24 or 4.27a, it is clear that the reactor operation diagram, similar to the case of consecutive reactions, consists of three distinct operation regions: HSO, PAO, and runaway. The situation is different for the optimal curve, which in this case exists only for $B > 39.8$. For $B < 39.8$, in fact, the outlet yield increases continuously as Da increases, so that optimality corresponds to an infinitely large Da value or reactor length. The optimal values of the outlet yield in this region, as shown in Fig. 4.29b, have been computed by using a Da value sufficiently large to yield complete conversion, which is also associated with a temperature value equal to that of the cooling medium, i.e., $x = 1$ and $\theta^o = 0$.

The performance corresponding to optimally designed reactors is summarized in Fig. 4.29b. It is seen that as long as the optimally designed reactor operates in the HSO region, i.e., for $B < 39.8$, the outlet yield decreases only slightly as B increases. However, at $B = 39.8$, the outlet yield undergoes a sharp decrease, and for larger values of B, the optimally designed reactor operates at the boundary between the PAO and the runaway regions. This is of course an operation regime that cannot be adopted in practice, since it is intrinsically sensitive. In conclusion, the value $B = 39.8$ can be regarded as a critical condition for runaway of optimally designed reactors. Thus, for the given set of values of the system physicochemical parameters, the optimally designed reactor should operate at $B < 39.8$, which corresponds to an inlet concentration value of oxygen:

$$C_A^i < 0.045 \text{ kmol/m}^3 \quad \text{or} \quad P_A^i < 186 \text{ kPa}$$

Nomenclature

B	$(-\Delta H) \cdot C^i \cdot \gamma / \rho \cdot c_p \cdot T^i$, heat-of-reaction parameter
Bi	$d_i \cdot h_w / 2 \cdot \lambda_r$, Biot number
c_p	Mean specific heat of reaction mixture, kJ/(K·kg)
$c_{p,co}$	Mean specific heat of coolant, kJ/(K·kg)
C	Concentration of reactant, kmol/m³
d_t	Diameter of the tubular reactor, m
D_r	Radial effective diffusivity, m²/s
Da	$\rho_B \cdot k(T^i) \cdot (C^i)^{n-1} \cdot L/v^o$, Damkohler number
E	Activation energy, kJ/kmol
f_i	$(i = 1, 2)$ functions, defined by Eqs. (4.7) and (4.8)
H_r	$\Delta H_2 / \Delta H_1$, heat-of-reaction ratio
k	Reaction rate constant, $(\text{kmol/m}^3)^{1-n}$/s
l	Axial coordinate of the tubular reactor, m
L	Reactor length, m
M_w	Molecular weight, kg/kmol
n	Reaction order

P	Pressure, kPa
Pe_h	$v^o \cdot d_t \cdot \rho \cdot c_p / 2 \cdot \lambda_r$, radial-heat Peclet number
Pe_m	$v^o \cdot d_t / 2 \cdot D_r$, radial-mass Peclet number
r	Radial coordinate, m; reaction rate, kmol/(m³·s)
R_1	$\exp[\gamma_1 \cdot \theta / (\gamma_1 + \theta)] \cdot u_A^{n_1}$, dimensionless rate expression for the first reaction in consecutive or parallel reactions
R_2	$\exp[\gamma_2 \cdot \theta / (\gamma_1 + \theta)] \cdot u_B^{n_2}$ or $\exp[\gamma_2 \cdot \theta / (\gamma_1 + \theta)] \cdot u_A^{n_2}$, dimensionless rate expression for the second reaction in consecutive reactions or in parallel reactions
R_g	Ideal-gas constant, kJ/(K·kmol)
R_r^i	$k_2(T^i) \cdot C_A^{n_2 - n_1} / k_1(T^i)$, reaction-rate ratio at inlet conditions
$s(y; \phi)$	$\partial y / \partial \phi$, local sensitivity of the output variable y to the input parameter ϕ
S	$u_B / (1 - u_A)$, selectivity
$S(\theta^*; \phi)$	$\partial \ln \theta^* / \partial \ln \phi$, normalized sensitivity of the temperature maximum θ^* to the input parameter ϕ
St	$4 \cdot U \cdot L / d_t \cdot v^o \cdot \rho \cdot c_p$, Stanton number
t_n	Number of reactor tubes
T	Temperature, K
u_A	C_A / C_A^i, dimensionless concentration of A
u_B	C_B / C_A^i, dimensionless concentration of B
U	Overall heat transfer coefficient, kJ/(m²·s·K)
v^o	Superficial velocity, m/s
w_{co}	Coolant flow rate, kg/s
x	$1 - C/C^i$, reactant conversion
y	$2 \cdot r / d_t$
Y	u_B, yield of B
z	l/L

Greek Symbols

α	Heat-loss parameter corresponding to the α model, defined implicitly by Eq. (4.42)
ΔH	Heat of reaction, kJ/kmol
ϕ	Generic model input parameter
ϕ	Vector of model input parameters
γ	$E / R_g \cdot T^i$, dimensionless activation energy
κ	$2 \cdot L / d_t$
λ_r	Radial effective thermal conductivity, kJ/(m·s·K)
θ	$(T - T^i) \cdot \gamma / T^i$, dimensionless temperature in the reactor
θ^*	Maximum of the dimensionless temperature θ along the tubular reactor
$\bar{\theta}$	Average value of radial temperature corresponding to the α model [Eqs. (4.39) to (4.42)]

ρ Density, kg/m^3

τ $\pi \cdot d_t^2 \cdot t_n \cdot v^o \cdot \rho \cdot c_p / 4 \cdot c_{p,co} \cdot w_{co}$, ratio of reactant to coolant heat capacity

Subscripts

c Critical condition
co Coolant side
m Value corresponding to the temperature maximum
opt Optimal condition

Superscripts

$*$ Maximum of the quantity
i Reactor inlet
o Reactor outlet

Acronyms

AE   Adler and Enig (1964)
DC   Dente and Collina (1964)
HP   Henning and Perez (1986)
HSO   Hot-spot operation, *i.e.*, temperature profile exhibiting a maximum along the reactor
MV   Morbidelli and Varma (1988)
PAO   Pseudo-adiabatic operation, *i.e.*, no temperature maximum along the reactor
VF   van Welsenaere and Froment (1970)
VR   Vajda and Rabitz (1992)

References

Adler, J., and Enig, J. W. 1964. The critical conditions in thermal explosion theory with reactant consumption. *Comb. Flame* **8**, 97.

Akella, L. M., Hong, J.-C., and Lee, H. H. 1985. Variable cross-section reactors for highly exothermic reactions. *Chem. Eng. Sci.* **40**, 1011.

Akella, L. M., and Lee, H. H. 1983. A design approach based on phase plane analysis: counter-current reactor/heat exchanger with parametric sensitivity. *A.I.Ch.E. J.* **29**, 87.

Barkelew, C. H. 1959. Stability of chemical reactors. *Chem. Eng. Prog. Symp. Ser.* **25**, 37.

Bauman, E. G., Varma, A., Lorusso, J., Dente, M., and Morbidelli, M. 1990. Parametric sensitivity in tubular reactors with co-current external cooling. *Chem. Eng. Sci.* **45**, 1301.

Bilous, O., and Amundson, N. R. 1956. Chemical reactor stability and sensitivity II. Effect of parameters on sensitivity of empty tubular reactors. *A.I.Ch.E. J.* **2**, 117.

Carberry, J. J., and Wendel, M. M. 1963. A computer model of the fixed bed catalytic reactor: the adiabatic and quasi-adiabatic cases. *A.I.Ch.E. J.* **9**, 129.

Carberry, J. J., and White, D. 1969. On the role of transport phenomena in catalytic reactor behavior. *Ind. Eng. Chem.* **61**, 27.

Degnan, T. F., and Wei, J. 1979. The co-current reactor heat exchanger: part I. Theory. *A.I.Ch.E. J.* **25**, 338.

De Lasa, H. 1983. Application of the pseudo-adiabatic operation to catalytic fixed bed reactors. *Can. J. Chem. Eng.* **61**, 710.

De Maria, F., Longfield, J. E., and Butler, G. 1961. Catalytic reactor design. *Ind. Eng. Chem.* **53**, 259.

Dente, M., Buzzi Ferraris, G., Collina, A., and Cappelli, A. 1966a. Sensitivity behavior of tubular chemical reactors, II. Sensitivity criteria in the case of parallel reactions. *Quad. Ing. Chim. Ital.* **48**, 47.

Dente, M., and Collina, A. 1964. Il comportamento dei reattori chimici a flusso longitudinale nei riguardi della sensitività. *Chim. Ind.* **46**, 752.

Dente, M., Collina, A., Cappelli, A., and Buzzi Ferraris, G. 1966b. Sensitivity behavior of tubular chemical reactors, III. Sensitivity criteria in the case of consecutive reactions. *Quad. Ing. Chim. Ital.* **48**, 55.

Finlayson, B. A. 1971. Packed bed reactor analysis by orthogonal collocation. *Chem. Eng. Sci.* **26**, 1081.

Froment, G. F. 1967. Fixed bed catalytic reactors: current design status. *Ind. Eng. Chem.* **59**, 18.

Froment, G. F. 1984. Progress in the fundamental design of fixed bed reactors. In *Frontiers in Chemical Reaction Engineering*, L. K. Doraiswarmy and R. A. Mashelkar, eds. Vol. 1, p. 12. New Delhi: Wiley Eastern.

Hagan, P. S., Herskowitz, M., and Pirkle, C. 1988a. A simple approach to highly sensitive tubular reactors. *SIAM J. Appl. Math.* **48**, 1083.

Hagan, P. S., Herskowitz, M., and Pirkle, C. 1988b. Runaway in highly sensitive tubular reactors. *SIAM J. Appl. Math.* **48**, 1437.

Henning, G. P., and Perez, G. A. 1986. Parametric sensitivity in fixed-bed catalytic reactors. *Chem. Eng. Sci.* **41**, 83.

Hlavacek, V. 1970. Packed catalytic reactors. *Ind. Eng. Chem.* **62**, 8.

Hlavacek, V., Marek, M., and John, T. M. 1969. Modeling of chemical reactors – XII. *Coll. Czech. Chem. Commun.* **34**, 3868.

Hosten, L. H., and Froment, G. F. 1986. Parametric sensitivity in co-current cooled tubular reactors. *Chem. Eng. Sci.* **41**, 1073.

Juncu Gh., and Floarea, O. 1995. Sensitivity analysis of tubular packed bed reactor by pseudo-homogeneous 2-D model. *A.I.Ch.E. J.* **41**, 2625.

Luss, D., and Medellin, P. 1972. Steady state multiplicity and stability in a countercurrently cooled tubular reactor. In *Proceedings of the Fifth European/Second*

International Symposium on Chemical Reaction Engineering, p. B4-47. Amsterdam: Elsevier.

Morbidelli, M., and Varma, A. 1985. On parametric sensitivity and runaway criteria of pseudo-homogeneous tubular reactors. *Chem. Eng. Sci.* **40**, 2165.

Morbidelli, M., and Varma, A. 1988. A generalized criterion for parametric sensitivity: application to thermal explosion theory. *Chem. Eng. Sci.* **43**, 91.

Soria Lopez, A., De Lasa, H. I., and Porras, J. A. 1981. Parametric sensitivity of a fixed bed catalytic reactor. *Chem. Eng. Sci.* **36**, 285.

Thomas, P. H., and Bowes, P. C. 1961. Some aspects of the self-heating and ignition of solid cellulosic materials. *Br. J. Appl. Phys.* **12**, 222.

Vajda, S., and Rabitz, H. 1992. Parametric sensitivity and self-similarity in thermal explosion theory. *Chem. Eng. Sci.* **47**, 1063.

van Welsenaere, R. J., and Froment, G. F. 1970. Parametric sensitivity and runaway in fixed bed catalytic reactors. *Chem. Eng. Sci.* **25**, 1503.

Varma, A., and Aris, R. 1977. Stirred pots and empty tubes. In *Chemical Reactor Theory: A Review*, L. Lapidus and N. Amundson, eds. Englewood Cliffs, NJ: Prentice-Hall.

Voge, H. H., and Adams, Ch. R. 1967. Catalytic oxidation of olefins. *Adv. Catal.* **17**, 151.

Westerterp, K. R. 1962. Maximum allowable temperatures in chemical reactors. *Chem. Eng. Sci.* **17**, 423.

Westerterp, K. R., and Overtoom, R. R. M. 1985. Safe design of cooled tubular reactors for exothermic, multiple reactions; consecutive reactions. *Chem. Eng. Sci.* **40**, 155.

Westerterp, K. R., and Ptasinski, K. L. 1984. Safe design of cooled tubular reactors for exothermic, multiple reactions; parallel reactions – I. Development of criteria. II. The design and operation of an ethylene oxide reactor. *Chem. Eng. Sci.* **39**, 235.

Wohlfahrt, K., and Emig, G. 1980. Compare maleic anhydride routes. *Hydrocarbon Process.* **59**, 83.

Wu, H., Morbidelli, M., and Varma, A. 1998. Pseudo-adiabatic operation and runaway in tubular reactors. *A.I.Ch.E. J.* **44**, 1157.

5

Parametric Sensitivity in Continuous-Flow Stirred Tank Reactors

S TIRRED VESSELS, with inlet and outlet fluid streams, are widely used as chemical reactors in practice. Their behavior can be approximated by an ideal model: the continuous-flow stirred tank reactor (CSTR), also called the perfectly mixed flow reactor, where temperature and concentration are uniform in the entire vessel and equal to those of the outlet stream. This device is particularly suited for processes where temperature and composition should be controlled and a significant amount of heat of reaction removed. Examples include nitration of aromatic hydrocarbons or glycerin, production of ethylene glycol, copolymerization of butadiene and styrene, polymerization of ethylene using a Ziegler catalyst, hydrogenation of α-methylstyrene to cumene, and air oxidation of cumene to acetone and phenol (Froment and Bischoff, 1990).

Although CSTRs are simple devices, they can exhibit a parametrically sensitive behavior when exothermic reactions are carried out. In this case, conversion and temperature in the reactor undergo large variations in response to small variations of one or more of the reactor operating conditions. Therefore, in practice, we need to determine operating conditions that avoid the parametrically sensitive region.

The parametric sensitivity behavior of CSTRs has been investigated only recently. The main reason is that in this case there is neither a temperature profile nor a hot spot, as in the case of batch or tubular reactors, so that all earlier parametric sensitivity criteria based on some geometric feature of these temperature profiles cannot be applied. It is only through the calculation of sensitivity of model outputs with respect to model inputs that the sensitivity behavior of such lumped systems can be investigated. This was done by Chemburkar *et al.* (1986) and subsequently by Vajda and Rabitz (1993), who used the generalized criteria to identify the parametrically sensitive region of CSTRs. This chapter summarizes these results, and also draws a connection between the sensitivity behavior of two ideal models: CSTRs and plug-flow reactors (PFRs), which represent two opposite extremes in modeling real reactors. The comparison between the parametric sensitivity behavior of these two ideal reactors provides an insight into the role of axial mixing on the behavior of tubular reactors. In addition, the analysis of this simple reacting system offers a unique opportunity to clearly state the relationship between two related but distinct phenomena: *steady-state multiplicity* and *parametric sensitivity*, which are often confused in the literature.

5.1 Sensitivity Analysis

For a nonadiabatic CSTR, where a single irreversible nth-order exothermic reaction occurs, the steady-state concentration and temperature in the outlet stream are given by (Froment and Bischoff, 1990)

$$q \cdot (C_f - C) - V \cdot k \cdot C^n = 0 \tag{5.1}$$

$$\rho \cdot c_p \cdot q \cdot (T_f - T) + V \cdot (-\Delta H) \cdot k \cdot C^n - A \cdot U \cdot (T - T_{co}) = 0 \tag{5.2}$$

In dimensionless form the above equations become

$$F(\theta, \phi) = \frac{B^{n-1}}{Da} \cdot [\theta + St \cdot (\theta - \theta_{co})] - \exp\left(\frac{\theta}{1+\theta/\gamma}\right) \cdot [B - \theta - St \cdot (\theta - \theta_{co})]^n$$
$$= 0 \tag{5.3}$$

$$B \cdot x - \theta - St \cdot (\theta - \theta_{co}) = 0 \tag{5.4}$$

where the following dimensionless quantities have been introduced:

$$\theta = \frac{T - T_f}{T_f} \cdot \gamma; \qquad \theta_{co} = \frac{T_{co} - T_f}{T_f} \cdot \gamma; \qquad x = \frac{C_f - C}{C_f}; \qquad \gamma = \frac{E}{R_g \cdot T_f};$$

$$Da = \frac{V \cdot k(T_f) \cdot C_f^{n-1}}{q}; \qquad B = \frac{(-\Delta H) \cdot C_f}{\rho \cdot c_p \cdot T_f} \cdot \gamma; \qquad St = \frac{A \cdot U}{\rho \cdot c_p \cdot q}$$

$$\tag{5.5}$$

and as usual we indicate with ϕ the vector of independent, model input parameters, i.e., in this case $\phi = [Da \; B \; St \; \gamma \; n \; \theta_{co}]^T$.

Since Eq. (5.4) is merely a linear algebraic relation between reactant conversion, x, and dimensionless temperature, θ, all the distinguishing features of this system can be determined by investigating the single nonlinear algebraic equation: $F(\theta, \phi) = 0$, given by Eq. (5.3), where conversion has been replaced by temperature using Eq. (5.4).

In order to apply the generalized parametric sensitivity criterion developed by Morbidelli and Varma (1988), we need to define an appropriate objective sensitivity. In the present case, a natural choice is to take the reactor (or outlet stream) temperature as the objective, so that

$$s(\theta; \phi) = \frac{d\theta}{d\phi} \tag{5.6}$$

where ϕ may represent any one of the six model input parameters in ϕ. As discussed in previous chapters, a more appropriate quantity in sensitivity analysis is the normalized

Table 5.1. Analytical expressions for the various partial derivatives of $F(\theta, \phi)$ needed to compute the objective sensitivities $S(\theta; \phi)$ according to Eq. (5.8)

$$\frac{\partial F}{\partial \theta} = \frac{B^{n-1}}{Da} \cdot (1 + St) - \exp\left(\frac{\theta}{1 + \theta/\gamma}\right) \cdot [B - \theta - St \cdot (\theta - \theta_{co})]^n$$

$$\cdot \left[\frac{1}{(1 + \theta/\gamma)^2} - \frac{n \cdot (1 + St)}{B - \theta - St \cdot (\theta - \theta_{co})}\right]$$

$$\frac{\partial F}{\partial B} = \frac{(n-1) \cdot B^{n-2}}{Da} \cdot [\theta + St \cdot (\theta - \theta_{co})] - n \cdot \exp\left(\frac{\theta}{1 + \theta/\gamma}\right) \cdot [B - \theta - St \cdot (\theta - \theta_{co})]^{n-1}$$

$$\frac{\partial F}{\partial Da} = -\frac{B^{n-1}}{Da^2} \cdot [\theta + St \cdot (\theta - \theta_{co})]$$

$$\frac{\partial F}{\partial St} = \frac{B^{n-1}}{Da} \cdot (\theta - \theta_{co}) + n \cdot \exp\left(\frac{\theta}{1 + \theta/\gamma}\right) \cdot [B - \theta - St \cdot (\theta - \theta_{co})]^{n-1} \cdot (\theta - \theta_{co})$$

$$\frac{\partial F}{\partial \gamma} = -\exp\left(\frac{\theta}{1 + \theta/\gamma}\right) \cdot [B - \theta - St \cdot (\theta - \theta_{co})]^n \cdot \left(\frac{\theta}{\gamma + \theta}\right)^2$$

$$\frac{\partial F}{\partial n} = \frac{B^{n-1}}{Da} \cdot [\theta + St \cdot (\theta - \theta_{co})] \cdot \ln(B) - \exp\left(\frac{\theta}{1 + \theta/\gamma}\right) \cdot [B - \theta - St \cdot (\theta - \theta_{co})]^n$$

$$\cdot \ln[B - \theta - St \cdot (\theta - \theta_{co})]$$

objective sensitivity,

$$S(\theta; \phi) = \frac{\phi}{\theta} \cdot s(\theta; \phi) = \frac{d(\ln \theta)}{d(\ln \phi)} \tag{5.7}$$

which has a clearer physical meaning since it serves to normalize the magnitudes of the dimensionless temperature θ and the parameter ϕ. An analytical expression of the normalized sensitivity can be derived by differentiating both sides of Eq. (5.3) with respect to ϕ, leading to

$$\frac{dF}{d\phi} = \frac{\partial F}{\partial \theta} \cdot \frac{\partial \theta}{\partial \phi} + \frac{\partial F}{\partial \phi} = 0 \tag{5.8}$$

which, substituted into Eq. (5.7), yields

$$S(\theta; \phi) = -\frac{\phi}{\theta} \cdot \frac{\partial F / \partial \phi}{\partial F / \partial \theta} \tag{5.9}$$

where the expressions of all the partial derivatives for all possible choices of the input parameter ϕ are reported in Table 5.1. Thus, the general procedure for computing objective sensitivities is as follows: First solve the algebraic equation (5.3) for θ, and then substitute the obtained θ value into Eq. (5.9) to compute the corresponding sensitivity value.

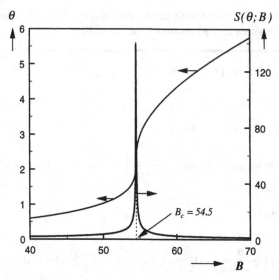

Figure 5.1. Reactor temperature θ and its normalized sensitivity $S(\theta; B)$ as functions of the heat-of-reaction parameter B for a CSTR operating in the region of unique steady states. $Da = 0.11$; $\gamma = 20$; $St = 10$; $n = 1$; $\theta_{co} = 0$. From Chemburkar *et al.* (1986).

In Fig. 5.1, the steady-state value of the reactor temperature θ is shown as a function of the heat-of-reaction parameter B, together with the corresponding normalized sensitivity of the reactor temperature with respect to B, $S(\theta; B)$. It is seen that the sensitivity exhibits a sharp maximum at a critical value of $B = B_c$, indicating a situation in which the reactor temperature is most sensitive to changes in B. As it is also apparent from the temperature values in Fig. 5.1, for B values greater than this critical value, *i.e.*, $B > B_c$, the reactor is in the ignited state. Thus, the reactor has a much higher temperature, which is undesirable in most cases, and it may be considered to be in a runaway condition. Therefore, according to the generalized sensitivity criterion, B_c corresponds to the critical value that separates two distinct operation regions for the reactor: the safe ($B < B_c$) and the runaway ($B > B_c$).

It is well known that nonisothermal CSTRs may exhibit steady-state multiplicity. This is characterized by the classical S-shape of the curve representing the steady-state reactor temperature in the θ–B plane, as shown in Fig. 5.2. For sufficiently small values of B, the reactor exhibits a unique steady state. As B increases, the available steady states become three, and then again only one for further increasing B values. Accordingly, when increasing the heat-of-reaction parameter B starting from low values, we see that the reactor temperature θ increases continuously, as indicated by the lower branch of the S-shaped curve shown in Fig. 5.2. However, when the upper bifurcation point B^* is reached, the steady-state solution on the lower branch is no longer available and the reactor operation "jumps" to the upper branch. This implies a discontinuous, significant increase of the reactor temperature, which can indeed be

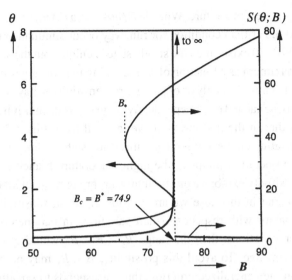

Figure 5.2. Reactor temperature, θ, and its normalized sensitivity, $S(\theta; B)$, as a function of the heat of reaction parameter, B, for a CSTR operating in the region of multiple steady states. $Da = 0.07$; $\gamma = 20$; $St = 10$; $n = 1$; $\theta_{co} = 0$. From Chemburkar, *et al.* (1986).

regarded as a strong sensitivity of the reactor temperature to small changes of the heat-of-reaction parameter B. Zeldovich (1941), Zeldovich and Zysin (1941), and Barkelew (1984) in fact used this bifurcation point to identify the critical conditions for parametric sensitivity. This corresponds to identifying runaway with ignition from a lower-temperature to a higher-temperature steady state. This behavior is confirmed by the values of the sensitivity $S(\theta; B)$ shown in Fig. 5.2 as a function of B. It is seen that as B approaches the upper bifurcation value B^*, the sensitivity increases progressively more rapidly, eventually becoming infinite. The generalized sensitivity criterion also identifies this behavior as the occurrence of reactor runaway, and the upper bifurcation value B^* as the critical condition for runaway to occur, since this is the location where the sensitivity exhibits its maximum.

It is important to note that even though for the conditions considered in Fig. 5.2 parametric sensitivity and ignition from one to another steady state occur simultaneously, they remain two distinct concepts. This clear for the conditions examined in Fig. 5.1, where no multiple steady states are present. Accordingly, no bifurcation is present, and the Barkelew (1984) criterion would then predict no parametric sensitivity. However, the sensitivity $S(\theta; B)$ exhibits a well-defined maximum, so that the generalized sensitivity criterion predicts two regions: safe and runaway. Thus we can conclude that *sensitivity is a phenomenon independent of multiplicity.*

Recently, Vajda and Rabitz (1993) have further emphasized this point by noting that steady-state bifurcation and the criticality to runaway correspond to singularity ($|\partial F/\partial \theta| = 0$) and near-singularity ($|\partial F/\partial \theta| \approx 0$), respectively, of Eq. (5.3), which determines the steady-state temperature. When $|\partial F/\partial\theta| = 0$, as in Fig. 5.2, then steady-state bifurcation and the critical condition for runaway occur simultaneously. If only a point where $|\partial F/\partial\theta| \approx 0$ exists, then the steady state is unique, but the sensitivity may exhibit a sharp maximum as a function of B, indicating reactor runaway, as for the case shown in Fig. 5.1. Thus, the analysis of parametric sensitivity behavior of a given system can, and should, be carried out independently of steady-state multiplicity.

It is worth pointing out that in the discussion above, as well as in Fig. 5.2, we always refer to a reactor operating on the low-temperature steady state, as long as this state is available. When the operation jumps to the high-temperature branch at the bifurcation point B^*, the sensitivity exhibits a discontinuity, dropping from an infinite to a finite value. It is clear that in this case we are considering local sensitivities, i.e., changes in system behavior with respect to small variations in parameters. If we include large parameter variations, then the range $B_* < B < B^*$ would also be considered parametrically sensitive. To avoid this possibility, $B < B_*$ may be used for safe operation. Based on the generalized criterion, this corresponds to sensitivity approaching infinity at $B = B_*$, as B decreases from larger values following the high-temperature steady-state branch. In the present work, when multiple steady states exist, we characterize runaway by the upper bifurcation point B^* since even small parameter changes near this point can lead to sharp temperature increases.

In Chapters 3 and 4, we have seen that a characteristic feature of the generalized runaway criterion is that the predicted critical conditions are the same, independent of the model input parameter selected for the sensitivity analysis. This feature is now verified for the specific case of a CSTR. Figure 5.3 shows the sensitivities of the

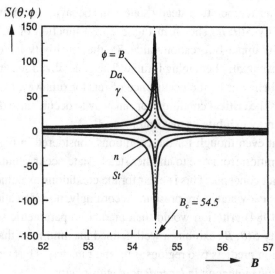

Figure 5.3. Normalized sensitivities $S(\theta; \phi)$ as a function of the heat-of-reaction parameter B for various choices of the model input parameter ϕ. $Da = 0.11$; $St = 10$; $\gamma = 20$; $n = 1$; $\theta_{co} = 0$.

Table 5.2. Critical values ψ_c ($=Da \cdot B_c/St$) corresponding to the maximum of $S(\theta; \phi)$ with respect to various choices of ϕ; $St = 10$; $\gamma = 20$; $n = 1$[a]

	ψ_c ($=Da \cdot B_c/St$)				
Da	$\phi = B$	$\phi = Da$	$\phi = St$	$\phi = \gamma$	$\phi = n$
0.01	0.4390	0.4390	0.4390	0.4390	0.4390
0.05	0.4935	0.4935	0.4935	0.4935	0.4935
0.11	0.5997	0.5997	0.5997	0.5997	0.5997
0.14	0.6677	0.6629	0.6677	0.6730	0.6702
0.17	0.7286	0.7113	0.7286	0.7464	0.7353
0.20	0.7845	0.7477	0.7845	0.8197	0.7954
0.30	0.9489	0.8063	0.9489	1.0641	0.9641
0.40	1.0944	0.79618	1.0944	1.3086	1.0944

[a] For $Da \leq 0.05$ the reactor may exhibit steady-state multiplicity.

reactor temperature with respect to each of the five model input parameters, B, Da, St, γ, and n. It appears that the critical value of B, where the sensitivity maximum is located, is the same to four significant figures ($B_c = 54.52$) for *any* choice of ϕ, thus indicating that the reactor temperature becomes sensitive to all the model parameters simultaneously. This finding is further substantiated by the results shown in Tables 5.2, 5.3, and 5.4, where, for three different values of γ, the critical values of ψ (defined as $Da \cdot B/St$) that maximize $S(\theta; \phi)$ for all possible choices of the input parameter ϕ are reported. Note that the quantity ψ has the same physical meaning as the Semenov number used in Chapters 3 and 4.

Two clear trends emerge from the results reported in Tables 5.2, 5.3, and 5.4:

(1) For a fixed value of γ, when Da is sufficiently small, all the ψ_c values predicted for different choices of ϕ are the same. However, the differences between the ψ_c values increase as Da increases.
(2) As γ increases, the differences between the values of ψ_c predicted for various choices of ϕ decrease.

Since, for a fixed ψ_c, an increase of Da implies a lower value of the critical heat-of-reaction parameter B_c, these findings indicate that in the region of low heat-of-reaction B and activation energy γ, the critical values of ψ predicted by the generalized criterion become dependent on the particular choice of ϕ. Thus, similarly to batch and tubular reactors examined in previous chapters, in this case the validity of a generalized runaway region is lost. This is because for low values of the heat-of-reaction parameter B, the low overall energy available to the system results in a mild nature of all temperature-related phenomena. Such a mild nature is evidenced by the relatively low values of the $S(\theta; \phi)$ maxima shown in Fig. 5.4, where it is seen that the sensitivity peaks reduce as Da increases (*i.e.*, as B_c decreases). In the same figure

Table 5.3. Critical values ψ_c ($=Da \cdot B_c/St$) corresponding to the maximum of $S(\theta; \phi)$ with respect to various choices of ϕ; $St = 10$; $\gamma = 20$; $n = 1^a$

Da	ψ_c ($=Da \cdot B_c/St$)				
	$\phi = B$	$\phi = Da$	$\phi = St$	$\phi = \gamma$	$\phi = n$
0.01	0.4202	0.4202	0.4202	0.4202	0.4202
0.05	0.4693	0.4693	0.4693	0.4693	0.4693
0.11	0.5581	0.5581	0.5581	0.5581	0.5581
0.14	0.6177	0.6172	0.6177	0.6181	0.6178
0.17	0.6794	0.6729	0.6794	0.6854	0.6815
0.20	0.7362	0.7166	0.7362	0.7528	0.7408
0.30	0.9034	0.7977	0.9034	0.9773	0.9101
0.40	1.0511	0.8049	1.0511	1.2018	1.0437

^{a}For $Da \leq 0.11$ the reactor may exhibit steady-state multiplicity.

Table 5.4. Critical values ψ_c ($=Da \cdot B_c/St$) corresponding to the maximum of $S(\theta; \phi)$ with respect to various choices of ϕ; $St = 10$; $\gamma = 1000$; $n = 1^a$

Da	ψ_c ($=Da \cdot B_c/St$)				
	$\phi = B$	$\phi = Da$	$\phi = St$	$\phi = \gamma$	$\phi = n$
0.01	0.4162	0.4162	0.4162	0.4162	0.4162
0.05	0.4643	0.4643	0.4643	0.4643	0.4643
0.11	0.5505	0.5505	0.5505	0.5505	0.5505
0.14	0.6069	0.6068	0.6069	0.6070	0.6069
0.17	0.6686	0.6637	0.6686	0.6731	0.6702
0.20	0.7255	0.7090	0.7255	0.7392	0.7293
0.30	0.8931	0.7949	0.8931	0.9596	0.8985
0.40	1.0411	0.8062	1.0411	1.1801	1.0326

^{a}For $Da \leq 0.11$ the reactor may exhibit steady-state multiplicity.

it may also be noted that for the smallest value of Da considered, i.e., $Da = 0.07$, the sensitivity curve exhibits a discontinuity. This is because in this case steady-state multiplicity is possible, and then as discussed above, the critical value B_c for runaway coincides with the upper bifurcation point B^*. Accordingly, the sensitivity of the reactor temperature undergoes a discontinuity when passing from the lower to the higher branch of the steady-state temperature curve.

Thus summarizing, for large B values, where thermal runaway is a significant phenomenon, the criterion for runaway retains its generalized nature. For smaller B values, where the obtained critical condition is no longer generalized, the runaway itself is a much less significant phenomenon.

In all instances discussed above, the critical value ψ_c ($=Da \cdot B_c/St$) has been obtained by maximizing the sensitivity $S(\theta; \phi)$ with respect to B. Let us now examine

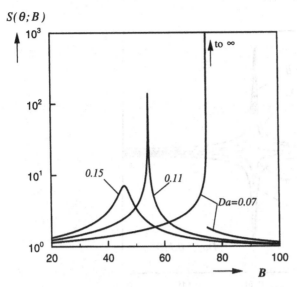

Figure 5.4. Normalized sensitivity as a function of the heat-of-reaction parameter B for various values of the Damkohler number, Da. $St = 10$; $\gamma = 20$; $n = 1$; $\theta_{co} = 0$. From Chemburkar *et al.* (1986).

the results obtained when the sensitivity maximization is performed with respect to some other parameter. For this we refer to the case shown in Fig. 5.3, where for $Da = 0.11$, $St = 10$, $\gamma = 20$, and $n = 1$, it is seen that all the sensitivity curves $S(\theta; \phi)$–B reach their maximum at $B_c = 54.52$, i.e., B_c is the critical B value for the reactor runaway, which corresponds to $\psi_c = 0.5997$. Now, let us fix $B = 54.52$, $St = 10$, $\gamma = 20$, and $n = 1$, and compute the values of the sensitivities $S(\theta; \phi)$ as a function of Da, as shown in Fig. 5.5. It appears that all the sensitivity curves again exhibit their maximum at the same location, which identifies the critical Da value for reactor runaway as $Da_c = 0.110$. This critical Da value is identical to that considered in Fig. 5.3, and implies again that $\psi_c = 0.5997$. This confirms the generalized character of the adopted runaway criterion, since it defines criticality as the condition where the sensitivity $S(\theta; \phi)$, regardless of the choice of ϕ, is maximized with respect to any of the involved input parameters. Thus, *a critical value found for the designated parameter, together with the fixed values of all the remaining parameters, determines the critical conditions for reactor runaway, i.e., all the system parameters are critical.*

Finally, a similar sensitivity analysis of the CSTR was performed by Vajda and Rabitz (1993), using their parametric sensitivity criterion discussed in Chapter 3. The basic idea is to introduce a small, unstructured perturbation of the parameters appearing in the steady-state equation (5.3) and examine its effect on the reactor temperature. The obtained results are in good agreement with those presented above and thus confirm the reliability of the two parametric sensitivity criteria. The details of the analysis performed by Vajda and Rabitz are not discussed here, and the interested reader may refer directly to their original paper.

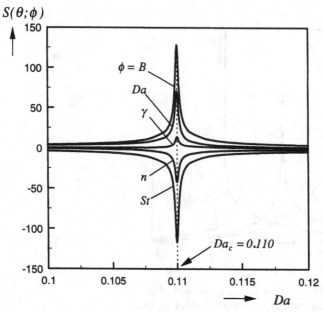

Figure 5.5. Normalized sensitivities $S(\theta; \phi)$ as a function of the Damkohler number Da for various choices of the model input parameter ϕ. $B = 54.5$; $St = 10$; $\gamma = 20$; $n = 1$; $\theta_{co} = 0$.

5.2 Regions of Parametrically Sensitive Behavior

5.2.1 Role of the Involved Physicochemical Parameters

After having established the existence of a generalized runaway region for a CSTR, let us now compute its boundaries in the $St/(DaB)$–B, *i.e.*, $1/\psi$–B, parameter plane, which is the same one considered earlier for batch (see Fig. 3.8) and tubular (see Fig. 4.2) reactors. In the case of a first-order reaction with $\gamma = 20$ and $\theta_{co} = 0$ (*i.e.*, coolant temperature equal to feed temperature), the obtained results are shown in Fig. 5.6 (solid curves) for various Da values. It appears that as Da increases, the runaway region for small $1/\psi$ values enlarges with respect to B, while for large B values it shrinks with respect to $1/\psi$. Moreover, as $B \to \infty$, the critical value of the Semenov number ψ approaches an asymptotic value, which is a function of Da.

In order to understand this behavior, let us rewrite Eq. (5.3) in the equivalent form:

$$F(\theta, \phi) = \frac{1}{Da} \cdot \left[\frac{St + 1}{B}(\theta - \theta_{co}) + \frac{\theta_{co}}{B} \right] - \exp\left(\frac{\theta}{1 + \theta/\gamma} \right)$$

$$\cdot \left[1 - \frac{St + 1}{B}(\theta - \theta_{co}) - \frac{\theta_{co}}{B} \right]^n = 0 \tag{5.10a}$$

$St/(DaB)$ and $(St+1)/(DaB)$

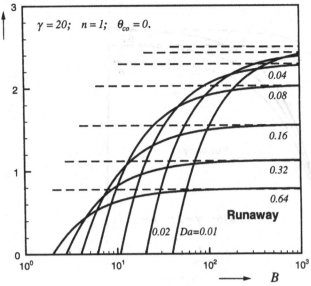

Figure 5.6. Boundaries of the runaway region for various Da values in the $St/(DaB)$–B plane (solid curves) and in the $(St + 1)/(DaB)$–B plane (broken curves).

For large values of $B(\rightarrow \infty)$, this reduces to

$$F(\theta, \phi) \approx \frac{St + 1}{Da \cdot B} \cdot (\theta - \theta_{co}) - \exp\left(\frac{\theta}{1 + \theta/\gamma}\right) \cdot \left[1 - \frac{St + 1}{B}(\theta - \theta_{co})\right]^n$$

$$= 0 \tag{5.10b}$$

It is seen that the heat-of-reaction parameter B and the heat-transfer parameter St are always coupled in the group $(St + 1)/B$, which may then be regarded as a single parameter. Thus, in the region of large B values, the critical value of the parameter $(St + 1)/B$ for runaway depends only on the Damkohler number, for fixed γ, n, and θ_{co}. This is consistent with the results shown in Fig. 5.6, indicating that the predicted critical values of $St/(DaB)$, for large B, depend on Da but not on B (or St). On physical grounds, this implies that, for highly exothermic reactions, in order to prevent reactor runaway, the ratio between heat removal and heat generation, $St/(DaB)$, cannot fall below a minimum value, which is lower the larger is the reactor conversion, *i.e.*, the Damkohler number.

A similar behavior is found in the case where the coolant temperature equals the feed temperature, *i.e.*, $\theta_{co} = 0$, so that Eq. (5.10a) reduces to

$$F(\theta, \phi) = \frac{St + 1}{Da \cdot B} \cdot \theta - \exp\left(\frac{\theta}{1 + \theta/\gamma}\right) \cdot \left[1 - \frac{St + 1}{B} \cdot \theta\right]^n = 0, \quad \text{for } \theta_{co} = 0 \tag{5.11}$$

St/(DaB) and (St+1)/(DaB)

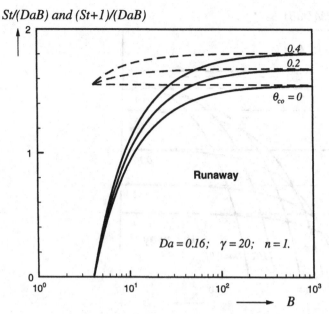

Figure 5.7. Boundaries of the runaway region for various values of the coolant temperature θ_{co} in the $St/(DaB)$–B plane (solid curves) and in the $(St + 1)/(DaB)$–B plane (broken curves).

Thus, for fixed γ and n, the reactor behavior is determined by only two parameters: Da and $(St + 1)/B$. In this case the runaway boundaries in the $(St + 1)/(DaB)$–B parameter plane are given by the broken horizontal straight lines shown in Fig. 5.6, indicating critical values of $(St + 1)/(DaB)$ depending on Da but not on B (or St). It is worth noting that in Fig. 5.6 each critical horizontal line cannot continue below a minimum B value, which depends on Da. This arises because of the physical constraint, $St \geq 0$. The minimum B value for each line corresponds to $St = 0$, i.e., adiabatic conditions. Note that a horizontal straight line as the runaway boundary is found in the $(St + 1)/(DaB)$–B parameter plane only in the case where the coolant and feed temperatures are equal. Otherwise, the runaway region is defined by curved boundaries as shown by the broken curves in Fig. 5.7.

Figures 5.8a–d show the runaway boundaries in the ψ–Da parameter plane, for various values of dimensionless activation energy γ, reaction order n, Stanton number St, and dimensionless coolant temperature θ_{co}, respectively. The regions above these boundaries correspond to runaway operations, while the safe operation regions are located below the boundaries. It can be seen that runaway is more likely as activation energy γ and coolant temperature θ_{co} increase, or as reaction order n and wall heat-transfer coefficient St decrease. Some anomalies at large Da values in Fig. 5.8d are due to the usual loss of the generalized nature of the runaway criterion, as discussed in several instances earlier. These plots can be used conveniently in the early stages of reactor design to avoid runaway operations.

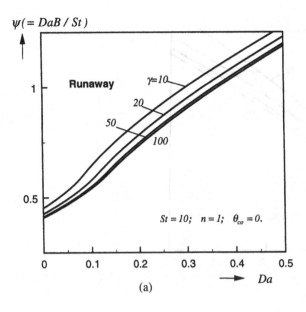

(a)

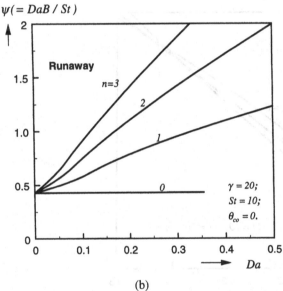

(b)

Figure 5.8. Runaway regions in the ψ–Da parameter plane for various values of (a) dimensionless activation energy γ, (b) reaction order n, (c) Stanton number St, and (d) dimensionless coolant temperature θ_{co}.

With reference to Fig. 5.8, we can investigate the limiting behavior of the critical value ψ_c, for low reaction rates, *i.e.*, as $Da \to 0$. Since ψ_c is finite (Fig. 5.8), it follows that, at criticality, $B \to \infty$ as $Da \to 0$. Since for low Da values, the reactor may exhibit steady-state multiplicity, the critical condition coincides with the upper bifurcation

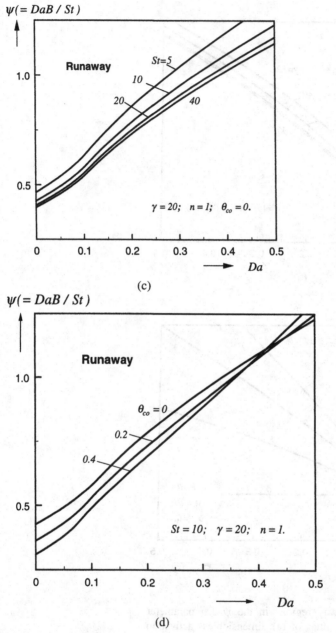

Figure 5.8. (cont.)

point of Eq. (5.3), where $\partial F / \partial \theta = 0$. Using the expressions reported in Table 5.1, this leads to

$$\frac{1 + St}{(1 + St) \cdot \theta - St \cdot \theta_{co}} - \frac{1}{(1 + \theta/\gamma)^2} + \frac{n \cdot (1 + St)}{B - (1 + St) \cdot \theta + St \cdot \theta_{co}} = 0 \quad (5.12)$$

which, since $B \to \infty$, reduces to

$$\frac{1 + St}{(1 + St) \cdot \theta - St \cdot \theta_{co}} - \frac{1}{(1 + \theta/\gamma)^2} = 0 \qquad (5.13)$$

This gives

$$\theta^* = \frac{\gamma}{2} \cdot [(\gamma - 2) - \sqrt{\gamma \cdot (\gamma - 4) - 4 \cdot \theta_{co} \cdot St/(1 + St)}] \qquad (5.14a)$$

and

$$\theta_* = \frac{\gamma}{2} \cdot [(\gamma - 2) + \sqrt{\gamma \cdot (\gamma - 4) - 4 \cdot \theta_{co} \cdot St/(1 + St)}] \qquad (5.14b)$$

which, when real, represent the critical temperatures corresponding to upper and lower bifurcation, respectively. Accordingly, the critical value ψ_c can be obtained by considering $\theta_c = \theta^*$ along with $B \to \infty$ in Eq. (5.3):

$$\psi = Da \cdot B_c/St = \left(1 + \frac{1}{St}\right)\left[(\theta_c - \theta_{co}) + \frac{\theta_{co}}{St + 1}\right] \cdot \exp\left(-\frac{\theta_c}{1 + \theta_c/\gamma}\right) \qquad (5.15)$$

This result confirms that the asymptotic value of ψ_c as $Da \to 0$ depends on γ, St, and θ_{co} but not on n, as indicated by the numerical computations shown in Figs. 5.8a–d. It is worth noting that in the particular case of zeroth-order reaction ($n = 0$), from Eq. (5.12) it is seen that the θ_c value given by Eq. (5.14a) is valid for *all* values of Da. Accordingly, in this case, the critical value ψ_c is given by Eq. (5.15) and is independent of Da, as clearly appears from the results shown in Fig. 5.8b for $n = 0$.

5.2.2 Relation between Multiplicity and Sensitivity Behavior

An important feature of the dynamics of reacting systems is the relation between multiplicity and sensitivity behavior. This is best addressed in the case of CSTRs, since the shapes of these two regions are particularly simple to analyze. In order to illustrate this aspect, let us consider the sensitivity behavior of a CSTR that can exhibit multiple steady states. We do this with reference to the B–Da parameter plane shown in Fig. 5.9, for fixed values of γ, n, and St, where runaway (I) and multiplicity (II) regions are indicated. Note that the multiplicity region (II) is bounded by the lower (B_*) and upper (B^*) bifurcation points shown in Fig. 5.2. In particular, we consider the reactor response to an increase of the B value, for various fixed values of Da. In the case where the reactor operates in the nonrunaway region and B increases below the lower bifurcation point, e.g., transition $P_1 Q_1$ in Fig. 5.9, the reactor operates always in the low conversion branch and no runaway occurs. The same holds true for the transition $P_2 Q_2$, although during this excursion the multiplicity region is encountered, since B is increased from a value less than B_* to a

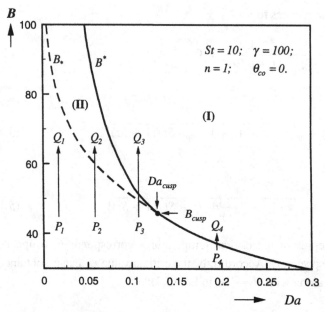

Figure 5.9. Regions of steady-state multiplicity and runaway in the B–Da parameter plane. (I) runaway region; (II) multiplicity region. From Chemburkar *et al.* (1986).

value lying between B_* and B^*. It is evident from Fig. 5.2 that the sensitivity of the low-temperature steady state does not exhibit a maximum, and hence these excursions are safe with respect to small perturbations. On the other hand, it is clear that in these conditions the reactor is asymptotically but not globally stable, since a sufficiently large perturbation in the operating conditions (*e.g.*, the transients during startup or shutdown) can lead to a transition from the low-temperature steady state to the ignited state. This can be regarded as a sensitivity behavior for the reactor. Using this approach, Bilous and Amundson (1955) investigated the sensitivity of CSTRs based entirely on multiplicity considerations. In the cases of excursions like P_3Q_3 and P_4Q_4, the sensitivity exhibits a maximum when the runaway boundary (*i.e.*, the solid curve in Fig. 5.9) is crossed, and then the reactor undergoes runaway. In the first case this occurs because the B value exceeds the upper bifurcation point B^* where sensitivity becomes infinite (Fig. 5.2), while in the second one the reactor steady state is always unique and the sensitivity goes through a maximum (Fig. 5.1). These two types of behavior have been discussed in detail in Section 5.1.

As a general conclusion, we can observe in Fig. 5.9 that steady-state multiplicity behavior is a subset of the parametrically sensitive behavior. If multiplicity exists for a given Da value, parametrically sensitive behavior also exists for the same Da. However, the converse is not true, as in the case of $Da > Da_{cusp}$, where although multiplicity is not possible, runaway may occur.

5.3 Role of Mixing on Reactor Parametric Sensitivity

The PFR and CSTR are two ideal models that represent two extreme conditions of mixing. In the first, no mixing occurs in the direction of flow and the fluid proceeds as an ideal piston, while in the second, mixing is so intense that composition and temperature are uniform throughout the entire reactor. Intermediate situations may be described by the axial dispersion model, which represents a tubular reactor with dispersion of mass and heat in the direction of flow, *i.e.*, the reactor axis (Varma and Aris, 1977). The magnitude of such dispersion processes is described by the Peclet numbers, one for mass and one for heat, defined as the ratio between the characteristic time of axial mixing and the average reactor residence time. It can be shown that the axial dispersion model approaches the CSTR as the Peclet numbers decrease toward zero, while the PFR model is obtained in the limit of large Peclet numbers. Thus, the axial dispersion model provides a more realistic description of reactors, where usually a finite degree of mixing is present. From the discussion above, it may be expected that the runaway behavior of tubular reactors with axial dispersion is intermediate between those of PFR and CSTR. In the sequel, we discuss the effect of mixing on parametric sensitivity by comparing the sensitivity behavior of CSTR and PFR.

Let us first consider typical runaway boundaries for a PFR in the $1/\psi$–B parameter plane, shown in Fig. 5.10. It is seen that the runaway region enlarges asymptotically as the Da value increases and approaches a limiting region when Da is greater than a certain value, equal to about 0.3 for the parameters considered in Fig. 5.10. This can

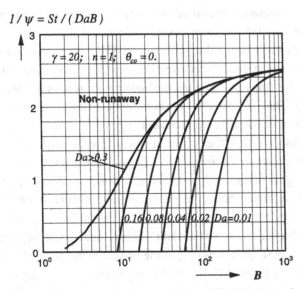

Figure 5.10. Runaway boundaries for a PFR in the $1/\psi$–B parameter plane for various values of Da.

be understood by noting that large Da values imply that the reactor is very long, or more precisely, the characteristic time of the chemical reaction is much smaller than the reactor residence time. Thus the reaction is completed inside the reactor and the maximum in the temperature profile occurs well before the reactor outlet and hence it is not affected by further increase of the reactor length (*i.e.*, Da).

A limiting behavior that is worth discussing is that corresponding to large values of the heat-of-reaction parameter, *i.e.*, $B \to \infty$, which is, for example, the typical situation for combustion reactions. In Fig. 5.10, it is seen that as $B \to \infty$ the critical Semenov number ψ approaches a unique asymptotic value for all Da. This is because for large values of the heat of reaction, small conversions are sufficient to cause large temperature increases so that runaway develops soon at the reactor entrance, and so the reactor length becomes irrelevant. This is the physical situation where Semenov analysis, which neglects reactant consumption, becomes valid. The results of this analysis have been discussed earlier in the context of batch reactors (see Section 3.2.1), but they also apply to tubular plug-flow reactors. It is in fact easy to see that the asymptotic value of ψ as $B \to \infty$, shown in Fig. 5.10, coincides with the value given by Eqs. (3.16) and (3.18) in Chapter 3, based on the Semenov criterion. In particular, in the case of large activation energy, *i.e.*, $\gamma = \infty$ (instead of $\gamma = 20$ in Fig. 5.10), the classical critical Semenov number, $\psi_c = 1/e$ is obtained.

The runaway behavior for PFRs described above is qualitatively different from that of CSTRs discussed in the previous section with reference to Fig. 5.6, where the same physicochemical parameter values as in Fig. 5.10 were considered. In particular, no asymptotic runaway region can be found as Da increases, and the asymptotic value of the critical Semenov number as $B \to \infty$ is a function of Da. This is due to the different objective in the definition of sensitivity adopted in the two cases. For PFRs, runaway is based on the temperature maximum along the reactor, which, when the reactor operates in the hot-spot regime, as discussed in detail in Chapter 4, becomes independent of Da for sufficiently large values of Da. For CSTRs, runaway is based on the reactor temperature, which, as can be seen from Eq. (5.3), depends always on Da, *i.e.*, the larger is the Da value (the reaction rate), the higher is the reactor temperature (the reactant conversion). Thus, for CSTRs, no asymptotic runaway region in the $1/\psi - B$ parameter plane can be found as Da increases.

It is worth noting that for CSTRs, in the particular case where the residence time is much shorter than the characteristic reaction time, *i.e.*, $Da \to 0$, the reactant conversion is negligible, and we are again in the conditions where Semenov analysis applies. In the previous section, we have shown that in this particular case the value of the critical Semenov number is given by Eqs. (5.14a) and (5.15), which can be compared with Eqs. (3.18) and (3.16) in Chapter 3, derived based on Semenov analysis. It may be seen that for $\theta_{co} = 0$ and $\theta_a = 0$, the expressions of θ_c in the two cases are identical. The ψ_c values are also equal, since in Eq. (5.15), $1 + (1/St)$ approaches 1 in the asymptotic region. Moreover, for $\gamma \to \infty$, the critical values of the Semenov number for both CSTR and PFR approach the classical value e^{-1}. This is an interesting result, which, in view of the different reacting systems examined, indicates the intrinsic nature of the parametric sensitivity concept.

About the effect of mixing on reactor runaway, by comparing Figs. 5.6 and 5.10 we see that when the characteristic time for the chemical reaction is smaller than the residence time, *i.e., for large Da values, runaway is less likely for a CSTR than for a PFR*, which indicates that mixing reduces the possibility of reactor runaway. This is because, for large *Da* values, the reactant conversion at the reactor outlet is high and in a tubular reactor the hot spot is located inside the reactor, as discussed in Chapter 4. Thus, mixing reduces the temperature at the hot spot, making runaway less likely. On the other hand, in the case where the characteristic time for the chemical reaction is large with respect to residence time, *i.e., for low Da values, runaway is more likely for a CSTR than for a PFR* (see for instance the runaway boundaries for $Da = 0.01$ in Figs. 5.6 and 5.10). This can be explained by recalling that the PFR at low *Da* values operates in the pseudo-adiabatic regime, where the reactor outlet conversion is low and the temperature maximum is located at the reactor outlet. Thus, axial mixing may increase the temperature toward the reactor inlet, increasing the possibility of reactor runaway.

Although the runaway regions are different for PFRs and CSTRs, depending on the *Da* value, there exists a similarity in behavior. In both Figs. 5.6 and 5.10, it is possible to identify a boundary that envelopes all the possible runaway regions for different *Da* values. In the region above this boundary, the reactor operation is safe for *all Da* values. In the sequel, we refer to this region as the *intrinsic nonrunaway region*. Knowledge of this region is very useful in the early stages of reactor design, because $1/\psi = St/(DaB)$ and B are parameters independent of the reactor size and feed flow rate. Thus, since ψ represents the ratio between the intrinsic heat generation rate and the heat removal rate, it is possible to determine for a given heat generation rate how large the heat removal rate should be, so as *to avoid reactor runaway, without the need to specify the reactor dimension or the feed flow rate*. In addition, it can be understood that, in the case of PFRs, the intrinsic nonrunaway boundary corresponds to the runaway boundary predicted by the MV generalized criterion when using the reactant conversion as the independent variable (since, in this case, the hot spot is guaranteed to occur within the reactor). In the case of CSTRs, this intrinsic nonrunaway boundary corresponds to the explicit criterion for CSTR runaway developed by Balakotaiah *et al.* (1995), discussed in the following section.

Figure 5.11 compares the two intrinsic nonrunaway boundaries indicated in Figs. 5.6 and 5.10. It is seen that the difference between the intrinsic nonrunaway regions of PFR and CSTR is relatively small. This indicates that, *for the intrinsic nonrunaway region, the influence of axial mixing on reactor runaway is not significant*. Figures 5.12a and b show the influence of the dimensionless activation energy γ on the intrinsic nonrunaway region for CSTRs and PFRs, respectively. As expected, the influence of γ is noticeable for relatively low activation energies, while for $\gamma > 50$ it can be neglected.

$1/\psi = St/(DaB)$

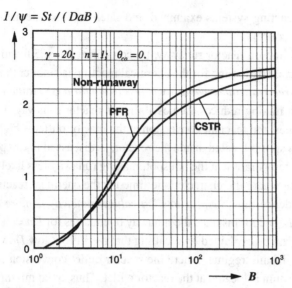

Figure 5.11. The intrinsic nonrunaway region in the $1/\psi$–B parameter plane for a CSTR and a PFR.

$1/\psi = St/(DaB)$

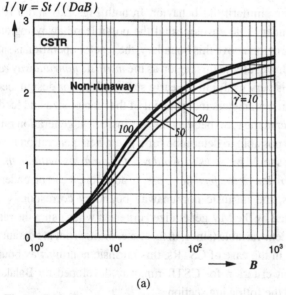

(a)

Figure 5.12. The intrinsic nonrunaway regions for various values of the dimensionless activation energy γ in the $1/\psi$–B parameter plane for (a) CSTR and (b) PFR. $n = 1$; $\theta_{co} = 0$.

Thus summarizing, by comparing the runaway behavior of CSTRs and PFRs, it can be concluded that mixing can either increase or decrease the possibility of reactor runaway, depending on the value of the Damkohler number. In particular, when the reactor residence time is much smaller than the characteristic reaction time, mixing

$1/\psi = St/(DaB)$

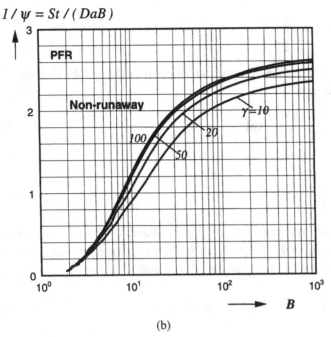

(b)

Figure 5.12. (cont.)

has a detrimental effect on runaway, while it has the opposite effect when the reactor residence time is much larger than the characteristic reaction time. However, when the intrinsic nonrunaway region is considered, the differences between the runaway behavior of CSTRs and PFRs are relatively small, indicating that mixing plays only a minor role in this characteristic.

5.4   Explicit Criteria for Parametric Sensitivity

Balakotaiah *et al.* (1995) derived explicit expressions for the runaway boundary in the case of a CSTR, based on the analysis of reaction paths in the temperature-conversion plane. The adopted runaway criterion is related to occurrence of reactor multiplicity, since it is assumed that runaway occurs when the reactor operates in the multiplicity region, as a consequence of large perturbations in the reactor operation that could lead to reactor ignition. Here, we discuss only the final results of this analysis, without going into the details of their derivation.

In the case of finite γ values, the explicit expression for the runaway boundary was derived for first-order reactions ($n = 1$), for $\theta_{co} = 0$. The runaway boundary is composed of two segments. The first segment is given by

$$\frac{1}{\psi} = \frac{St}{Da \cdot B} = \left[1 - \frac{4 \cdot \gamma}{B \cdot (\gamma - 4)} \right] \cdot \frac{e^2}{4}, \qquad (5.16a)$$

which is valid for

$$\frac{4 \cdot \gamma}{\gamma - 4} < B \le \frac{8 \cdot \gamma \cdot (\gamma - 2)}{(\gamma - 4)^2} \tag{5.16b}$$

When the B value is larger than the upper bound given by Eq. (5.16b), then the second segment applies, given by

$$\frac{1}{\psi} = \frac{St}{Da \cdot B} = \frac{(1-x)^2}{B \cdot x^2} \cdot \exp\left(\frac{B \cdot x^2}{1 + B \cdot x^2 / \gamma}\right) \tag{5.17a}$$

with

$$0 < x < \frac{\gamma - 4}{2 \cdot (\gamma - 2)} \tag{5.17b}$$

where a typographical error in the original paper of Balakotaiah *et al.* (1995) has been corrected, *i.e.*, B in Eq. (5.17a) replaces their B^2. Note that B and the conversion x in Eq. (5.17a) satisfy the following algebraic equation:

$$\frac{x^4}{\gamma^2} \cdot B^2 - x^2 \cdot \left(1 - x - \frac{2}{\gamma}\right) \cdot B + 1 = 0 \tag{5.17c}$$

Thus, in order to compute the second segment, two procedures can be followed: (1) for a given B value larger than the upper bound in Eq. (5.16b), we determine the x value from Eq. (5.17c), which should satisfy Eq. (5.17b), and then compute the critical ψ value from Eq. (5.17a); (2) for a given x value in the range given by Eq. (5.17b), we determine the B value as the smaller root of Eq. (5.17c), and then calculate the critical ψ value from Eq. (5.17a).

The runaway boundaries predicted by this explicit criterion are shown by the solid curves in Fig. 5.13 for various γ values, together with the intrinsic nonrunaway boundaries defined in the previous section (broken curves). It may be noted that the runaway boundaries given by Eqs. (5.16) and (5.17) are practically identical to the intrinsic nonrunaway boundaries. Differences arise only in the region of very low B values, where parametric sensitivity is no longer a significant phenomenon. Thus, it appears that the explicit criterion developed by Balakotaiah *et al.* (1995) predicts only the intrinsic nonrunaway boundaries, and for a specific Da, it is expected to provide conservative runaway boundaries. This is confirmed by the results shown in Fig. 5.14, where the critical conditions for runaway predicted by the explicit criterion and the MV generalized criterion in the case of $Da = 0.16$, $B = 100$, and $n = 1$ are compared. In this case, the predicted critical values, $1/\psi_c$ for the explicit and generalized criteria, are 2.05 and 1.49, respectively. It is evident that the value $1/\psi_c = 1.49$, given by the generalized criterion, represents the true transition between runaway and safe operation, because the temperature profile exhibits a sharp increase as $1/\psi$ decreases below $1/\psi_c = 1.49$. Instead, $1/\psi_c = 2.05$ is a conservative estimate, since there is no sharp temperature increase with the decrease of $1/\psi$ and the temperature values remain relatively low.

$1/\psi = St/(DaB)$

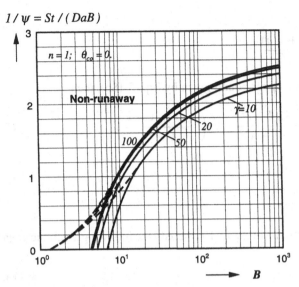

Figure 5.13. The intrinsic nonrunaway regions in the $1/\psi-$ B plane for first-order reactions in a CSTR: generalized (broken curves) and explicit (solid curves) criteria.

$S(\theta; B)$ and θ

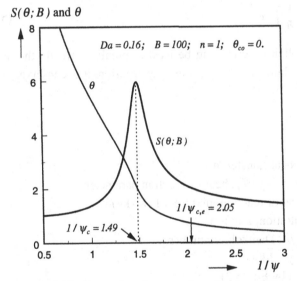

Figure 5.14. Comparison between the critical conditions for runaway in the case of a CSTR, predicted by the generalized ($1/\psi_c$) and the explicit ($1/\psi_{c,e}$) criteria.

In the case of $\gamma \to \infty$, a similar explicit criterion has been derived by Balakotaiah *et al.* (1995) for all positive-order reactions, *i.e.*, $n \geq 0$

$$\frac{1}{\psi} = \frac{St}{Da \cdot B} = \left[1 - \frac{(1 + \sqrt{n})^2}{B}\right] \cdot (1 + \sqrt{n})^{-1-n} \cdot n^{n/2} \cdot \exp(1 + \sqrt{n}) \quad (5.18a)$$

with

$$(1 + \sqrt{n})^2 < B \le \frac{(1 + \sqrt{n})^3}{\sqrt{n}} \tag{5.18b}$$

When B is greater than the upper bound given by Eq. (5.18b), then

$$\frac{1}{\psi} = \frac{St}{Da \cdot B} = \frac{(1 - x)^{n+2}}{(n \cdot x + 1 - x)^2} \cdot \exp\left(\frac{n \cdot x + 1 - x}{1 - x}\right) \tag{5.19a}$$

with

$$0 < x < \frac{1}{1 + \sqrt{n}} \tag{5.19b}$$

In this case, B and conversion x in Eq. (5.19a) satisfy the following algebraic equation:

$$B = \frac{(n \cdot x + 1 - x)^2}{n \cdot x^2 \cdot (1 - x)} \tag{5.19c}$$

For the special case of a zeroth-order reaction, the entire runaway boundary is defined by Eq. (5.18), which reduces to

$$\frac{1}{\psi} = \frac{St}{Da \cdot B} = \left(1 - \frac{1}{B}\right) \cdot e \tag{5.20}$$

Also in this case, the explicit criterion can be used to predict correctly the intrinsic nonrunaway boundary, but, as a general runaway criterion, it is conservative for practical applications.

Nomenclature

A	Surface area for heat transfer, m^2
B	$(-\Delta H) \cdot C_f \cdot \gamma / \rho \cdot c_p \cdot T_f$, heat-of-reaction parameter
c_p	Mean specific heat of reaction mixture, kJ/(K $\cdot$ kg)
C	Reactant concentration, kmol/m^3
Da	$V \cdot k(T_f) \cdot C_f^{n-1}/q$, Damkohler number
E	Activation energy, kJ/kmol
F	Function, defined by Eq. (5.3)
k	Reaction rate constant, (kmol/m^3)$^{1-n}$/s
n	Reaction order
q	Volumetric flow rate, m^3/s
R_g	Ideal-gas constant, kJ/(K $\cdot$ kmol)
$s(\theta; \phi)$	$d\theta/d\phi$, local sensitivity of θ with respect to the independent parameter ϕ
$S(\theta; \phi)$	$d \ln \theta / d \ln \phi$, normalized sensitivity of θ with respect to the independent parameter ϕ
St	$A \cdot U / \rho \cdot c_p \cdot q$, Stanton number

T	Temperature, K
U	Overall heat transfer coefficient, kJ/(m$^2 \cdot$ s $\cdot$ K)
V	Volume of CSTR, m^3
x	$1 - C/C_f$, reactant conversion

Greek Symbols

ΔH	Heat of reaction, kJ/kmol
γ	$E/R_g \cdot T_f$, dimensionless activation energy
ϕ	Generic model input parameter
ϕ	Vector of model input parameters
θ	$\gamma \cdot (T - T_f)/T_f$, dimensionless reactor temperature
ρ	Density, kg/m^3
ψ	DaB/St, Semenov number

Subscripts

c	Critical value
co	Coolant
$cusp$	Cusp point
f	Feed
$*$	Lower bifurcation point

Superscript

| $*$ | Higher bifurcation point |

References

Balakotaiah, V., Kodra, D., and Nguyen, D. 1995. Runaway limits for homogeneous and catalytic reactors. *Chem. Eng. Sci.* **50**, 1149.

Barkelew, C. H. 1984. Stability of adiabatic reactors. *Am. Chem. Soc. Ser.* **237**, 337.

Bilous, O., and Amundson, N. R. 1955. Chemical reactor stability and sensitivity. *A.I.Ch.E. J.* **1**, 513.

Chemburkar, R. M., Morbidelli, M., and Varma, A. 1986. Parametric sensitivity of a CSTR. *Chem. Eng. Sci.* **41**, 1647.

Froment, G. F., and Bischoff, K. B. 1990. *Chemical Reactor Analysis and Design*, 2nd ed. New York: Wiley.

Morbidelli, M., and Varma, A. 1988. A generalized criterion for parametric sensitivity: application to thermal explosion theory. *Chem. Eng. Sci.* **43**, 91.

Vajda, S., and Rabitz, H. 1993. Generalized parametric sensitivity: application to a CSTR. *Chem. Eng. Sci.* **48**, 2453.

Varma, A., and Aris, R. 1977. Stirred pots and empty tubes. In *Chemical Reactor Theory: A Review*, L. Lapidus and N. Amundson, eds. Englewood Cliffs, NJ: Prentice-Hall.

Zeldovich, Ya. B. 1941. On the theory of heat-release from chemical reactions in streams. Part I. *Zh. Tekh. Fiz.* **11**, 493.

Zeldovich, Ya. B., and Zysin, Yu. A. 1941. On the theory of heat-release from chemical reactions in streams. Part II. Analysis of heat losses during the reaction. *Zh. Tekh. Fiz.* **11**, 501.

6

Runaway in Fixed-Bed
Catalytic Reactors

IXED-BED CATALYTIC REACTORS consist of single tubes or bundles of
tubes, packed with catalyst particles. They can be simulated by using the *pseudo-homogeneous model* where interparticle and intraparticle mass and heat transport resistances are neglected. The reactor runaway behavior predicted by this model has been discussed in Chapter 4. However, this simple model can be confidently applied to simulate catalytic reactors only for slow reactions. In most cases of practical interest, catalytic reactions are relatively fast and the roles of interparticle and intraparticle transport resistances have to be considered in the simulations. Thus, in the present chapter, the parametric sensitivity behavior of fixed-bed catalytic reactors is investigated by using a *heterogeneous model*, where both interparticle and intraparticle mass and heat transport resistances are included.

In a heterogeneous catalytic reactor, the temperature inside the catalyst particle is the key variable to be controlled, since it affects the reaction rate as well as the catalyst activity, selectivity, and life. Runaway of the particle temperature may occur because the fluid temperature is running away and the particle temperature simply follows it. This is the same phenomenon as in homogeneous tubular reactors discussed in Chapter 4. In this chapter we also account for the runaway of the particle temperature, which is strictly related to the heterogeneous nature of the system and is governed by the interaction between interparticle and intraparticle mass and heat transport and the chemical reactions.

Two approaches have been typically adopted to investigate the runaway behavior of the particle temperature. In the first, proposed by McGreavy and Adderley (1973), a single catalyst particle is extracted from the reactor and the temperature runaway behavior of this isolated particle is investigated. Since this approach deals only with a single particle located somewhere along the reactor, it is usually referred to as the *local runaway* problem. The study of local runaway indicates that the particle temperature may run away even though the external fluid temperature lies within the safe regime.

The second approach considers the entire particle temperature profile along the reactor length and, in particular, it focuses on the particle temperature maximum to identify the operating conditions that lead to reactor runaway. This may be referred to as the *global runaway* problem. It should be noted that the reactor runaway discussed in Chapter 4 in the context of homogeneous reactors is sometimes also referred to as global runaway (Rajadhyaksha *et al.*, 1975). The difference between the global runaway of this chapter and the one discussed in Chapter 4 is that the former refers to the hot spot of the particle temperature profile, while the latter refers to the hot spot of the fluid temperature profile.

The local runaway is first discussed in Section 6.2, through applications of the Morbidelli and Varma (1988) generalized criterion. Then, in Section 6.3, the global runaway is investigated. Maps of the runaway regions in the space of the reactor operating conditions are then presented, where both local and global runaway are accounted for. These maps can be used in practice to avoid the operating conditions leading to reactor runaway at the early stages of reactor design. Some examples to illustrate these ideas are discussed. Finally, some approximate, explicit runaway criteria are discussed and their predictions are compared with those of the exact criteria. However, before proceeding further, let us discuss in detail the main features of the heterogeneous model to be used in the parametric sensitivity analysis.

6.1   The Heterogeneous Model of a Fixed-Bed Catalytic Reactor

The heterogeneous model of a fixed-bed catalytic reactor involves the mass and energy balances in the fluid and solid phases, which for a single reaction can be written for the fluid phase as

$$-v^o \cdot \frac{dC}{dl} = k_g \cdot a_v \cdot (C - C_s) \tag{6.1}$$

$$\rho_f \cdot c_p \cdot v^o \cdot \frac{dT}{dl} = h \cdot a_v \cdot (T_s - T) - \frac{4 \cdot U}{d_t} \cdot (T - T_{co}) \tag{6.2}$$

and for the solid phase as

$$\rho_B \cdot r = k_g \cdot a_v \cdot (C - C_s) \tag{6.3}$$

$$(-\Delta H) \cdot \rho_B \cdot r = h \cdot a_v \cdot (T_s - T) \tag{6.4}$$

with inlet conditions

$$C = C^i, \qquad T = T^i, \quad \text{at } l = 0 \tag{6.5}$$

Deriving the above equations, we have assumed plug flow in the axial direction and perfect mixing in the radial direction. It is worth noting that the first assumption is probably acceptable in practice (Carberry and Wendel, 1963), while the second is less realistic, particularly for nonadiabatic reactors. However, when the radial mass and heat dispersions are included, the fluid-phase model becomes two dimensional and thus cumbersome to deal with. In practical applications, if the effect of the radial dispersions needs to be included, the approximate methods proposed by Finlayson (1971) or Hagan *et al.* (1988a,b) may be used. Finlayson proposed to estimate the radial temperature profile through a one-point collocation approximation, thus reducing the original two-dimensional model to a one-dimensional model, while Hagan *et al.* derived a one-dimensional model, referred to as the α model, to approximate the two-dimensional model through an axial location-dependent parameter α. As shown in Section 4.3, the α model is relatively more accurate and thus is recommended for studies of reactor parametric sensitivity. In the following, we will consider only the one-dimensional plug-flow model for the fluid phase.

In the case of an irreversible nth-order reaction, Eqs. (6.1) and (6.2) for the fluid phase, when combined with those for the solid phase, become, in dimensionless form,

$$\frac{dx}{dz} = Da \cdot \exp\left(\frac{\theta_s}{1 + \theta_s/\gamma}\right) \cdot (1 - x_s)^n \cdot \eta \tag{6.6}$$

$$\frac{d\theta}{dz} = B \cdot Da \cdot \exp\left(\frac{\theta_s}{1 + \theta_s/\gamma}\right) \cdot (1 - x_s)^n \cdot \eta - St \cdot (\theta - \theta_{co}) \tag{6.7}$$

with inlet conditions

$$x = 0, \qquad \theta = 0 \quad \text{at } z = 0 \tag{6.8}$$

where

$$x = 1 - \frac{C}{C^i}; \qquad \theta = \gamma \cdot \frac{T - T^i}{T^i}; \qquad z = \frac{l}{L}; \qquad Da = \frac{\rho_B \cdot k(T^i) \cdot (C^i)^{n-1} \cdot L}{v^o};$$

$$B = \frac{(-\Delta H) \cdot C^i}{\rho_f \cdot c_p \cdot T^i} \cdot \gamma; \qquad St = \frac{4 \cdot U \cdot L}{v^o \cdot \rho_f \cdot c_p \cdot d_t}; \qquad \gamma = \frac{E}{R_g \cdot T^i} \tag{6.9}$$

The quantities x_s and θ_s in Eqs. (6.6) and (6.7) are the reactant conversion and the dimensionless temperature at the external surface of the particle, respectively, and are determined by the solid-phase equations (6.3) and (6.4), which take the dimensionless forms

$$x = x_s - Da_p \cdot \exp\left[\frac{\theta_s}{1 + \theta_s/\gamma}\right] \cdot (1 - x_s)^n \cdot \eta \tag{6.10}$$

$$\theta = \theta_s - \frac{B \cdot Da_p}{Le} \cdot \exp\left[\frac{\theta_s}{1 + \theta_s/\gamma}\right] \cdot (1 - x_s)^n \cdot \eta \tag{6.11}$$

where

$$Da_p = \frac{\rho_B \cdot k(T^i) \cdot (C^i)^{n-1}}{k_g \cdot a_v}; \qquad Le = \frac{h}{k_g \cdot \rho_f \cdot c_p} \tag{6.12}$$

The parameter η is the *effectiveness factor*, which takes into account the effect of concentration and temperature gradients inside the catalyst particles on the global reaction rate. It is defined as the ratio between the actual reaction rate in the catalyst particle and the reaction rate computed at the concentration and temperature values at the external surface of the particle. Its evaluation requires the solution of the detailed mass and energy balances inside the catalyst particle, which generally involves appropriate numerical techniques. However, it has been shown (Carberry, 1975; Pereira *et al.*, 1979) that in most cases of interest in applications, the temperature gradient within the catalyst particle is much smaller than the temperature difference between the external surface of the particle and the fluid phase. This leads to the so-called *internal isothermal model*, where the temperature is assumed to be constant throughout the catalyst particle but different from the corresponding value in the fluid phase. With this assumption, the solution of the detailed mass and energy balances inside the catalyst particle can be approximated by the following expression of the effectiveness factor η, which applies to any particle geometry and reaction order $n > -1$ (Aris, 1965; Bischoff, 1965; Petersen, 1965):

$$\eta = \frac{3\Phi \coth(3\Phi) - 1}{3\Phi^2} \tag{6.13}$$

where

$$\Phi^2 = (\Phi^i)^2 \cdot (1 - x_s)^{n-1} \cdot \exp[\theta_s/(1 + \theta_s/\gamma)] \tag{6.14}$$

and

$$(\Phi^i)^2 = \left(\frac{V_p}{S_p}\right)^2 \cdot \frac{n_p + 1}{2} \cdot \frac{\rho_p \cdot k(T^i) \cdot (C^i)^{n-1}}{D_e} \tag{6.15}$$

The parameter Φ, usually referred to as the normalized Thiele modulus, represents the ratio between the chemical reaction rate and the mass diffusion rate in the catalyst particle. Note that in this model the temperature at the external surface, θ_s, is equal to the temperature throughout the entire catalyst particle.

6.2   Runaway of a Single Catalyst Particle: Local Runaway

By investigating the local runaway behavior, McGreavy and Adderley (1973) showed that, even though the fluid temperature lies in the safe operation region, the particle temperature may undergo thermal runaway. A technique was developed for determining the critical conditions for particle runaway, which, however, as will be shown later, is conservative. Thus, the runaway behavior of a single catalyst particle is discussed here through applications of the MV generalized criterion.

When considering a single catalyst particle, we need to focus only on the mass and energy balance equations (6.10) and (6.11) for the solid phase. It is readily seen that these equations imply the following relationship between conversion and temperature on the particle surface:

$$x_s = x + \frac{Le}{B} \cdot (\theta_p - \theta) \tag{6.16}$$

Substituting this into Eq. (6.11) gives

$$F(\theta_p) = \theta_p - \theta - \frac{B \cdot Da_p}{Le} \cdot \exp\left(\frac{\theta_p}{1 + \theta_p/\gamma}\right) \cdot \left(1 - x - \frac{Le}{B} \cdot (\theta_p - \theta)\right)^n \cdot \eta = 0 \tag{6.17}$$

Note that since the internal isothermal model has been used here, the temperature at the external surface equals the temperature in the entire catalyst particle. Thus, θ_s in Eqs. (6.10) and (6.11) has been replaced by θ_p in Eqs. (6.16) and (6.17). The distinguishing features of a single catalyst particle behavior are now investigated through Eq. (6.17), together with the expression (6.13) for the effectiveness factor η. Note that in this context we assume the values of the conversion x and temperature θ in the fluid phase as fixed, depending on the specific location of the considered catalyst particle along the reactor.

6.2.1 Critical Conditions for Local Runaway of Particle Temperature

In the following, we apply the generalized runaway criterion (Morbidelli and Varma, 1988) to a single catalyst particle. For this, we need to define the objective sensitivity, which in this case is given naturally by the sensitivity of the particle temperature, defined as

$$s(\theta_p; \phi) = \frac{d\theta_p}{d\phi} \tag{6.18}$$

where ϕ represents any of the six model input parameters, B, Da_p, Le, γ, n, and Φ^i. As discussed in previous chapters, a more appropriate quantity in sensitivity analysis is the normalized objective sensitivity, $S(\theta_p; \phi)$, defined as

$$S(\theta_p; \phi) = \frac{\phi}{\theta_p} \cdot s(\theta_p; \phi) \tag{6.19}$$

which has a clearer physical meaning since it serves to normalize the magnitudes of the parameter, ϕ, and the dimensionless catalyst temperature, θ_p. The expression of the normalized sensitivity in this case can be readily obtained by differentiating Eq. (6.17) with respect to ϕ, leading to the following relationship:

$$S(\theta_p; \phi) = -\frac{\phi}{\theta_p} \cdot \frac{\partial F/\partial \phi}{\partial F/\partial \theta_p} \tag{6.20}$$

Table 6.1. Analytical expressions for partial derivatives of F given by Eq. (6.17)

$$\frac{\partial F}{\partial \theta_p} = 1 - (\theta_p - \theta) \cdot \left[\frac{1}{(1 + \theta_p/\gamma)^2} + \frac{1}{\eta} \cdot \frac{\partial \eta}{\partial \theta_p} - \frac{n \cdot Le/B}{1 - x - Le \cdot (\theta_p - \theta)/B} \right]$$

$$\frac{\partial F}{\partial B} = -(\theta_p - \theta) \cdot \left[\frac{1}{B} + \frac{1}{\eta} \frac{\partial \eta}{\partial B} + \frac{n \cdot Le \cdot (\theta_p - \theta)/B^2}{1 - x - Le \cdot (\theta_p - \theta)/B} \right]$$

$$\frac{\partial F}{\partial Da_p} = -\frac{(\theta_p - \theta)}{Da_p}.$$

$$\frac{\partial F}{\partial Le} = -(\theta_p - \theta) \cdot \left[-\frac{1}{Le} + \frac{1}{\eta} \cdot \frac{\partial \eta}{\partial Le} - \frac{n \cdot (\theta_p - \theta)/B}{1 - x - Le \cdot (\theta_p - \theta)/B} \right]$$

$$\frac{\partial F}{\partial \gamma} = -(\theta_p - \theta) \cdot \left[\left(\frac{\theta_p}{\gamma + \theta_p} \right)^2 + \frac{1}{\eta} \cdot \frac{\partial \eta}{\partial \gamma} \right]$$

$$\frac{\partial F}{\partial n} = -(\theta_p - \theta) \cdot \left[\ln[1 - x - Le \cdot (\theta_p - \theta)/B] + \frac{1}{\eta} \cdot \frac{\partial \eta}{\partial n} \right]$$

$$\frac{\partial F}{\partial \Phi^i} = -(\theta_p - \theta) \cdot \frac{1}{\eta} \cdot \frac{\partial \eta}{\partial \Phi^i}$$

where the two partial derivatives, for all possible choices of the model input parameter, ϕ, can be expressed analytically as reported in Tables 6.1 and 6.2. The general procedure for computing the sensitivity value involves first solving the algebraic equation (6.17) to obtain the θ_p value, and then substituting it into Eq. (6.20) to evaluate the corresponding sensitivity. Note that the θ_p value used corresponds to the smallest root of Eq. (6.17), since, as discussed later, we consider here the sensitivity of a catalyst particle operating on the low conversion branch, when multiple steady states are available for the catalyst particle.

Figure 6.1 shows the values of the normalized sensitivity, $S(\theta_p; B)$, as a function of B, together with the corresponding catalyst temperature, θ_p. It is seen that the sensitivity exhibits a sharp maximum at $B = B_c$, which, according to the generalized runaway criterion, is defined as the critical B value that separates the safe from the runaway operation region. As it clearly appears from the catalyst temperature values, the safe operation region is for $B < B_c$, while for $B > B_c$ the catalyst particle has undergone runaway since its temperature is much higher.

It is well known (Aris, 1975) that in the case of nonisothermal reactions (or sufficiently nonlinear isothermal reaction rate expressions), a catalyst particle may exhibit steady-state multiplicity. For example, in the case of a nonisothermal first-order reaction, this feature is given by the classical S-shape curve shown in Fig. 6.2, representing the solution θ_p of Eq. (6.17) as a function of the heat-of-reaction parameter B. This leads to the characteristic multiplicity pattern 1-3-1, where three steady states are possible for B values between the lower and the upper bifurcation points, i.e.,

Table 6.2. Analytical expressions for partial derivatives of η given by Eq. (6.13)

$$\frac{\partial \eta}{\partial \Phi} = \frac{9\Phi^2 + 2 - 3\Phi \coth(3\Phi)[1 + 3\Phi \coth(3\Phi)]}{3\Phi^3}$$

$$\frac{\partial \eta}{\partial \theta_p} = \frac{\partial \eta}{\partial \Phi} \cdot \frac{\Phi}{2} \cdot \left[\frac{1}{(1 + \theta_p/\gamma)^2} - \frac{(n-1) \cdot Le/B}{1 - x - Le \cdot (\theta_p - \theta)/B} \right]$$

$$\frac{\partial \eta}{\partial B} = \frac{\partial \eta}{\partial \Phi} \cdot \frac{\Phi}{2} \cdot \frac{(n-1) \cdot Le \cdot (\theta_p - \theta)/B^2}{1 - x - Le \cdot (\theta_p - \theta)/B}$$

$$\frac{\partial \eta}{\partial Le} = -\frac{\partial \eta}{\partial \Phi} \cdot \frac{\Phi}{2} \cdot \frac{(n-1) \cdot (\theta_p - \theta)/B}{1 - x - Le \cdot (\theta_p - \theta)/B}$$

$$\frac{\partial \eta}{\partial \gamma} = \frac{\partial \eta}{\partial \Phi} \cdot \frac{\Phi}{2} \cdot \left(\frac{\theta_p}{\gamma + \theta_p} \right)^2$$

$$\frac{\partial \eta}{\partial n} = \frac{\partial \eta}{\partial \Phi} \cdot \frac{\Phi}{2} \cdot \ln[1 - x - Le \cdot (\theta_p - \theta)/B]$$

$$\frac{\partial \eta}{\partial \Phi^i} = \frac{\partial \eta}{\partial \Phi} \cdot \frac{\Phi}{\Phi^i}$$

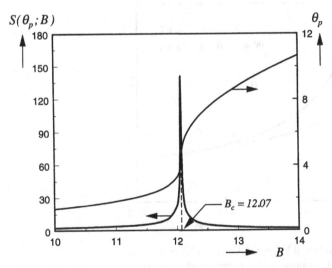

Figure 6.1. Temperature of the catalyst particle θ_p and its normalized sensitivity with respect to the heat-of-reaction parameter $S(\theta_p; B)$ as a function of B. $Da_p = 0.1$; $Le = 1$; $\gamma = 20$; $n = 1$; $\Phi^i = 1$; $\theta = 0$; $x = 0$.

$B_* < B < B^*$, while a unique steady state exists for all other B values. The values of the bifurcation points are shown in Fig. 6.3 as a function of Da_p. They identify two regions in the B–Da_p parameter plane: region I, where a unique steady state exists for the catalyst particle; and region II, where instead three steady states are simultaneously available.

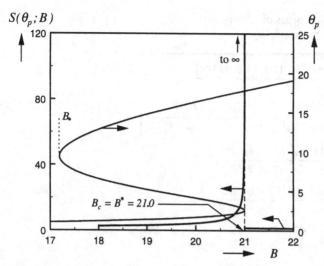

Figure 6.2. Temperature of the catalyst particle θ_p and its normalized sensitivity with respect to the heat-of-reaction parameter $S(\theta_p; B)$ as a function of B, in the multiplicity region. $Da_p = 0.05$; $Le = 1$; $\gamma = 20$; $n = 1$; $\Phi^i = 1$; $\theta = 0$; $x = 0$.

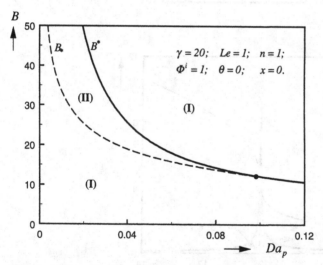

Figure 6.3. Upper (B^*) and lower (B_*) bifurcation points and critical value for reactor runaway (B_c) represented by the solid curve, as functions of the Damkohler number Da_p. (I) unique steady state; (II) three steady states; ($\bullet$) cusp point where the two bifurcation points merge.

In the case where multiple steady states exist, the normalized sensitivity $S(\theta_p; B)$ for a catalyst particle operating on the lower conversion branch increases with B and becomes infinite at the upper bifurcation point $B = B^*$, as shown in Fig. 6.2. This arises because, at any bifurcation point, we have $\partial F/\partial\theta_p = 0$, which from Eq. (6.20) leads to $|S(\theta_p; \phi)| \to \infty$ for any ϕ. Beyond the upper bifurcation point the particle operates on the high conversion branch, where its sensitivity is again finite and actually rather small. This leads to the discontinuity in the sensitivity value at B^* shown in Fig. 6.2. Thus, according to the generalized sensitivity criterion (as discussed in Section 5.1 in the context of CSTRs), this bifurcation point represents the critical condition for the catalyst runaway, $i.e.$, $B_c = B^*$. It is worth noting that, when a catalyst particle exhibits multiple steady states, its runaway boundary is always coincident with the upper bifurcation point of the multiple steady-state region. This clearly appears in Fig. 6.3, where in the B–Da_p parameter plane the lower and upper bifurcation points, $i.e.$, B_* and B^*, are shown together with the critical value for reactor runaway, B_c, predicted by the generalized sensitivity criterion. However, sensitivity and multiplicity are two independent phenomena, and, as shown in Fig. 6.1, runaway may occur in the region where the catalyst particle exhibits a unique steady state. This corresponds to the portion of the solid curve in Fig. 6.3, to the right of the cusp point, where only B_c is shown since the bifurcation points do not exist. Note that all the above features are similar to those discussed previously in Section 5.2.2 in the context of CSTR.

It is worth stressing that, in our runaway considerations, we always refer to a catalyst particle operating on the low conversion branch. This is because even small parameter variations near B^* can lead to sharp temperature increases. A notable exception is given by a reactor operating with periodic flow reversal (Matros and Bunimovich, 1996), where the catalyst particles are forced to operate periodically in ignited conditions, $i.e.$, on the high conversion branch. In this case, the absolute value of the normalized sensitivity of the catalyst temperature, $S(\theta_p; B)$, increases when B decreases. Hence it becomes infinite at the lower bifurcation point $B = B_*$, and then drops to a finite value for $B < B_*$, where the catalyst temperature jumps to the low conversion branch. Thus, B_* can be regarded as the critical condition for extinction of the catalyst particle operating in ignited conditions (Wu et $al.$, 1998). It is clear geometrically that the critical B value for catalyst ignition (runaway) is larger than that for catalyst extinction in the multiple steady-state region, while the two critical B values become equal in the unique steady-state region.

In Figs. 6.1 and 6.2, the critical value of B for the catalyst particle runaway has been obtained by analyzing the behavior of the normalized sensitivity of the particle temperature with respect to the heat-of-reaction parameter, $S(\theta_p; B)$. In previous chapters, we have seen that the generalized runaway criterion predicts the occurrence of runaway for the same conditions, independently of the particular choice of the model input parameter ϕ used in the definition of the sensitivity, $S(\theta_p; \phi)$. When this is not the case, the system is essentially parametrically insensitive, and the boundary that indicates the transition between runaway and nonrunaway behavior is not well defined. This conclusion applies also to the case of a catalyst particle.

As an example, two different situations are illustrated in Figs. 6.4a and 6.4b. In the first case, the sensitivity of the particle temperature with respect to each of the six model input parameters, $i.e.$, B, Da_p, Le, γ, n, and Φ^i, is shown as a function of B. It is seen that this exhibits a maximum (or a minimum) at the same value of B ($=12.07$), which

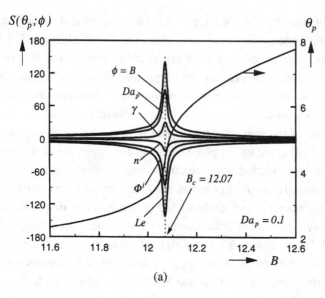

(a)

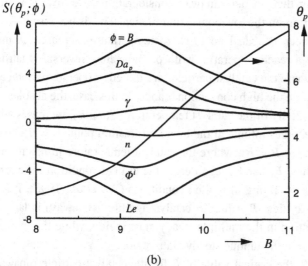

(b)

Figure 6.4. Temperature of the catalyst particle θ_p and its normalized sensitivity $S(\theta_p; B)$ with respect to each of the six independent parameters, $B, Da_p, Le, \gamma, n,$ and Φ^i, as a function of the heat-of-reaction parameter B. $Le = 1; \gamma = 20; n = 1;$ $\Phi^i = 1; \theta = 0; x = 0$. (a) $Da_p = 0.1$; (b) $Da_p = 0.15$.

is then defined as the critical value $B_c = 12.07$ that separates safe from runaway operation. In the second figure, the location of the maximum (or the minimum) of the normalized sensitivity $S(\theta_p; \phi)$ depends on the particular choice of the model input parameter ϕ. In this case, a generalized runaway boundary cannot be defined, and this system may be classified as parametrically insensitive. The above arguments are confirmed by the computed temperature values shown in the same figures. The catalyst particle temperature in Fig. 6.4a increases sharply when the parameter B crosses the critical value

$B_c = 12.07$, while in Fig. 6.4b it changes smoothly with B over the entire range of B values.

It is evident that the generalized nature of the adopted runaway criterion always holds when the catalyst particle operates in the multiple steady-state region, since in this case $\partial F/\partial \theta_p$ in Eq. (6.20) equals zero at the bifurcation point, leading to $|S(\theta_p; \phi)| \to \infty$ for any choice of the input parameter ϕ.

6.2.2 Runaway Regions

Role of interparticle transport resistances

Let us first consider the case where only interparticle mass and heat transfer resistances are present, while intraparticle transport resistances are negligible, i.e., $\Phi = 0$ and $\eta = 1$ in Eq. (6.13). This situation is of direct interest in many applications, such as those where the catalytically active element is present only on the external surface of the catalyst particle. We discuss the parametric sensitivity behavior of this system with reference to the $1/\psi_p$–B parameter plane, where

$$\psi_p = Da_p \cdot B/Le \tag{6.21}$$

may be referred to as the Semenov number for a catalyst particle and represents the ratio between the heat-production rate of the reaction and the heat-removal rate from the external surface of the particle. Figure 6.5 shows the runaway regions predicted by the generalized runaway criterion for various values of the reaction order, the activation energy, and the temperature and conversion in the fluid phase. As expected from the results discussed in previous chapters, it is seen that runaway becomes more likely as the activation energy, γ, increases or as the reaction order, n, decreases. In addition, the runaway region enlarges as the temperature in the fluid phase, θ, increases, whereas it shrinks as the reactant conversion in the fluid phase, x, increases.

From the same figures it appears that in the limiting case of very large heat of reaction, i.e., as $B \to \infty$, the critical ψ_p value, $\psi_{p,c}$, approaches an asymptotic value that is a function of γ, θ, and x but is independent of the reaction order n. This can be proven by analyzing the steady-state multiplicity behavior of the catalyst particle. The boundaries (or bifurcation points) of the multiple steady-state region are determined as solutions of the equation $\partial F/\partial \theta_p = 0$, which from Table 6.1 implies

$$\frac{1}{\theta_p - \theta} = \frac{1}{(1 + \theta_p/\gamma)^2} + \frac{1}{\eta} \cdot \frac{\partial \eta}{\partial \theta_p} - \frac{n \cdot Le/B}{1 - x - Le \cdot (\theta_p - \theta)/B} \tag{6.22}$$

In the case where $B \to \infty$, with the expressions for η [Eq. (6.13)] and $\partial \eta/\partial \theta_p$ reported in Table 6.2, the above equation reduces to

$$\frac{1}{\theta_p - \theta} = \frac{1}{(1 + \theta_p/\gamma)^2} \left[1 + \frac{9\Phi^2 + 2 - 3\Phi \coth(3\Phi)[1 + 3\Phi \coth(3\Phi)]}{2[3\Phi \coth(3\Phi) - 1]} \right] \tag{6.23}$$

where Φ is given by Eq. (6.14).

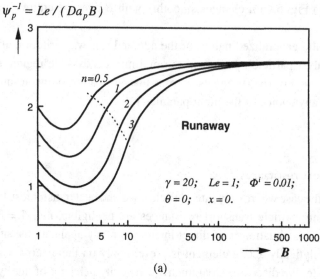

(a)

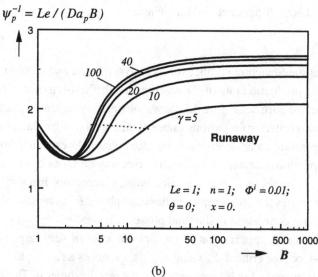

(b)

Figure 6.5. Role of various parameters on the runaway regions of a single catalyst particle in the ψ_p^{-1}–B parameter plane, in the case where only interparticle transport resistances are present. (a) Reaction order n; (b) activation energy γ; (c) temperature in the fluid phase θ; (d) conversion in the fluid phase x. For B values greater than those indicated by the dashed line, the runaway boundary is generalized.

Now, for the case under examination, where intraparticle mass-transfer resistances are negligible, *i.e.*, $\Phi^i \to 0$, Eq. (6.23) takes the following simpler form:

$$\theta_p - \theta = \left(1 + \frac{\theta_p}{\gamma}\right)^2 \tag{6.24}$$

$$\psi_p^{-1} = Le / (Da_p B)$$

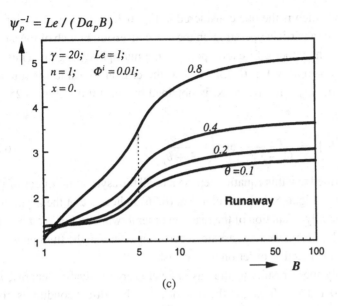

(c)

$$\psi_p^{-1} = Le / (Da_p B)$$

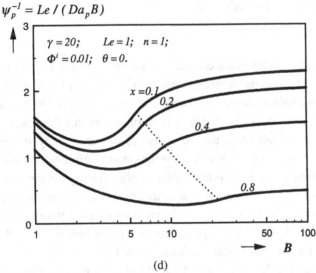

(d)

Figure 6.5. (cont.)

which gives

$$\theta_p = \frac{\gamma}{2}[(\gamma - 2) - \sqrt{\gamma(\gamma - 4) - 4\theta}] \tag{6.25a}$$

and

$$\theta_p = \frac{\gamma}{2}[(\gamma - 2) + \sqrt{\gamma(\gamma - 4) - 4\theta}] \tag{6.25b}$$

The equations above indicate that, for $B \to \infty$, when $\gamma > 2 \cdot (1 + \sqrt{1 + \theta})$, the roots above are real and the catalyst particle operates in the multiple steady-state region. In this case, which is the one considered in Fig. 6.2, as discussed earlier, the runaway boundary for a particle operating on the low conversion branch of the steady state corresponds to the upper bifurcation point of the multiplicity region. Thus, the particle temperature given by Eq. (6.25a) is also the critical θ_p value for runaway and the critical value, $\psi_{p,c}$, for $B \rightarrow \infty$ is obtained by substituting Eq. (6.25a) in Eq. (6.17):

$$\psi_{p,c} = \left(\frac{B \cdot Da_p}{Le} \right)_c = \frac{\theta_p - \theta}{(1 - x)^n} \cdot \exp\left(-\frac{\theta_p}{1 + \theta_p/\gamma} \right) \tag{6.26}$$

It can be readily verified that this equation reproduces all the asymptotic values of $\psi_{p,c}$ for $B \rightarrow \infty$ shown in Fig. 6.5. In particular, Eq. (6.26) indicates that the asymptotic value of $\psi_{p,c}$ is generally a function of the reaction order n, while in Fig. 6.5a it is not. This is because, in Fig. 6.5a, we have assumed zero conversion in the fluid phase, for which the effect of the reaction order on $\psi_{p,c}$ vanishes.

Let us now apply these results to the case of very large activation energy, *i.e.*, $\gamma \rightarrow \infty$. For $\theta = 0$ and $x = 0$, *i.e.*, at the reactor inlet, the critical conditions given by Eqs. (6.24) and (6.26) further reduce to

$$\theta_{p,c} = 1, \qquad \psi_{p,c} = e^{-1} \tag{6.27}$$

These are identical to the well-known critical conditions for thermal runaway, first derived by Semenov (1928) for batch reactors and discussed in detail in Section 3.2.1. Note that in this case we refer to the catalyst particle temperature θ_p and to the particle Semenov number ψ_p defined by Eq. (6.21). It is worth noting that this result arises because, for large heat-of-reaction values, runaway occurs immediately at very low reactant conversion, in agreement with the assumption of negligible reactant consumption in the Semenov analysis. This conclusion applies to many different reacting systems, as shown in Table 6.3, where it is seen that the critical conditions for runaway for all the reacting systems examined in the previous chapters have very similar asymptotic behavior for $B \rightarrow \infty$. In the particular case where also $\gamma \rightarrow \infty$, they all approach the classical Semenov critical condition, $\psi_c = 1/e$.

The runaway boundaries shown in Fig. 6.5 have been obtained by maximizing the normalized sensitivity, $S(\theta_p; B)$, as a function of Da_p. Since it is expected that the peak value of $S(\theta_p; B)$ reduces as B decreases, we investigate whether the predicted runaway boundaries maintain the generalized character as the value of B is reduced. For this, we consider the critical conditions for runaway as predicted by the generalized runaway criterion using the normalized sensitivity of the particle temperature $S(\theta_p; \phi)$ for different choices of the model input parameter ϕ. An example is shown in Fig. 6.6, where for a given set of operating conditions and physicochemical parameters the runaway boundary in the ψ_p^{-1}–B parameter plane is shown based on the normalized sensitivity $S(\theta_p; \phi)$ for $\phi = Da_p$, B, and γ. It is seen that, in all three cases, the predicted runaway boundaries are identical in the region of high B values, while they tend to deviate from each other as B decreases below about $B = 5$. According to the

Table 6.3. Asymptotic values of the critical temperature and the Semenov number for $B \to \infty$ for various reacting systems. The Semenov number is defined as the ratio between heat production and heat removal rates for each specific case

Reacting System	Critical Conditions	
	For Finite γ Values	For $\gamma \to \infty$*
Batch reactors	$\theta_c = \dfrac{\gamma}{2}[(\gamma - 2) - \sqrt{\gamma(\gamma - 4) - 4\theta_a}]$	$\theta_c = 1$
	$\psi_c = (\theta_c - \theta_a) \cdot \exp\left(-\dfrac{\theta_c}{1 + \theta_c/\gamma}\right)$	$\psi_c = e^{-1}$
Pseudo-homogeneous plug-flow reactors	$\theta_c = \dfrac{\gamma}{2}[(\gamma - 2) - \sqrt{\gamma(\gamma - 4) - 4\theta_{co}}]$	$\theta_c = 1$
	$\psi_c = \left(\dfrac{B \cdot Da}{St}\right)_c = (\theta_c - \theta_{co}) \cdot \exp\left(-\dfrac{\theta_c}{1 + \theta_c/\gamma}\right)$	$\psi_c = e^{-1}$
Continuous-flow well stirred tank reactors (CSTRs)	$\theta_c = \dfrac{\gamma}{2}[(\gamma - 2) - \sqrt{\gamma(\gamma - 4) - 4\theta_{co} \cdot St/(1 + St)]}$	$\theta_c = 1$
	$\psi_c = \left(\dfrac{B \cdot Da}{St}\right)_c = \theta_c \cdot \left(1 + \dfrac{1}{St}\right) \cdot \exp\left(-\dfrac{\theta_c}{1 + \theta_c/\gamma}\right)$	$\psi_c = e^{-1}$
Single catalyst particle with negligible intraparticle transport limitations	$\theta_{p,c} = \dfrac{\gamma}{2}[(\gamma - 2) - \sqrt{\gamma(\gamma - 4) - 4\theta}]$	$\theta_{p,c} = 1$
	$\psi_{p,c} = \left(\dfrac{B \cdot Da_p}{Le}\right)_c = \dfrac{\theta_{p,c} - \theta}{(1 - x)^n} \cdot \exp\left(-\dfrac{\theta_{p,c}}{1 + \theta_{p,c}/\gamma}\right)$	$\psi_{p,c} = e^{-1}$

*For batch reactor at $\theta_a = 0$; for PFA and CSTR at $\theta_{co} = 0$; for a single particle at $\theta = x = 0$.

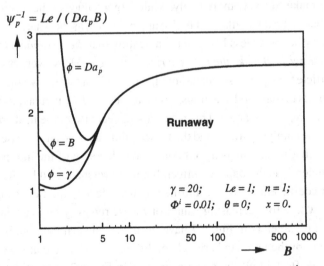

Figure 6.6. Catalyst particle runaway regions in the ψ_p^{-1}–B parameter plane, predicted by the generalized runaway criterion based on the normalized sensitivity $S(\theta_p; \phi)$ for different choices of the input model parameter ϕ.

generalized runaway criterion, in the region where the predicted runaway boundary depends on the particular choice of the model input parameter in the sensitivity definition, the system is essentially parametrically insensitive. Computations of this type have been repeated for all the cases shown in Fig. 6.5, and the parametrically sensitive and insensitive regions have been separated from each other by dotted curves. For a B value lower than that at the intersection point between the runaway boundary and the dotted curve, the runaway boundary is no longer generalized, and the system is parametrically insensitive. It can be seen that the parametrically insensitive region enlarges as n increases (Fig. 6.5a) and γ decreases (Fig. 6.5b). This coincides with the previous observation that runaway becomes less likely as the reaction order increases and the activation energy decreases. Because of the decrease of the available heat of reaction, as the reactant conversion in the fluid phase x increases, the parametrically insensitive region enlarges (Fig. 6.5d). However, results in Fig. 6.5c indicate that the parametrically insensitive region is independent of the fluid-phase temperature, θ, at least for the operating conditions considered in this example.

Role of intraparticle mass-transfer resistances

The influence of intraparticle mass transport resistance on the catalyst particle runaway has been investigated by computing the critical value of B for runaway as a function of the Thiele modulus Φ^i through the generalized runaway criterion. The obtained runaway boundaries in the B–Φ^i parameter plane are shown for various values of the involved parameters (which include the presence of some interparticle transport resistances as well) in Figs. 6.7a–d. It is seen that larger values of the activation energy and Damkohler number make runaway more likely, while larger values of the reaction order and Lewis number make it less likely. It is also apparent that, as the Thiele modulus Φ^i increases, runaway becomes less likely, thus indicating that *the intraparticle mass transport resistance always shrinks the particle runaway region*. However, this is not the case for interparticle transport resistances. In Fig. 6.7c, it is in fact shown that as the interparticle resistances to mass and heat transport increase, *i.e.*, Da_p increases with Le constant, the runaway region enlarges. This is because a decrease of the heat-transfer coefficient between the catalyst particle and the flowing fluid leads to an increase in the particle temperature that favors runaway. Of course, the decrease of the interparticle mass-transfer coefficient has the opposite effect, but it is overshadowed by the effect of the heat-transfer coefficient. This is confirmed by the effect of the Lewis number shown in Fig. 6.7d, where for decreasing values of Le the runaway region enlarges.

The values of the parameters in Fig. 6.7 have been chosen such that, in all cases, the generalized nature of the predicted critical conditions holds. Note that the critical values of B for $\Phi^i = 0.01$ in all the cases examined in Fig. 6.7 correspond to the critical B values shown in Fig. 6.5. It is seen from Fig. 6.7 that, in general, when $\Phi^i < 0.1$, the influence of intraparticle mass-transfer resistance on the critical conditions is negligible. For large values of the Thiele modulus, *i.e.*, $\Phi^i \gg 0.1$, the critical B value for runaway increases as Φ^i increases, and can be computed analytically as

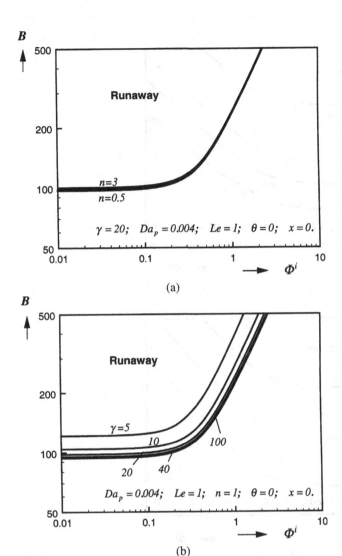

Figure 6.7. Role of various parameters on the runaway regions of a single catalyst particle in the ψ_p^{-1}–B parameter plane, in the case where both inter- and intraparticle transport resistances are present. (a) Reaction order n; (b) activation energy γ; (c) interparticle mass transfer resistance Da_p; (d) Lewis number Le.

follows: From Eq. (6.14), we see that $\Phi \to \infty$ as $\Phi^i \to \infty$, so that from Eq. (6.13)

$$\eta \to 1/\Phi \tag{6.28}$$

which, substituted into Eq. (6.17), gives

$$\psi_{p,c} = \left(\frac{B \cdot Da_p}{Le}\right)_c = \Phi^i \cdot \frac{\theta_p - \theta}{[1 - x - Le(\theta_p - \theta)/B]^{(n+1)/2}} \cdot \exp\left(-\frac{\theta_p/2}{1 + \theta_p/\gamma}\right) \tag{6.29}$$

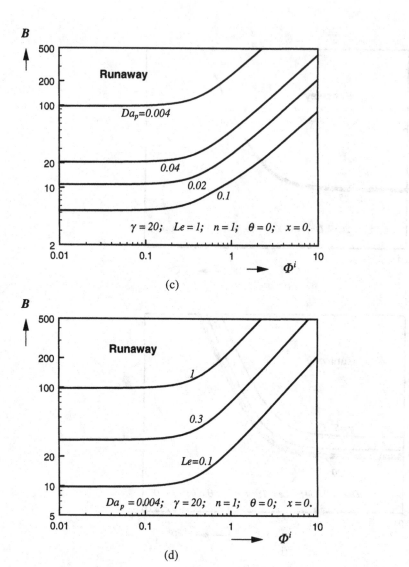

Figure 6.7. (cont.)

From Fig. 6.7d, we see that at criticality, when $\Phi^i \to \infty$, also $B \to \infty$; thus we can expect, based on the same arguments discussed in the previous section, that the runaway boundary coincides with the upper bifurcation point of the steady-state multiplicity region. Then, using Eq. (6.23) for calculating the particle temperature for $\Phi \to \infty$ yields

$$(1 + \theta_p/\gamma)^2 = (\theta_p - \theta)/2 \tag{6.30}$$

which leads to

$$\frac{\theta_p}{2} = \frac{1}{2} \cdot \frac{\gamma}{2}\left[\left(\frac{\gamma}{2} - 2\right) - \sqrt{\frac{\gamma}{2}\left(\frac{\gamma}{2} - 4\right) - 4 \cdot \frac{\theta}{2}}\right] \tag{6.31a}$$

and

$$\frac{\theta_p}{2} = \frac{1}{2} \cdot \frac{\gamma}{2}\left[\left(\frac{\gamma}{2} - 2\right) + \sqrt{\frac{\gamma}{2}\left(\frac{\gamma}{2} - 4\right) - 4 \cdot \frac{\theta}{2}}\right]$$ (6.31b)

Note that since $B \to \infty$, Eq. (6.29) can also be simplified as

$$\psi_{p,c} = \left(\frac{B \cdot Da_p}{Le}\right)_c = \Phi^i \cdot \frac{\theta_p - \theta}{(1 - x)^{(n+1)/2}} \cdot \exp\left(-\frac{\theta_p/2}{1 + \theta_p/\gamma}\right)$$ (6.32)

which, in combination with Eq. (6.31a), allows one to reproduce analytically the runaway boundaries for $\Phi^i \to \infty$ computed numerically in Fig. 6.7.

The results shown in Fig. 6.7a indicate that the effect of the reaction order on the runaway region in the case where intraparticle transport resistances are present is much smaller than when only interparticle transport resistances are present, shown in Fig. 6.5a. This can be explained by noting that, for a constant value of the *normalized* Thiele modulus Φ^i defined by Eq. (6.15), the effectiveness factor η in Eq. (6.13) depends only slightly on the reaction order, and then only in the region of intermediate Φ^i values (Aris, 1975).

An interesting feature of the behavior for large values of the Thiele modulus can be uncovered by noting that Eqs. (6.31) and (6.32) can be rewritten in the following equivalent form:

$$\theta_p' = \frac{\gamma'}{2} \cdot [(\gamma' - 2) - \sqrt{\gamma'(\gamma' - 4) - 4 \cdot \theta'}]$$ (6.33a)

$$\theta_p' = \frac{\gamma'}{2} \cdot [(\gamma' - 2) + \sqrt{\gamma'(\gamma' - 4) - 4 \cdot \theta'}]$$ (6.33b)

$$\psi_{p,c}' = \left(\frac{B' \cdot Da_p'}{Le'}\right)_c = \frac{\theta_p' - \theta'}{(1 - x)^{n'}} \cdot \exp\left(-\frac{\theta_p'}{1 + \theta_p'/\gamma'}\right)$$ (6.34)

where

$$\gamma' = \gamma/2; \qquad \theta_p' = \theta_p/2; \qquad \theta' = \theta/2; \qquad n' = (n + 1)/2;$$
$$B' = B/2; \qquad Da_p' = Da_p/\Phi^i; \qquad Le' = Le$$ (6.35)

Now, by comparing Eqs. (6.33) and (6.34) with Eqs. (6.25) and (6.26) for the case of negligible intraparticle mass-transfer resistance ($\Phi^i \to 0$), it appears that the limiting ($B \to \infty$) critical conditions for particle runaway in the case of *severe* ($\Phi^i \to \infty$) intraparticle mass-transfer resistance are identical to those for *negligible* ($\Phi^i \to 0$) intraparticle mass-transfer resistance, but with the original parameters replaced by those defined by Eq. (6.35). This finding agrees with the well-known fact that, under severe intraparticle transport resistances, the reaction rate exhibits an apparent activation energy that is one-half the true value and an apparent reaction order that is equal to the true one plus one divided by two (Aris, 1975). From the

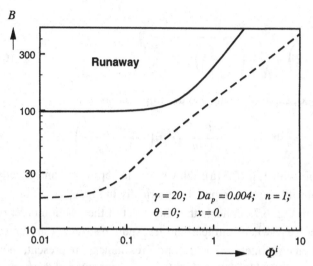

Figure 6.8. Comparison between the catalyst particle runaway regions predicted by the generalized runaway (solid curve) and the McGreavy and Adderley (1973) criteria (broken curve).

expression of B in Eq. (6.9), since $\gamma' = \gamma/2$, it follows that the apparent heat-of-reaction parameter B' also becomes one-half the true one, B. Further, the original parameter Da_p, representing the ratio between chemical reaction and interparticle mass transfer rates, is replaced by $Da'_p = Da_p/\Phi^i$, which represents the original parameter modified by intraparticle diffusion. This result is useful in applications, since it allows one to obtain an approximate solution of the runaway problem in the case of severe intraparticle diffusion resistance from the corresponding solution in the case of negligible intraparticle resistance.

Finally, let us compare the runaway regions predicted by the generalized and the McGreavy and Adderley (1973) runaway criteria. The latter identifies the occurrence of runaway with the inflection point on the θ_p–θ versus θ curve. Without going into the details, we compare the predictions of the two criteria for a given set of parameter values in Fig. 6.8. It is apparent that the runaway boundaries predicted by the McGreavy and Adderley criterion are conservative with respect to those predicted by the generalized one. It should be noted that for the conditions considered in Fig. 6.8, the reactor operates in the multiple steady-state region. When the reactor operates in the region of unique steady states, the difference between the two criteria decreases.

6.3 Runaway of Fixed-Bed Reactors: Global Runaway

In order to investigate the runaway behavior of a fixed-bed reactor, we consider the catalyst particle temperature profile along the reactor axis, since, as discussed earlier, this is the temperature that controls the process. In particular, we will identify the runaway region for this reactor by applying the generalized runaway criterion using the maximum in the particle temperature profile (*i.e.*, hot spot) as the objective.

In order to reduce the number of independent parameters in the model equations, we use here reactant conversion, x, rather than the reactor length, z, as the independent variable. As discussed in Chapter 4 in the context of homogeneous tubular reactors, this choice makes the results of this analysis conservative. We will return to this point later (Section 6.3.4) when discussing the pseudo-adiabatic operation for fixed-bed reactors.

By dividing Eq. (6.7) by Eq. (6.6), we obtain the following equation describing the reactor behavior in the fluid temperature (θ)–fluid conversion (x) plane:

$$\frac{d\theta}{dx} = B\left(1 - \frac{1}{\psi} \cdot \frac{\theta - \theta_{co}}{\exp[\theta_p/(1 + \theta_p/\gamma)] \cdot [1 - x - Le \cdot (\theta_p - \theta)/B]^n \cdot \eta}\right) = F_1 \tag{6.36}$$

with the inlet condition

$$\theta = 0 \quad \text{at } x = 0 \tag{6.37}$$

The catalyst particle temperature, θ_p, is given by the relationship (6.17) derived earlier:

$$\theta_p = \theta + \psi_p \cdot \exp\left(\frac{\theta_p}{1 + \theta_p/\gamma}\right) \cdot \left[1 - x - \frac{Le}{B} \cdot (\theta_p - \theta)\right]^n \cdot \eta = F_2 \tag{6.38}$$

Note that, in the equations above, in addition to the *particle* Semenov number, ψ_p, we also have the parameter

$$\psi = \frac{Da \cdot B}{St} \tag{6.39}$$

referred to as the *reactor* Semenov number, as defined first in Chapter 4 for homogeneous systems. All the remaining quantities have been defined earlier in Section 6.2, including the effectiveness factor, η, given by Eq. (6.13).

6.3.1 Critical Conditions for Global Runaway of Particle Temperature

The normalized sensitivity of the particle temperature maximum, θ_p^*, along the reactor length is given by

$$S(\theta_p^*; \phi) = \frac{\phi}{\theta_p^*} \cdot s(\theta_p^*; \phi) = \frac{\phi}{\theta_p^*} \cdot \frac{d\theta_p^*}{d\phi} \tag{6.40}$$

where ϕ represents a generic model input parameter such as B, ψ, γ, n, θ_{co}, Da_p, Le, or Φ^i. The critical conditions for reactor runaway, according to the generalized sensitivity criterion, are then identified as the situation in which the normalized objective sensitivity, $S(\theta_p^*; \phi)$, is maximized (Morbidelli and Varma, 1986b).

The evaluation of the normalized objective sensitivity, $S(\theta_p^*; \phi)$, requires the solution of the local fluid temperature sensitivity equations obtained by direct differentiation of Eq. (6.36) with respect to the generic model input parameter, ϕ, as follows:

$$\frac{ds(\theta; \phi)}{dx} = \frac{\partial F_1}{\partial \phi} + \frac{\partial F_1}{\partial \theta} \cdot s(\theta; \phi) + \frac{\partial F_1}{\partial \theta_p} \cdot s(\theta_p; \phi) \tag{6.41}$$

with IC

$$s(\theta; \phi) = 0 \quad \text{at } x = 0 \tag{6.42}$$

The local particle temperature sensitivity, $s(\theta_p; \phi)$, is obtained by differentiating both sides of Eq. (6.38) with respect to ϕ:

$$s(\theta_p; \phi) = \frac{(\partial F_2/\partial \phi) + (\partial F_2/\partial \theta) \cdot s(\theta; \phi)}{1 - (\partial F_2/\partial \theta_p)} \tag{6.43}$$

Thus, the general procedure for computing the normalized objective sensitivity, $S(\theta_p^*; \phi)$, is to first integrate simultaneously Eqs. (6.36) and (6.41) from their respective ICs up to the temperature maximum, θ_p^*, where the local sensitivity, $s(\theta; \phi)$, is also obtained. The particle temperature local sensitivity, $s(\theta_p^*; \phi)$, is then computed from Eq. (6.43), followed by $S(\theta_p^*; \phi)$ from Eq. (6.40).

Typical curves of the normalized objective sensitivity, $S(\theta_p^*; \phi)$, are shown in Fig. 6.9 as a function of the heat-of-reaction parameter B for various choices of the input parameter, ϕ ($=B$, St and Da_p). It is seen that the critical value, B_c, characterized by the maximum value of $S(\theta_p^*; \phi)$, is identical (up to the first four significant

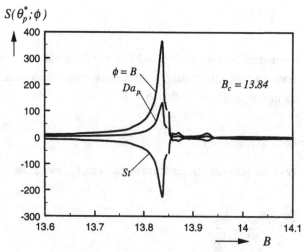

Figure 6.9. Normalized sensitivity of the catalyst particle temperature maximum, $S(\theta_p^*; \phi)$, as a function of the heat-of-reaction parameter B for various choices of the input model parameter ϕ. $\gamma = 20$; $n = 1$; $Da_p = 0.01$; $Le = 1$; $\Phi^i = 0$; $\beta = 30$; $\theta_{co} = 0$.

Table 6.4. Critical values of the heat-of-reaction parameter B_c, predicted by the generalized criterion based on the maximum of (a) the $S(\theta_p^*; \phi)$–B curve and (b) the $S(\theta^*; \phi)$–B curve, for various choices of the independent parameter ϕ

	$\phi = n$		$\phi = \theta_{co}$		$\phi = \gamma$	
β	(a)	(b)	(a)	(b)	(a)	(b)
0.1	2.46	2.30	1.99	1.98	3.50	3.38
5	6.93	6.71	6.93	6.83	7.07	6.90
10	8.83	8.72	8.84	8.77	8.88	8.78
30	13.8	13.8	13.8	13.8	13.8	13.8
50	17.5	17.5	17.5	17.5	17.5	17.5

$n = 1; \gamma = 20; \theta_{co} = 0; Le = 1; Da_p = 0.01; \Phi^i = 0.$
From Morbidelli and Varma (1986b).

digits) for *all* curves. This finding, which holds also for all the other input parameters (*i.e.*, ψ, γ, n, Le, or Φ^i), allows one to define a generalized runaway boundary in the parameter space, where the reactor becomes simultaneously sensitive to small changes of *all* the involved physicochemical parameters.

It should be noted that in Fig. 6.9 we obtain the critical value B_c (=13.84) for runaway by maximizing $S(\theta_p^*; \phi)$ with respect to B for fixed values of the remaining parameters. As discussed in previous chapters, when this is done, all the fixed values of the remaining parameters are also critical. For example, in the case of Fig. 6.9, if one would fix $B = 13.84$ and maximize $S(\theta_p^*; \phi)$ with respect to γ, one would obtain the critical γ value for runaway as $\gamma_c = 20$. Indeed, once the critical value of a given parameter is computed for fixed values of all the remaining parameters, they all together constitute a critical point in the reactor parameter space.

A set of critical B values, computed based on the maximum of $S(\theta_p^*; \phi)$ versus B for $\phi = n$, θ_{co}, and γ, are reported in Table 6.4, Column (a), as a function of β (=St/Da). It appears that B_c decreases as β decreases and, for small β values, B_c becomes dependent on the particular choice of the input parameter, ϕ. This occurs because the runaway boundary is located in the region of low B values, where the total thermal energy that can be produced by the reaction is low. In this case, the hot-spot magnitude of the particle temperature is modest, and so is the numerical value of the normalized objective sensitivity. Thus, as discussed in previous chapters, we cannot define a generalized boundary that separates the runaway and nonrunaway regions. If a situation of this type needs to be examined, then one should analyze the maximum of the normalized objective sensitivity, $S(\theta_p^*; \phi)$, for the particular ϕ of interest.

In Table 6.4, Column (b), the critical B values computed based on the normalized sensitivity of the hot spot in the *fluid* temperature, $S(\theta^*; \phi)$, are reported. In the region of large β values, the predictions are identical to those obtained based on $S(\theta_p^*; \phi)$. Deviations occur only in the cases noted above, where a generalized boundary separating the runaway from the nonrunaway region cannot be defined. This finding indicates that *when runaway of the particle temperature occurs, it is always accompanied by runaway of the fluid temperature. Thus, in order to investigate the global runaway of a fixed-bed catalytic reactor, we can use either the hot spot of the particle temperature or that of the fluid temperature, as the objective to define the normalized objective sensitivity.*

6.3.2 Runaway Regions

Role of interparticle mass- and heat-transfer resistances

Let us first consider the case of a heterogeneous reactor where only interparticle heat- and mass-transfer resistances are present, while intraparticle transport limitations are negligible. This implies a small Thiele modulus, Φ^i, in Eq. (6.13), which leads to the unity effectiveness factor in Eqs. (6.36) and (6.38). Another typical application of this model is in the case of externally coated catalyst particles, which are frequently used in industrial practice.

In order to obtain a complete picture of runaway behavior in the reactor, it is necessary to combine information derived separately from both local and global runaway analyses, as shown in Fig. 6.10. Curve (a) represents the boundary for local runaway of the particle temperature at the reactor inlet conditions; curve (b) represents the

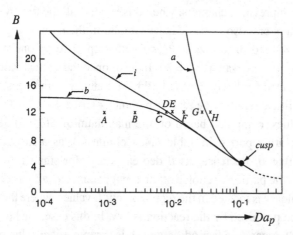

Figure 6.10. Runaway boundaries in the B–Da_p parameter plane in the case of negligible intraparticle transport limitations: (a) local runaway boundary of particle temperature at the reactor inlet; (b) global runaway boundary of particle temperature along the reactor; (i) occurrence of ignition somewhere along the reactor. $n = 1$; $Le = 1$; $\gamma = 20$; $\theta_{co} = 0$; $\beta = 20$; $\Phi^i = 0$. From Morbidelli and Varma (1986a).

boundary for global runaway of the particle temperature along the reactor; curve (i) corresponds to situations where particle ignition (*i.e.*, local runaway) occurs somewhere along the reactor length.

The global runaway boundary, curve (b), indicates that, in the region of low Da_p values, the critical B value for global runaway becomes essentially independent of Da_p and, as $Da_p \to 0$, it approaches the value predicted by the pseudo-homogeneous model. This is expected since, having fixed the value of the Lewis number, Le, as $Da_p \to 0$, *both* mass and heat interparticle transfer resistances vanish, and the heterogeneous model reduces to the pseudo-homogeneous one. In this region, curve (i) is located at much higher B values relative to curve (b), indicating that the occurrence of local runaway (or ignition) somewhere along the reactor is impossible unless the reactor is already under global runaway conditions. As Da_p (*i.e.*, interparticle mass- and heat-transfer resistances) increases, the global and local runaway boundaries approach and eventually merge with each other. At this stage, the two phenomena occur simultaneously. Under these conditions, interparticle transport limitations play an important role in reactor runaway, and the predictions of the pseudo-homogeneous model become substantially wrong. It is worth reiterating that, since curve (i) in Fig. 6.10 is always either above or coincident with curve (b), local runaway along the reactor cannot occur independently, and it always follows global runaway. On the other hand, global runaway can occur independently of local runaway, as in the region of low Da_p values in Fig. 6.10. These results indicate that *the global runaway boundary, obtained through the sensitivity analysis of the particle temperature maximum along the reactor, defines a nonrunaway region (below curve b) that is safe from both local and global points of view.*

The global runaway boundary in Fig. 6.10 ends when it merges with curve (a) at high Da_p values. Since curve (a) represents the critical conditions for local runaway of the particle temperature at the reactor inlet, global runaway is now occurring immediately at the reactor inlet. In addition, since the point where curve (b) merges with curve (a) is also the cusp of the multiplicity region of the particle located at the reactor inlet, it follows that after this point the entire reactor operates in the unique steady-state region. As discussed in Section 6.2, in this case one generally needs to examine whether or not the predicted runaway boundary has the generalized character. In Fig. 6.10, in the region where the runaway boundary is given by the broken curve, the critical conditions depend on the particular choice of the input parameter in the definition of the normalized objective sensitivity. Thus, a sharp transition between the runaway and nonrunaway regions tends to vanish.

A detailed illustration of the reactor runaway behavior is given in Fig. 6.11, where the particle temperature versus conversion profiles are shown for increasing values of Da_p, along the path indicated in Fig. 6.10. It appears that the profiles A, B, and C correspond to safe operation, while profile D exhibits a much larger change in the hot-spot value, thus indicating runaway behavior. This occurs because, when the operation conditions move from point C to point D in Fig. 6.10, the global runaway

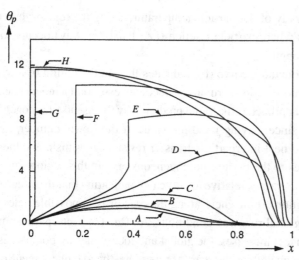

Figure 6.11. Particle temperature versus conversion profiles for various values of Da_p. Data as in Fig. 6.10; $Da_p = 0.001$(A), 0.003(B), 0.0065(C), 0.0085(D), 0.01(E), 0.015(F), 0.03(G), 0.04(H). From Morbidelli and Varma (1986a).

boundary (*i.e.*, curve b) is crossed. Further increase in Da_p leads to particle ignition (local runaway) at some location along the reactor length (profile E), which then moves toward the reactor inlet up to point H, where also the particle located at the reactor inlet is ignited. This is indicated by the fact that when operation conditions move from point G to point H in Fig. 6.10, the boundary for local runaway of particles at the reactor inlet [*i.e.*, curve (a)] is crossed. For larger values of Da_p, the entire bed operates in the ignited regime for the given B value.

The global runaway regions in the B–Da_p parameter plane are shown in Figs. 6.12a, b, c, and d, for various values of the reaction order n, the heat removal parameter β, the activation energy γ, and the coolant temperature θ_{co}, respectively. As expected on physical grounds, the runaway region shrinks for increasing values of the reaction order and of the heat removal parameter, while it enlarges for increasing values of the activation energy and of the coolant temperature. For large values of γ, roughly above 100, the runaway region becomes independent of γ.

In the region of low Da_p values, *i.e.*, when the interparticle transport limitations tend to vanish, the critical B value for runaway approaches that obtained by using the pseudo-homogeneous reactor model. In the region of high Da_p values, where all the catalyst particles operate in the unique steady-state region, in all cases of Fig. 6.12 there is a region (as in Fig. 6.10, but not shown for simplicity) where a generalized runaway boundary between the runaway- and nonrunaway-region behavior cannot be defined.

Finally, it is worth comparing the results obtained using the generalized runaway criterion with those obtained by Rajadhyaksha *et al.* (1975) using the runaway criterion developed by van Welsenaere and Froment (1970), in the context of pseudo-homogeneous reactors (see Section 3.2). Rajadhyaksha *et al.* considered four

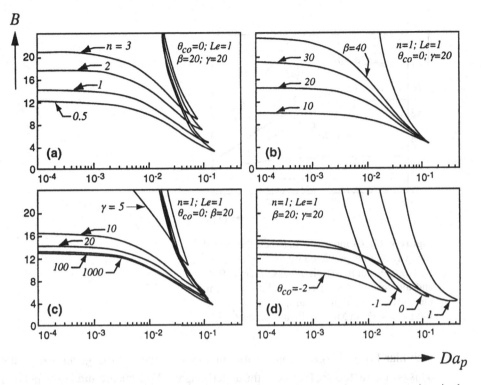

Figure 6.12. Influence of various physicochemical parameters on the runaway regions in the case of negligible intraparticle transport limitations. From Morbidelli and Varma (1986a,b). (a) reaction order, n; (b) wall heat-transfer parameter, β; (c) activation energy, γ; (d) coolant temperature, θ_{co}.

separate limiting regimes for heterogeneous catalytic reactors and developed separate procedures for calculating the runaway boundary for each. In the particular case where intraparticle mass- and heat-transfer resistances are small, the following expression for the critical B value was derived:

$$B_c = \frac{1}{2} \cdot \left\{ G + \bar{\theta} + \bar{\theta} \cdot \left[1 + \left(\frac{\beta \cdot \exp(-\bar{\theta}_p/(1 + \bar{\theta}_p/\gamma))}{1 + \beta \cdot Da_p/Le} \right)^{1/2} \right]^2 \right\} \tag{6.44}$$

where

$$G = \frac{\beta \cdot \bar{\theta}_p}{\exp[\bar{\theta}_p/(1 + \bar{\theta}_p/\gamma)] \cdot (1 + \beta \cdot Da_p/Le)} \tag{6.45}$$

$$\bar{\theta}_p = \frac{\gamma}{2} \cdot [\gamma - 2 - \sqrt{\gamma \cdot (\gamma - 4)}] \tag{6.46}$$

$$\bar{\theta} = \bar{\theta}_p - \frac{Da_p \cdot G}{Le \cdot \exp[-\bar{\theta}_p/(1 + \bar{\theta}_p/\gamma)]} \tag{6.47}$$

which according to the original work applies for $Da_p < 0.05$ and $\Phi^i = 0$. The runaway boundaries calculated by Eq. (6.44) (broken curves) are compared to those predicted by the generalized criterion (solid curves) in Fig. 6.13. Although the qualitative behavior

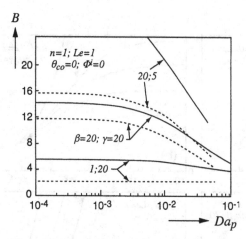

Figure 6.13. Comparison between the global runaway regions predicted by the explicit criterion (broken curves) (Rajadhyaksha *et al.*, 1975) and the generalized criterion (solid curves), in the case of negligible intraparticle transport limitations. From Morbidelli and Varma (1987).

is similar, Eq. (6.44) gives conservative predictions relative to the generalized criterion. This is a characteristic feature of the underlying van Welsenaere and Froment criterion, as discussed in Section 3.2 in the context of pseudo-homogeneous reactors.

Example 6.1 Experimental analysis of runaway in a fixed-bed reactor for vinyl acetate synthesis. Despite the relevance of the runaway problem in fixed-bed catalytic reactors, only a few experimental studies on this subject are available in the literature. The work by Emig *et al.* (1980) is one that presents a complete experimental picture of the runaway region. In particular, they utilized a stainless-steel reactor, 0.05 m in diameter, packed with zinc acetate catalyst supported on activated carbon, where vinyl acetate synthesis was conducted. The reactor was well equipped, in order to have close control of inlet flow rate and composition and to measure the fluid temperature at various positions along the reactor axis. In all experiments, the inlet temperature was set equal to the wall temperature (*i.e.*, $\theta^i = \theta_{co} = 0$). As experimental criteria for runaway, a temperature increase rate above 3 K/s or a temperature maximum along the reactor above 510 K (upper limit for stability of the catalyst) were utilized. The experimental results in the $1/\psi - B$ parameter plane are shown in Fig. 6.14, where open and filled circles indicate safe and runaway operating conditions, respectively. We now apply the generalized runaway criterion to predict the runaway boundary (*i.e.*, the boundary between open and filled circles in Fig. 6.14) using the heterogeneous reactor model accounting only for interparticle mass- and heat-transfer resistances.

A previous detailed kinetic study (El-Sawi *et al.*, 1977) showed that this reaction follows first-order kinetics, with dimensionless activation energy $\gamma \approx 20$. The values

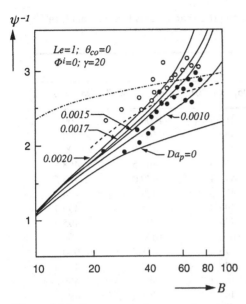

Figure 6.14. Runaway regions experimentally measured by Emig *et al.* (1980) and predicted by the generalized criterion with various reactor models. Experimental data: ○ = safe operation; ● = runaway. Calculated results: solid curves, heterogeneous model with interparticle transport limitations; dash-dot curves, pseudo-homogeneous model with $n = 0$; broken curve, two-dimensional homogeneous model. From Morbidelli and Varma (1986b).

used for all the other physicochemical parameters, selected according to the indications of Emig *et al.* (1980), are summarized at the bottom of Table 6.5. The same j factor has been used for heat and mass transfer, which implies $Le = 1$. The calculated runaway boundaries are shown by the solid curves in Fig. 6.14 for various values of the interparticle transport resistance, *i.e.*, Da_p. Specifically, the generalized runaway criterion based on the maximum of the normalized sensitivity of the particle temperature maximum to the heat-of-reaction parameter B, $S(\theta_p^*; B)$, was used. It is seen that the shape of the calculated boundary in the case of nonzero external resistances is very similar to that indicated by the experimental data. It is worth stressing that here we are not attempting a quantitative comparison. For example, the various experimental runs involve different values of the Reynolds number, yet they are compared in the same graph with curves calculated using fixed external transport coefficients. The comparison between the experimental and calculated runaway boundaries will be based on qualitative shape, which is physically more significant and less affected by the accuracy of the parameter values.

Figure 6.14 also shows the runaway boundaries calculated by the generalized criterion using the pseudo-homogeneous reactor model, *i.e.*, $Da_p = 0$. In particular,

Table 6.5. Temperature maximum along the reactor axis for vinyl acetate synthesis

Run	T^i, K	T^*_{exp}, K	$Da_p = 0.001$	$Da_p = 0.0015$	$Da_p = 0.0017$	$Da_p = 0.002$
				T^*_{cal}, K		
3	448	473.9	462.7	468.5	>1000	>1000
4	448	464.4	461.0	463.6	465.8	>1000
7	458	474.8	468.2	470.8	472.8	>1000
8	453	469.4	465.6	467.0	467.5	468.8
10	458	475.9	475.0	477.4	479.0	483.5
12	463	494.3	483.3	486.2	488.2	494.7
14	468	495.5	490.9	493.8	495.5	499.7
16	473	485.7	493.0	494.0	494.5	495.3
18	453	477.0	469.1	>1000	>1000	>1000
21	458	477.9	473.4	476.3	478.7	>1000
24	463	485.6	477.8	479.2	479.9	481.4
26	468	493.0	485.2	486.7	487.5	489.0
28	473	496.2	491.8	493.0	493.6	494.6
31	458	485.0	473.3	476.9	481.3	>1000
33	463	487.5	480.6	484.2	487.5	>1000
35	468	494.3	487.4	490.4	492.5	502.4
37	473	501.8	495.7	499.2	501.7	513.2

Experimental results and physicochemical parameters are from Emig *et al.* (1980). The calculated results are from Morbidelli (1987). $T_{co} = T^i$; $\Delta H = -1.057 \times 10^5$ kJ/kmol; $d_t = 0.05$ m; $\rho_B \cdot A = 4.915 \times 10^6$ 1/s; $G = \rho \cdot v^o = 0.05 - 0.15$ kg/m²/s; $U = 0.045 - 0.0701$ kJ/m²/s/K; $E = 7.630 \times 10^7$ kJ/kmol; $n = 1$.

two cases are considered: first-order reaction (solid curve) and zeroth-order reaction (dash-dot curve). It is seen that the shape of both boundaries is different from that indicated by the experimental data. In addition, the dashed curve in the same figure represents the predictions of the pseudo-homogeneous model accounting also for radial concentration and temperature gradients used in the original work, with parameter values determined by direct fitting of the experimental data. Again, the shape of this curve does not match the experimental evidence and is of the same type as all the others obtained using pseudo-homogeneous models. Thus, although radial gradients are certainly significant, they do not seem to be responsible for the change in the *shape* of the runaway boundary.

From the above comparison, it can be concluded that interparticle transport resistances are the only phenomena that can alter the typical shape of the pseudo-homogeneous runaway boundary to that indicated by the experimental data. Accordingly, *care must be taken when dealing with runaway in fixed-bed reactors to use heterogeneous models rather than simplified pseudo-homogeneous models.*

It should be stressed that the qualitative change in the shape of the runaway boundary discussed above occurs for relatively small interparticle resistances, *i.e.*, Da_p of the order of 10^{-3}. According to the Mears (1971) criterion, which is widely used to establish the importance of transport intrusions on reaction rate measurements, interparticle temperature gradients can be neglected when

$$\frac{B \cdot Da_p}{Le} < 0.05 \qquad (6.48)$$

At the reactor inlet, which is the only location where the operating conditions are known *a priori*, the Da_p values used in Fig. 6.14 satisfy this condition, thus indicating that interparticle temperature gradients should be absent. This is confirmed by the simulations, which indicate that in all cases temperature gradients are no larger than about 1–2 K at the reactor inlet, and then increase to no more than 6 K at the hot spot. Thus, it can be concluded that, although interparticle temperature gradients never exceed 1.5% of the hot-spot magnitude, they have a strong effect on the shape and location of the runaway boundary. This result is a further manifestation of the parametrically sensitive reactor behavior, such that even slight changes in parameters yield significantly different reactor behavior.

A cross-check of the reliability of the results discussed above can be obtained by comparing the calculated temperature and concentration profiles along the reactor length with the experimental data. Without attempting a detailed comparison, some useful insights can be obtained by simply comparing the measured temperature maximum with that predicted by the heterogeneous one-dimensional plug-flow model. The results are reported in Table 6.5 for various values of the external transport resistance parameter, Da_p. Each run is labeled by a number according to the nomenclature of the original paper (Emig *et al.*, 1980), where the detailed experimental conditions (*i.e.*, T_{co}, B and β) are also reported. Only those experimental runs that do not lead to runaway are shown in Table 6.5. This is because, when runaway occurs, very large temperature values are attained, where the adopted models for transport as well as the kinetic rate expressions may no longer be valid. Moreover, at such high temperatures new physicochemical phenomena may arise, such as catalyst deactivation, which are not included in the model. It is remarkable that whenever runaway occurs, the model always predicts hot-spot values greater than 1000 K (due to the occurrence of ignition or local runaway somewhere along the reactor), which do not have a realistic physical meaning but serve to indicate that runaway has occurred.

From the comparison shown in Table 6.5, it can be observed that the predicted maximum temperature values closest to the experimental data are obtained for Da_p equal to about 0.0017, which is approximately the same value that produces runaway boundaries closest to those observed experimentally (see Fig. 6.14). This consistency in the model predictions supports the conclusions drawn above.

Another indication of the reliability of the results described above can be obtained by comparing the Da_p values used in Fig. 6.14 and Table 6.5, with those predicted *a priori* from semiempirical relationships reported in the literature. In particular, by applying the correlations recommended by Doraiswamy and Sharma (1984) to the various experimental conditions considered by Emig *et al.* (1980), Da_p values in the range of 0.001–0.004 are calculated. Note that the variation in the values of the Reynolds number used in the experiments leads to Da_p changes by a factor of about 1.7. Thus the Da_p values obtained from the correlations can be regarded as in good agreement with the Da_p values used in Fig. 6.14 and Table 6.5.

Role of intraparticle mass-transfer resistance

Let us now extend the analysis performed above to cases including intraparticle mass-transfer resistance. Since, as discussed in Section 6.1, the internal isothermal model is used, the intraparticle heat-transfer resistance has been ignored.

The steady-state behavior of this heterogeneous model is fully characterized by eight dimensionless parameters: reaction order n, dimensionless activation energy γ, heat-of-reaction parameter B, heat-removal parameter β, cooling temperature θ_{co}, Damkohler number Da_p, Lewis number Le, and Thiele modulus Φ^i. In Figs. 6.15a–f the regions of global runaway of the particle temperature are shown in the heat-of-reaction B versus the Thiele modulus Φ^i plane for various values of each of the other six dimensionless parameters. In all cases, in order to obtain a clear graphical representation, we have omitted the boundaries for the local runaway of the particle temperature either at the inlet or at some location along the reactor. As discussed in the previous subsection, the global runaway boundary, obtained through analysis of the parametric sensitivity of the particle temperature at the hot spot, defines a nonrunaway region (below each curve) that is safe from both local and global points of view.

As expected on physical grounds, runaway occurs for larger B values as the Thiele modulus increases. For sufficiently large Φ^i, the reactor becomes controlled by interparticle mass transport, and runaway does not occur. The runaway region enlarges for lower reaction order n (Fig. 6.15a), as well as for lower Lewis number Le (Fig. 6.15e). The latter behavior may be explained physically by noting that lower Le values, with fixed Da_p and Φ^i, lead to lower interparticle heat-transfer coefficients. Therefore, temperature gradients between the catalyst particle and the fluid phase increase, enhancing the possibility for the particle temperature runaway. A similar behavior is exhibited when, for fixed Le and Φ^i, the value of the interparticle mass-transfer resistance parameter Da_p is increased (Fig. 6.15f), making runaway more likely. As already established using simpler reactor models, it is found that runaway is more likely for smaller heat-removal parameter β (Fig. 6.15b), larger activation energy γ (Fig. 6.15c), and smaller external cooling temperature θ_{co} (Fig. 6.15d).

The findings above are in agreement with the results discussed earlier in the context of only interparticle transport limitations, illustrated in the $B–Da_p$ parameter plane in Fig. 6.12. A similar representation is shown in Fig. 6.16, indicating the effect of the Thiele modulus on the runaway region. It is seen that, as the Thiele modulus Φ^i increases, the runaway region progressively shrinks, and the boundary of the global runaway region approaches and eventually merges with the local runaway boundary at the inlet conditions. This indicates that, for sufficiently large Φ^i values, the reactor runaway is determined by the local runaway of the particles at the reactor inlet. On the

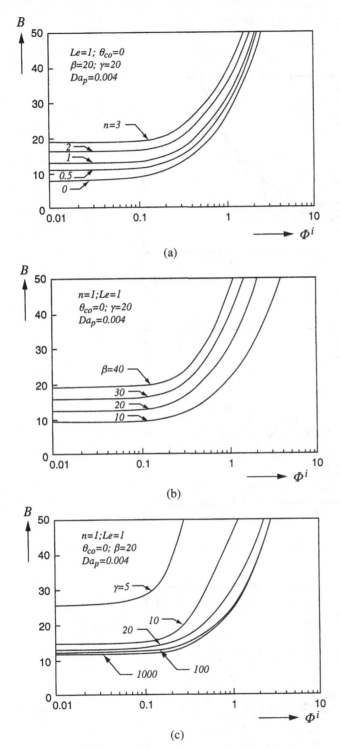

Figure 6.15. Influence of various physicochemical parameters on runaway regions in the case of inter- and intraparticle transport limitations. From Morbidelli and Varma (1987). (a) reaction order n; (b) wall heat transfer β; (c) activation energy γ; (d) coolant temperature θ_{co}; (e) Lewis number Le; (f) interparticle mass-transfer resistance Da_p.

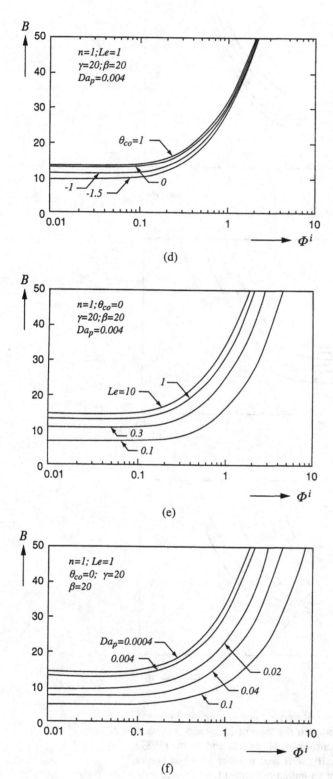

Figure 6.15. (cont.)

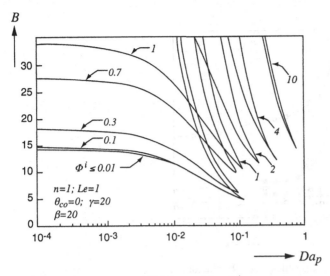

Figure 6.16. Influence of Thiele modulus Φ^i on runaway regions in the B–Da_p parameter plane. From Morbidelli and Varma (1987).

other hand, for fixed Φ^i and sufficiently large Da_p values, the reactor becomes controlled by interparticle mass and heat transfer, and the transition between runway and nonrunaway regions tends to vanish. In each case, there is a region (as in Fig. 6.10, but not shown for simplicity) where a generalized boundary that separates the runaway and nonrunaway behavior cannot be defined.

In summary, by examining the runaway regions shown in Figs. 6.15 and 6.16, it can be concluded that both interparticle and intraparticle transport limitations have a significant impact on reactor runaway, which cannot be ignored as is done when pseudo-homogeneous models are used. It also appears that interparticle and intraparticle transport limitations have a qualitatively different effect on the runaway region. *As interparticle mass- and heat-transfer resistances increase, runaway becomes more likely,* since the critical B value for runaway decreases continuously (Fig. 6.16). *On the other hand, as intraparticle mass-transfer resistance increases, runaway becomes less likely* (Fig. 6.15).

Example 6.2 Experimental analysis of runaway in a fixed-bed reactor for carbon monoxide oxidation. The parametric sensitivity behavior of a fixed-bed reactor, where carbon monoxide oxidation on CuO catalyst supported on γ-alumina occurs, was investigated experimentally by Bauman and Varma (1990). In the case of excess oxygen, the kinetics of this reaction were expressed by the following power law:

$$r = 1.72 \times 10^5 \exp\left(-\frac{6.85 \times 10^3 \text{ K}}{T}\right) \cdot C^{0.737}, \quad \text{kmol/s/m}^3$$

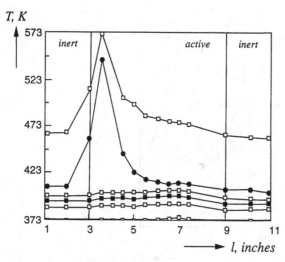

Figure 6.17. Temperature profiles along the reactor axis for various inlet temperatures in the case of CO oxidation on CuO/γ-Al_2O_3 catalyst. From Bauman and Varma (1990).

where C represents the concentration of carbon monoxide. A tubular fixed-bed reactor, with inert and catalytically active zones, was constructed. Several thermocouple ports were located along the reactor axis in order to measure the temperature profile within the reactor. A set of temperature profiles measured at different values of the inlet temperature is shown in Fig. 6.17. Based on the definition of sensitivity, the derivative of the temperature maximum with respect to the inlet temperature was estimated numerically to yield a plot of the experimentally measured normalized sensitivity, shown in Fig. 6.18. The experimental results were compared with the theoretical predictions obtained from the heterogeneous plug-flow reactor model, with both interparticle and intraparticle transport resistances.

A comparison of the experimental values of normalized sensitivity with theoretical predictions requires a complete specification of the reactor model. In addition to the intrinsic reaction kinetics, this involves the estimation of several transport parameters. The values of the interparticle transport coefficients were taken as the average values predicted by two commonly used assumptions: The Nusselt number equals the Sherwood number, and the Chilton-Colburn analogy, which states that the j factors for mass and heat transfer are equal. The reactor wall heat-transfer coefficient was estimated from the correlation of Dhalewadikar (1984), which leads to values relatively close to those predicted by Li and Finlayson (1977). The intraparticle mass diffusivity was measured experimentally using a Weisz (1957) diffusivity cell.

Figure 6.18 shows the normalized sensitivity of the fluid temperature maximum to the inlet temperature as a function of the inlet temperature, as predicted by the heterogeneous plug-flow model (solid curve) and measured experimentally ($\bullet$). It is seen that the theoretical predictions are in good agreement with the experimental

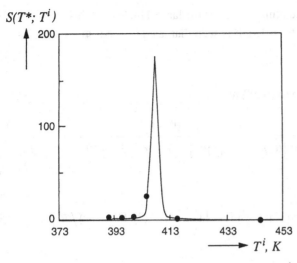

Figure 6.18. Values of the normalized sensitivity $S(T^*; T^i)$ as a function of the inlet temperature, measured experimentally (•) and predicted by the generalized criterion (solid curve). From Bauman and Varma (1990).

results. The solid curve exhibits a maximum at $T^i = 407$ K, which according to the generalized criterion provides the critical inlet temperature that separates the runaway region ($T^i > 407$ K) from the nonrunaway one ($T^i < 407$ K). Similarly, the experimental data indicate that the critical inlet temperature for runaway should be somewhat higher than 405 K (see also Fig. 6.17). Mainly because of the difficulty in obtaining reproducible data for changes in the inlet temperature lower than 2 K near the parametrically sensitive region, a better agreement between the experimental results and model simulations was not possible.

6.3.3 Limiting Behavior

Let us now consider the reactor behavior for some limiting values of the involved parameters, which are of interest in applications. From Eqs. (6.13) to (6.15), it follows that, for small values of the Thiele modulus, *i.e.*, $\Phi^i \to 0$, the effectiveness factor becomes unity, *i.e.*, $\eta \to 1$. In this case, the heterogeneous model given by Eqs. (6.36) to (6.39) reduces to the one examined in Section 6.3.1, which accounts only for interparticle mass- and heat-transfer resistances. Accordingly, it is readily seen that the critical B values shown in Figs. 6.15a–d approach, as $\Phi^i \to 0$, those shown in Figs. 6.12a–d. Moreover, when also $Da_p \to 0$ and Le is fixed (*i.e.*, when both interparticle mass- and heat-transfer resistances vanish), the heterogeneous model approaches the pseudo-homogeneous model analyzed in Chapter 4. Also in this case, the critical B values calculated using the heterogeneous model with $Da_p \to 0$ and $\Phi^i \to 0$ (see Figs. 6.15f and 6.16) approach those calculated with the pseudo-homogeneous model.

Let us now consider the limiting case of the large Thiele modulus, $\Phi^i \to \infty$. From inspection of Eqs. (6.13) to (6.15), it is seen that, as $\Phi^i \to \infty$, $\Phi \to \infty$; thus

$$\eta \to 1/\Phi \tag{6.49}$$

and the model Eqs. (6.36) to (6.38) reduce to

$$\frac{d\theta'}{dx} = B'\left(1 - \frac{1}{\psi'} \cdot \frac{\theta' - \theta'_{co}}{\exp[\theta'_p/(1 + \theta'_p/\gamma')] \cdot [1 - x - Le' \cdot (\theta'_p - \theta')/B']^{n'}}\right) \tag{6.50}$$

$$\theta' = \theta'^i = 0 \quad \text{at } x = 0 \tag{6.51}$$

$$\theta'_p = \theta' + \psi'_p \cdot \exp[\theta'_p/(1 + \theta'_p/\gamma')] \cdot [1 - x - Le' \cdot (\theta'_p - \theta')/B']^{n'} \tag{6.52}$$

where

$$
\begin{array}{lll}
\theta'_p = \theta_p/2; & \theta' = \theta/2; & \theta'_{co} = \theta_{co}/2; \\
\gamma' = \gamma/2; & n' = (n+1)/2; & B' = B/2; \\
Da'_p = Da_p/\Phi^i; & Le' = Le; & \psi'_p = B'Da'_p/Le'; \\
Da' = Da/\Phi^i; & St' = St; & \psi' = B'Da'/St'
\end{array} \tag{6.53}
$$

By comparing equations (6.50) and (6.52) with the model equations (6.36) and (6.38) for the case of negligible intraparticle mass-transfer resistance (*i.e.*, $\eta = 1$), it can be concluded that the model equations in the case of severe intraparticle mass-transfer resistance ($\Phi^i \to \infty$) are identical to those in the case of negligible intraparticle mass-transfer resistance ($\Phi^i \to 0$) with the original parameters replaced by those defined in Eq. (6.53). As discussed earlier (Section 6.2.2), this corresponds to well-known fact that under intraparticle diffusion control the apparent reaction rate exhibits an activation energy equal to one-half of the true value and a reaction order equal to the intrinsic order plus one and divided by two.

This finding allows one to evaluate the critical B value for global runaway in the case of the large Thicle modulus using the model equations for the case of negligible intraparticle mass-transfer resistance. The latter are easier to solve from the numerical viewpoint.

Example 6.3 Runaway regions in the case of severe intraparticle mass-transfer resistance.
Let us now compare the predictions given by the model equations including both interparticle and intraparticle transport resistances [*i.e.*, Eqs. (6.36) to (6.39) with Eqs. (6.13) to (6.15)] and by the model equations including only the interparticle transport resistance [*i.e.*, Eqs. (6.50) to (6.52)] but with modified parameters given by Eq. (6.53).

Typical results in terms of runaway boundaries are shown in Fig. 6.19 in the B–Φ^i parameter plane for three different sets of parameter values. The solid curves represent the global runaway boundaries predicted by the model [Eqs. (6.36) to (6.39) and (6.13)

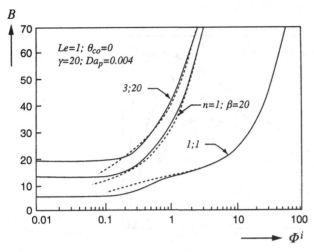

Figure 6.19. Comparison between the runaway regions predicted by the full heterogeneous reactor model [Eqs. (6.36) to (6.39) with (6.13) to (6.15), solid curves] and the simplified reactor model [Eqs. (6.50) to (6.53), broken curves]. From Morbidelli and Varma (1987).

to (6.15)], while the broken curves correspond to the simpler model with modified parameters [Eqs. (6.50) to (6.53)]. As expected, the broken curves merge into the solid ones as $\Phi^i \to \infty$.

However, it may be noted that the broken curve remains close to the solid one even at relatively small values of the Thiele modulus. In other words, the heterogeneous model including only interparticle transport resistances but with modified parameters given by Eq. (6.53) predicts essentially the same runaway region as the one including both interparticle and intraparticle transport resistances, for Φ^i values as low as about 0.2. This is surprising, since it is known that the approximation (6.49), on which Eq. (6.53) is based, provides accurate results only for Φ^i values larger than about 5 (Froment and Bischoff, 1990).

In order to investigate this point further, let us consider the particle temperature versus conversion profiles shown in Figs. 6.20a and b for two different sets of parameter values, both near criticality. The solid and broken curves correspond to the predictions given by Eqs. (6.36) to (6.39) with Eqs. (6.13) to (6.15), and by Eqs. (6.50) to (6.53), respectively. In Fig. 6.20a, since $\Phi^i = 20$, as expected, the two models predict very similar particle temperature versus conversion profiles, and consequently also very similar critical B values. On the other hand, in Fig. 6.20b, due to the small value of the Thiele modulus ($\Phi^i = 0.15$), the temperature profiles predicted by the two models are vastly different; however, they still predict very similar critical B values. This indicates that although the simpler model [Eqs. (6.50) to (6.53)] predicts quantitatively different reactor behavior, it preserves the intrinsic sensitivity character and gives almost the same prediction of the runaway region as the complete model. This feature could be

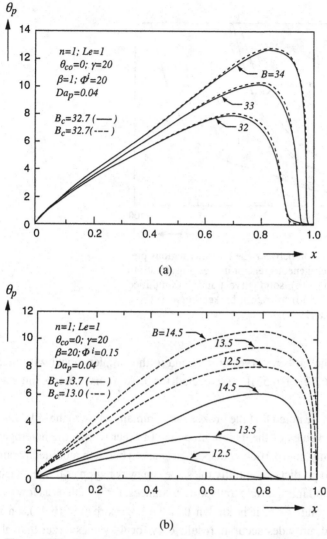

Figure 6.20. Particle temperature versus conversion profiles for various values of the heat-of-reaction parameter B around criticality. Solid curves, given by the full heterogeneous model [Eqs. (6.36) to (6.39) with (6.13) to (6.15)]; broken curves, given by the simplified model [Eqs. (6.50) to (6.53)]. (a) $\Phi^i = 20$; (b) $\Phi^i = 0.15$. From Morbidelli and Varma (1987).

ascribed at least in part to the intrinsic nature of the generalized runaway criterion that has been utilized.

6.3.4 Effect of Pseudo-Adiabatic Operation on Runaway Regions

It has been shown in Section 4.2.2 that a pseudo-homogeneous tubular reactor of finite length may operate in the pseudo-adiabatic operation (PAO) regime, where the

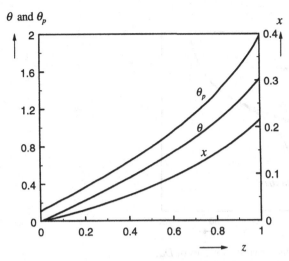

Figure 6.21. Profiles of conversion x, fluid temperature θ, and particle temperature θ_p, along the reactor axis z for a fixed-bed catalytic reactor in the PAO regime. $B = 10$; $Da = 0.1$; $St = 1$; $\gamma = 20$; $n = 1$; $Da_p = 0.01$; $Le = 1$; $\Phi^i = 0$; $\theta_{co} = 0$.

temperature along the reactor axis increases monotonically, without presenting a hot spot. The PAO regime may occur also in the case of fixed-bed catalytic reactors. An example is shown in Fig. 6.21, where conversion as well as fluid and particle temperatures are plotted against the reactor length. It is seen that both temperatures increase monotonically along the reactor axis, and their maxima would occur for reactor lengths longer than that considered in this example.

The PAO regime of a reactor cannot be predicted by a model using conversion as the independent variable. As can be seen from Eq. (6.36), for the typical case where the coolant temperature does not exceed the initial temperature, we have, when $x \to 1$, $d\theta/dx < 0$. This indicates that a temperature maximum occurs at a lower conversion value. It is clear that when the model equations (6.6) to (6.8) with reactor length as the independent variable are transformed into equations (6.36) to (6.38) with reactant conversion as the independent variable, the constraint of the reactor length on model predictions is lost; *i.e.*, an infinite value of the reactor length is implicitly assumed. Consequently, as discussed in detail in Section 4.2.2, using conversion as the independent variable, unrealistic runaway behavior is predicted for reactors in the PAO regime.

In Fig. 6.22, the global runaway boundary in the $\psi^{-1}-B$ parameter plane predicted by the generalized criterion using conversion as the independent variable (broken curve) is compared with those obtained using reactor length as the independent variable (solid curves) for various Da values. Note that, when taking conversion as independent variable, the critical ψ value is independent of Da, *i.e.*, reactor length. It is seen that the runaway boundary predicted in this case coincides with those predicted using reactor length as the independent variable only for long reactors ($Da \geq 1$). For shorter

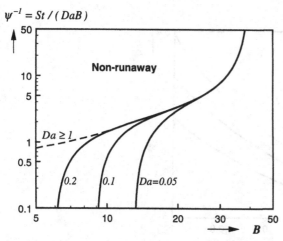

Figure 6.22. Runaway regions for various values of Da, predicted by the Morbidelli and Varma generalized criterion. Solid curves, reactor length as independent variable; broken curve, conversion (or reactor length for $Da \geq 1$) as independent variable. $\gamma = 20$; $n = 1$; $Da_p = 0.01$; $Le = 1$; $\Phi^i = 0$; $\theta_{co} = 0$.

reactors, the two boundaries deviate from each other in the region of low B values, and the differences increase for decreasing reactor lengths, *i.e.*, for lower Da values.

In order to have a better illustration of this point, let us now consider the runaway boundaries for a given reactor length, together with the numerically computed boundaries of the region where the reactor operates in the PAO regime. It is worth noting that, in the case of heterogeneous reactors, one may define PAO with respect to either the fluid or the particle temperature. For the case of $Da = 0.1$ shown in Fig. 6.22, the numerically computed PAO boundaries referring to fluid (curve 1) and particle temperature (curve 2) are shown in Fig. 6.23. The upper branches of the two PAO boundaries are identical, while the lower ones deviate from each other, and in particular the PAO boundary based on the particle temperature enters the runaway region slightly before that based on the fluid temperature. This arises because the particle at its hot spot is in the ignited state, with a temperature much higher than the fluid temperature. Continued heat transfer after the particle hot spot delays the fluid temperature hot spot.

Figure 6.23 also plots the runaway boundaries, curves 3 and 4 (from Fig. 6.22), corresponding to reactor length and conversion as independent variables, respectively. The top portions of these curves coincide and separate the hot-spot operation (HSO) and runaway regions, where in the former a particle hot spot occurs within the reactor but is not in runaway condition. However, as already seen in Fig. 6.22, in this case, the runaway boundary predicted using conversion as independent variable is conservative, and obviously unrealistic, because the temperature maximum, θ_p^* (or θ^*), in the normalized sensitivity, $S(\theta_p^*; \phi)$ [or $S(\theta^*; \phi)$], does not occur inside the reactor of given length. It is clear that *the runaway boundary predicted with conversion as the*

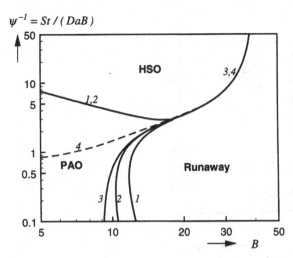

Figure 6.23. Reactor operation diagram in the ψ^{-1}–B parameter plane: curves 1 and 2 are the PAO boundaries referring respectively to particle and fluid temperatures; curves 3 and 4 are the runaway boundaries obtained with reactor length and conversion as independent variable, respectively. $\gamma = 20$; $n = 1$; $Da_p = 0.01$; $Le = 1$; $\Phi^i = 0$; $Da = 0.1$; $\theta_{co} = 0$.

independent variable cannot be used to describe the transition of the reactor behavior from the PAO to the runaway region, while it can be used for the transition from the HSO to the runaway region.

It should be noted that the runaway boundary is located inside the PAO region also when reactor length is used as the independent variable (see Fig. 6.23). In this case, runaway is investigated by considering the sensitivity of the reactor outlet temperature to changes of the independent parameter ϕ, $S(\theta_p^*; \phi)$ [or $S(\theta^*; \phi)$], since this is the temperature maximum in a reactor operating in the PAO regime. Figures 6.24a and b show the normalized sensitivities, $S(\theta_p^*; B)$ and $S(\theta^*; B)$, respectively, as a function of the heat-of-reaction parameter B, for a reactor whose runaway boundary is located inside the PAO region (*i.e.*, $St = 1$ in Fig. 6.23). The sensitivity of the particle temperature maximum $S(\theta_p^*; B)$ in Fig. 6.24a exhibits a sharp maximum at $B = 10.7$, where the particle temperature at the reactor outlet, θ_p^*, increases with B sharply, thus indicating a transition from safe to runaway operation. Similarly, in Fig. 6.24b, the sensitivity of the fluid temperature maximum, $S(\theta^*; B)$, exhibits a sharp maximum at practically the same B value. As discussed in Section 6.3.1, if the predicted runaway boundary has the generalized feature, the same global runaway boundary is predicted using either the particle or the fluid temperature maximum as objective of the normalized sensitivity. Accordingly, the results in Figs. 6.24a and b show that in this case the runaway boundary exhibits the generalized feature even when the reactor operates in the PAO regime.

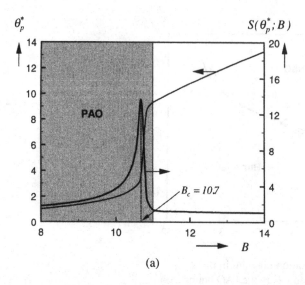

(a)

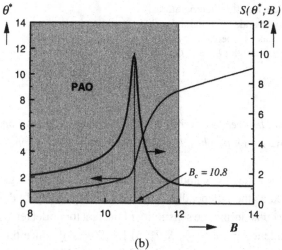

(b)

Figure 6.24. Temperature maximum and its normalized sensitivity to the heat-of-reaction parameter B as a function of B: (a) particle temperature; (b) fluid temperature. The sensitivity peaks indicate runaway occurrence inside the PAO region. $\gamma = 20$; $n = 1$; $Da_p = 0.01$; $Le = 1$; $\Phi^i = 0$; $St = 1$; $Da = 0.1$; $\theta_{co} = 0$.

The conclusion above is somewhat in contrast to that reached in Section 4.2 for pseudo-homogeneous reactors. Specifically, it was found that when the boundaries of the PAO and the runaway (predicted using reactor length as independent variable) regions are coincident, *e.g.*, the lower branch in the St–B parameter plane in Fig. 4.7, the latter has the generalized feature. On the other hand, when significant derivations between the runaway and the PAO boundaries arise, then the transition between safe and runaway operation tends to vanish, *i.e.*, a generalized boundary that separates the two does not exist. The explanation of this contrast is that in the case of fixed-bed reactors, the difference between PAO and runaway boundaries is not due to the loss of generalized validity of the latter as in homogeneous reactors but rather to the higher complexity of the involved physicochemical phenomena. In the case of homogeneous reactors, the occurrence of PAO depends only on the competition between heat-removal rate by external coolant and heat-generation rate by chemical reaction. The picture is complicated in the case of the fixed-bed reactors, by heat exchange between the stationary solid and the flowing fluid phases. These findings further indicate that the runaway behavior of fixed-bed catalytic reactors cannot be investigated using pseudo-homogeneous models.

It is worth stressing that, for $B \in [10.7, 11.0]$ in Fig. 6.24a and $B \in [10.8, 12.0]$ in Fig. 6.24b, since the reactor operates in the PAO regime, runaway is driven by the reactor outlet temperature, which is also the hot spot in the reactor.

One final comment is worth making. Studying reactor runaway behavior using conversion as the independent variable typically is chosen because PAO is relatively rare in industrial practice. In the PAO regime, the outlet conversion, as shown in Fig. 6.21, is generally low. Except for some cases of complex reactions, where large outlet conversion may lead to low selectivity toward the desired product, industrial reactors are generally designed sufficiently long to achieve significant conversion, and thus a hot spot is commonly present inside the reactor (*i.e.*, the HSO regime). Therefore, in most cases of practical relevance, one can apply the MV generalized criterion with conversion as the independent variable to predict the runaway behavior of fixed-bed catalytic reactors.

6.4 Explicit Criteria for Runaway

As discussed above, in order to correctly describe runaway in fixed-bed catalytic reactors it is necessary to account for interparticle and intraparticle transport resistances. This means that the heterogeneous model presented in Section 6.1 has to be utilized. This model is, however, too complicated for developing explicit criteria, and numerical methods are therefore required. Exceptions are two approximate explicit criteria, which are discussed in the following.

In the case of negligible intraparticle transport limitations, the analytical expressions (6.44) to (6.47) derived by Rajadhyaksha *et al.* (1975), although very conservative as discussed in the context of Fig. 6.13, may be used for a rough first estimation of the runaway boundary. Balakotaiah and Luss (1991) derived an explicit criterion based on the following two key assumptions:

(1) The temperature rise at ignition, which then leads to runaway, is small enough so that the Arrhenius temperature dependence can be replaced by the positive exponential approximation, *i.e.*, the Frank-Kamenetskii truncation.

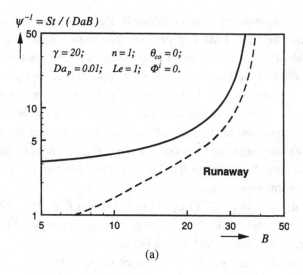

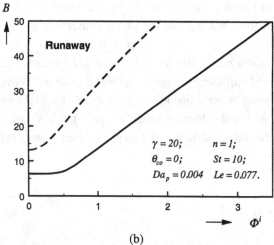

Figure 6.25. Runaway boundaries predicted by the explicit criterion developed by Balakotaiah and Luss (1991) (solid curves) and the generalized criterion (broken curves) in the case of (a) interparticle and (b) both inter- and intraparticle transport limitations.

(2) The reactant consumption in the fluid phase is negligible, although diffusional limitations within the catalyst particles are accounted for.

These two assumptions are the same ones used by Semenov (1928) when deriving the thermal runaway criterion in homogeneous batch reactors (see Section 3.2.1). Criticality for runaway is defined as the situation where ignition of the low-temperature steady state occurs. This arises where, while moving along the reactor axis, the high conversion branch of the particle multiplicity pattern becomes available (*i.e.*, the lower bifurcation point for the particle temperature). In this case, even though the particle temperature has not crossed the upper bifurcation point, it is possible that ignition occurs as a consequence of a large perturbation in the reactor that drives the particle temperature from the low to the high conversion branch. For a first-order reaction, the derived explicit criterion accounting for both interparticle and intraparticle transport limitations can be written as follows:

$$\frac{1}{\psi_c} = \left(\frac{St}{Da \cdot B}\right)_c = \frac{e}{[f(\Phi^i) - e \cdot \psi_p]} \tag{6.54}$$

where

$$f(\Phi^i) = \begin{cases} \exp(-\theta_{co}); & \Phi^i < 0.5\exp(-\theta_{co}/2) \\ 2\Phi^i \exp(-\theta_{co}/2); & \Phi^i > 0.5\exp(-\theta_{co}/2) \end{cases} \tag{6.55}$$

It is seen that, since Semenov's assumptions are used, in the case of negligible interparticle and intraparticle transport limitations (i.e., $\psi_p \to 0$ and $\Phi^i \to 0$) and $\theta_{co} = 0$, Eq. (6.54) reduces to the Semenov criterion (i.e., $\psi_c \to 1/e$).

Typical runaway boundaries (solid curves) calculated through Eq. (6.54) are shown in Fig. 6.25, in the cases where only interparticle (a) or both interparticle and intraparticle (b) transport limitations are present. In the same figure are also shown the results of the MV generalized criterion (broken curves), which differ significantly. Let us consider the case of $B = 40$ in Fig. 6.25b and compute the temperature maximum and its normalized sensitivity $S(\theta^*; \Phi^i)$ as a function of Φ^i. The obtained results are shown in Fig. 6.26. The normalized sensitivity reaches its maximum at $\Phi^i = 1.375$, which according to the generalized criterion is the critical Φ^i value that separates safe

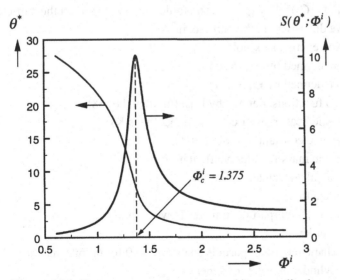

Figure 6.26. Temperature maximum, θ^*, and its normalized sensitivity to the Thiele modulus, $S(\theta^*; \Phi^i)$, as a function of the Thiele modulus. $\gamma = 20$; $n = 1$; $Da_p = 0.004$; $Le = 0.077$; $B = 40$; $Da = 0.5$; $St = 10$; $\theta_{co} = 0$.

($\Phi^i > 1.375$) from runaway ($\Phi^i < 1.375$) operation. This is indeed physically sound since around $\Phi^i = 1.375$ the temperature maximum increases sharply with the decrease of Φ^i. However, the explicit criterion locates the criticality at $\Phi^i = 2.788$, which is about twice as large as that predicted by the generalized criterion. From Fig. 6.26 it is seen that, at $\Phi^i = 2.788$, the temperature maximum in the reactor is still small and is far from the runaway region. Thus, the explicit criterion, Eq. (6.54), generally gives very conservative estimation of the runaway boundary. This is primarily due to the two assumptions noted earlier, as also confirmed by Balakotaiah *et al.* (1995). In any case, since the obtained criterion (6.54) is simple and predicts correctly the shape of the runaway boundary, it can be conveniently used in practice for a first estimation of runaway occurrence.

The above explicit criterion has also been extended to an nth-order reaction in the case of negligible interparticle mass- and heat-transfer resistances, by using the normalized Thiele modulus given by Eq. (6.15) (Balakotaiah and Luss, 1991).

Nomenclature

a_v	Particle surface per unit reactor volume, 1/m
B	$(-\Delta H) \cdot C^i \cdot \gamma / \rho_f \cdot c_p \cdot T^i$, heat-of-reaction parameter
c_p	Mean specific heat of reaction mixture, kJ/(K·kg)
C	Reactant concentration, kmol/m^3
d_t	Tubular reactor diameter, m
Da	$\rho_B \cdot k(T^i) \cdot (C^i)^{n-1} \cdot L/v^o$, Damkohler number
Da_p	$\rho_B \cdot k(T^i) \cdot (C^i)^{n-1} \cdot /k_g \cdot a_v$, Damkohler number based on the particle
D_e	Effective diffusivity in the particle, m^2/s
E	Activation energy, kJ/kmol
$f(\Phi^i)$	Function, defined by Eq. (6.55)
F	Function, defined by Eq. (6.17)
F_i	($i = 1, 2$) functions, defined by Eqs. (6.36) and (6.38)
h	Interparticle heat-transfer coefficient, kJ/(m^2·s·K)
k	Reaction rate constant, (kmol/m^3)$^{1-n}$/s
k_g	Interparticle mass-transfer coefficient, m/s
l	Reactor axial coordinate, m
L	Reactor length, m
Le	$h/k_g \cdot \rho_f \cdot c_p$, interparticle transfer Lewis number
n	Reaction order
n_p	Integer characteristic of particle shape: $n_p = 0$ for infinite slab, $n_p = 1$ for infinite cylinder, $n_p = 2$ for sphere
r	Reaction rate, kmol · (m^3/kmol)^n/(kg·s)
R_g	Ideal-gas constant, kJ/(K·kmol)
$s(y; \phi)$	$\partial y/\partial \phi$, local sensitivity of dependent variable y to independent parameter ϕ

$S(y; \phi)$ $s(y; \phi) \cdot \phi/y$, normalized sensitivity of independent variable y to independent parameter ϕ

St $4 \cdot U \cdot L/v^o \cdot \rho_f \cdot c_p \cdot d_t$, Stanton number

T Temperature, K

U Overall heat-transfer coefficient, kJ/(m^2·s·K)

v^o Velocity, m/s

x $1 - C/C^i$, reactant conversion

z l/L

Greek Symbols

β St/Da

ΔH Heat of reaction, kJ/kmol

ϕ Generic system parameter

Φ $(\Phi^i)^2 \cdot (1 - x_s)^{n-1} \cdot \exp[\theta_s/(1 + \theta_s/\gamma)]$, normalized Thiele modulus

Φ^i $(V_p/S_p)^2 \cdot (n_p + 1) \cdot \rho_p \cdot k(T^i) \cdot (C^i)^{n-1}/2 \cdot D_e$, normalized Thiele modulus at reactor inlet conditions

γ $E/R_g \cdot T^i$, dimensionless activation energy

η $[3 \cdot \Phi \cdot \coth(3 \cdot \Phi) - 1]/3 \cdot \Phi^2$, effectiveness factor

θ $\gamma \cdot (T - T^i)/T^i$, dimensionless temperature

$\theta*$ Maximum of the dimensionless temperature along the reactor axis

ρ Density, kg/m^3

ψ $Da \cdot B/St$, Semenov number for a tubular reactor

ψ_p $Da_p \cdot B/Le$, Semenov number for a catalyst particle

Subscripts

c Critical condition

co Coolant side

f Fluid phase

p Catalyst particle

s External surface of the particle

opt Optimal condition

Superscripts

i Reactor inlet

o Reactor outlet

Acronyms

HSO   Hot-spot operation

MV   Morbidelli and Varma

PAO   Pseudo-adiabatic operation

References

Aris, R. 1965. A normalization for the Thiele modulus. *Ind. Eng. Chem. Fund.* **4**, 227.

Aris, R. 1975. *The Mathematical Theory of Diffusion and Reaction in Permeable Catalysts*. Oxford: Clarendon Press.

Balakotaiah, V., Kodra, D., and Nguyen, D. 1995. Runaway limits for homogeneous and catalytic reactors. *Chem. Eng. Sci.* **50**, 1149.

Balakotaiah, V., and Luss, D. 1991. Explicit runaway criterion for catalytic reactors with transport limitations. *A.I.Ch.E. J.* **37**, 1780.

Bauman, E. G., and Varma, A. 1990. Parametric sensitivity and runaway in catalytic reactors: experiments and theory using carbon monoxide oxidation as an example. *Chem. Eng. Sci.* **45**, 2133.

Bischoff, K. B. 1965. Effectiveness factors for general reaction rate forms. *A.I.Ch.E. J.* **11**, 351.

Carberry, J. J. 1975. On the relative importance of external-internal temperature gradients in heterogeneous catalysis. *Ind. Eng. Chem. Fundam.* **14**, 129.

Carberry, J. J., and Wendel, M. M. 1963. A computer model of the fixed bed catalytic reactor: the adiabatic and quasi-adiabatic cases. *A.I.Ch.E. J.* **9**, 129.

Dhalewadikar, S. V. 1984. *Ethylene Oxidation on Supported Platinum Catalyst in a Non-Adiabatic Fixed-Bed Reactor: Experimental and Model*. Ph.D. Thesis: University of Notre Dame.

Doraiswamy, L. K., and Sharma, M. M. 1984. *Heterogeneous Reactions: Analysis, Examples, and Reactor Design*, Vol. 1. New York: Wiley.

El-Sawi, M., Emig, G., and Hofmann, H. 1977. A study of the kinetics of vinyl acetate synthesis. *Chem. Eng. J.* **13**, 201.

Emig, G., Hofmann, H., Hoffmann, U., and Fiand, U. 1980. Experimental studies on runaway of catalytic fixed-bed reactors (vinyl-acetate-synthesis). *Chem. Eng. Sci.* **35**, 249.

Finlayson, B. A. 1971. Packed-bed reactor analysis by orthogonal collocation. *Chem. Eng. Sci.* **26**, 1081.

Froment, G. F., and Bischoff, K. B. 1990. *Chemical Reactor Analysis and Design*. New York: John Wiley & Sons.

Hagan, P. S., Herskowitz, M., and Pirkle, C. 1988a. A simple approach to highly sensitive tubular reactors. *SIAM J. Appl. Math.* **48**, 1083.

Hagan, P. S., Herskowitz, M., and Pirkle, C. 1988b. Runaway in highly sensitive tubular reactors. *SIAM J. Appl. Math.* **48**, 1437.

Li, C. H., and Finlayson, B. A. 1977. Heat transfer in packed beds: a re-evaluation. *Chem. Eng. Sci.* **32**, 1055.

Matros, Yu. Sh., and Bunimovich, G. A. 1996. Reverse-flow operation in fixed bed catalytic reactors. *Catal. Rev. Sci. Eng.* **38**, 1.

McGreavy, C., and Adderley, C. I. 1973. Generalized criteria for parametric sensitivity and temperature runaway in catalytic reactors. *Chem. Eng. Sci.* **28**, 577.

Mears, D. E. 1971. Tests for transport limitations in experimental catalytic reactors. *Ind. Eng. Chem. Process Des. Dev.* **10**, 541.

Morbidelli, M. 1987. *Parametric Sensitivity and Runaway in Chemically Reacting Systems*. Ph.D. Thesis: University of Notre Dame.

Morbidelli, M., and Varma, A. 1986a. Parametric sensitivity in fixed-bed catalytic reactors: the role of interparticle transfer resistance. *A.I.Ch.E. J.* **32**, 297.

Morbidelli, M., and Varma, A. 1986b. Parametric sensitivity and runaway in fixed-bed catalytic reactors. *Chem. Eng. Sci.* **41**, 1063.

Morbidelli, M., and Varma, A. 1987. Parametric sensitivity in fixed-bed catalytic reactors: inter- and intraparticle resistance. *A.I.Ch.E. J.* **33**, 1949.

Morbidelli, M., and Varma, A. 1988. A generalized criterion for parametric sensitivity: application to thermal explosion theory. *Chem. Eng. Sci.* **43**, 91.

Pereira, C. J., Wang, J. B., and Varma, A. 1979. A justification of the internal isothermal model for gas-solid catalytic reactions. *A.I.Ch.E. J.* **25**, 1036.

Petersen, E. E. 1965. *Chemical Reaction Analysis*. Englewood Cliffs, NJ: Prentice-Hall.

Rajadhyaksha, R. A., Vasudeva, K., and Doraiswamy, L. K. 1975. Parametric sensitivity in fixed-bed reactors. *Chem. Eng. Sci.* **30**, 1399.

Semenov, N. N. 1928. Zur theorie des verbrennungsprozesses. *Z. Phys.* **48**, 571.

van Welsenaere, R. J., and Froment, G. F. 1970. Parametric sensitivity and runaway in fixed bed catalytic reactors. *Chem. Eng. Sci.* **25**, 1503.

Weisz, P. B. 1957. Diffusivity of porous particles. measurements and significance for internal reaction velocities. *Z. Phys. Chem.* **11**, 1.

Wu, H., Rota, R., Morbidelli, M., and Varma, V. 1998. Parametric sensitivity in fixed-bed catalytic reactors with reverse-flow operation. *Chem. Eng. Sci.* Submitted.

7

Parametric Sensitivity and Ignition Phenomena in Combustion Systems

COMBUSTION PROCESSES are of central importance in a variety of applications, such as engines, turbines, and furnaces. The *ignition* of a combustion process can be either endogenous (*i.e.*, self-ignition) or exogenous (*i.e.*, induced by an external agent such as a spark or a local temperature increase). The identification of the self-ignition conditions for a given chemical system is not only of practical interest, but it is also a challenging test for the validation of combustion kinetic models. The first fundamental question that we address in this chapter is the definition of a criterion to establish whether a system has been ignited or not. As we will see in the following, this can be done by using the concepts related to parametric sensitivity discussed in previous chapters in the context of chemical reactors. As a model system we will use hydrogen oxidation, since it constitutes a prototype for more complex combustion processes, and its kinetic behavior has been well studied both experimentally and theoretically.

Ignition can be considered as a transition region or a boundary that separates slow from fast combustion processes. For combustion occurring in a shock tube or in a closed vessel, it is often required to determine the so-called *ignition limits* in a parameter space (typically temperature, pressure, and composition) that identify where the system is ignited, *i.e.*, it undergoes a fast combustion process. For combustion induced by an external ignition source, a threshold needs to be defined to estimate the minimum energy required to ignite the system. Earlier studies on defining quantitatively such ignition limits were based on some geometric properties of the temperature profile, involving the assumptions of quasi steady state and negligible reactant consumption, as well as substantially simplified kinetic models. Results of these studies are discussed in several textbooks and review papers (*e.g.*, Semenov, 1959; Lewis and von Elbe, 1961; Dixon-Lewis and Williams, 1977; Zeldovich *et al.*, 1985; Kordylewski and Scott, 1984). In general, the features of the temperature profile used to identify an ignited system are the same as discussed in Chapter 3 in the context of runaway in batch reactors. On the other hand, the definition of ignition proposed by Gray and Yang (1965, 1967) is based on a different concept. Specifically, they defined the critical condition for ignition as the vanishing of the Jacobian of the system of equations describing the system behavior. Although the original investigation involved an algebraic system of equations obtained using the assumptions of quasi steady state and negligible reaction consumption, it is based on intrinsic system properties and can, as noted by Kordylewski and Scott (1984), be applied to more complex systems described by ordinary differential or partial differential equations.

Alternatively, since ignition limits separate two different types of system behavior, their identification can be pursued using the criteria for parametric sensitivity. Along these lines, Wu *et al.* (1993) proposed to describe the ignition phenomena in combustion processes through applications of the generalized criterion (Morbidelli and Varma, 1988) described in Chapter 3. It was shown that by using hydrogen oxidation in a closed vessel as an example along with a detailed kinetic model based on a number of elementary reactions, it is possible to describe correctly the three experimental explosion limits (the well-known inverse-S-shaped explosion boundaries) that separate the explosion from nonexplosion regions in the initial pressure-temperature plane.

This chapter, based primarily on the work of Wu *et al.* (1993) and Morbidelli and Wu (1992), illustrates the application of the generalized criterion to define the ignition limits quantitatively, using the example of hydrogen–oxygen mixtures in closed vessels. It will be shown that the generalized criterion provides an effective tool to predict the ignition dynamics in the initial pressure–initial temperature parameter plane.

7.1   General Definition of Ignition Limits

Let us consider the ignition or explosion limits shown in Fig. 7.1, as measured by Lewis and von Elbe (1961) for the stoichiometric H_2–O_2 reaction occurring in a KCl-coated vessel with radius $R_v = 3.7$ cm. These separate, in the initial pressure–initial temperature plane, the slow reaction or nonexplosion region (on the left-hand side of the boundary) from the fast reaction or explosion region (on the right-hand side of the boundary). The inverse-S-shaped explosion boundaries in Fig. 7.1 are divided by two bifurcation points, B_1 and B_2, into three branches, which from low to high initial pressure are usually referred to as the first, second, and third explosion limits, respectively. If we consider two H_2–O_2 mixtures with the same initial pressure, but with two slightly different initial temperature values, one to the left and the other to the right of the explosion boundary, they exhibit a distinctly different behavior in time, in terms of both temperature and conversion values. This indicates that near the explosion boundary the system behavior is sensitive to small changes in the initial temperature. On the other hand, if the initial temperature value is located far from the explosion boundary, then the system behavior is insensitive to changes in the initial temperature; *i.e.*, small changes in the initial temperature do not change the qualitative behavior of the system (*e.g.*, slow or fast reaction), but simply lead to correspondingly small differences in the transient temperature and conversion values.

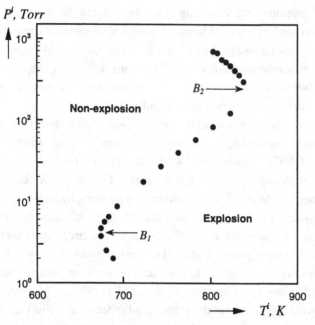

Figure 7.1. Explosion limits for the stoichiometric H_2–O_2 mixture in the initial pressure–initial temperature plane, measured experimentally by Lewis and von Elbe (1961) in a KCl-coated vessel of 7.4 cm in diameter.

This observation indicates that explosion phenomena can be regarded as instances of parametrically sensitive system behavior. Thus, we can determine the explosion limits through application of the MV generalized criterion based on the normalized objective sensitivity.

With respect to the thermal explosion or runaway in chemical reactors illustrated in Chapters 3 to 6, the explosion phenomena in combustion systems are more complex. For example, thermal explosion in a batch reactor is due to the rate of heat production by chemical reactions that is faster than the rate of heat removal by the cooling system. This leads to a continuous rise of the reactor temperature with a consequent acceleration of the chemical reactions, leading eventually to explosion. In this case, the key parameter describing the reactor behavior is the temperature maximum, which has been used in previous chapters to define the normalized objective sensitivity in the MV generalized criterion. In the case of flammable mixtures, the combustion process involves a complex network of chemical reactions involving radicals. This network arises from the classical chain mechanism consisting of initiation, propagation, branching, and termination reactions. In this case, in addition to the thermal processes described above, the chain branching reactions are also responsible for the ignition behavior. Chain branching reactions produce two or more active species or radicals from a single one, thus accelerating the process that under certain conditions may lead to explosion. Accordingly, in the following, we consider not only the temperature maximum but also the concentration maximum of specific radicals in defining the normalized objective sensitivities in the generalized criterion.

Let us consider a flammable mixture in a closed, well-mixed vessel. The changes in concentration of all chemical species in the vessel can be described by the following system of ordinary differential equations in vector form:

$$\frac{dy}{dt} = f(t, y, \phi), \tag{7.1}$$

with initial conditions (ICs)

$$y = y^i \quad \text{at } t = 0, \tag{7.2}$$

where t is time, $y = [y_1 \, y_2 \cdots y_{N_s}]^T$ is a vector including all the species concentrations, N_s is the number of chemical species, ϕ is the vector of independent parameters (e.g., initial temperature T^i; initial pressure P^i; kinetic constants, etc.) and f is the vector of functions representing the reaction rate of each species. The reactor energy balance is given by the following differential equation:

$$\frac{dT}{dt} = \left[\sum_{i=1}^{N_s} (h_i - R_g \cdot T) \cdot f_i - \frac{3 \cdot U}{R_v} \cdot (T - T^i) \right] \Big/ C_{vm}, \tag{7.3}$$

where

$$C_{vm} = \sum_{i=1}^{N_s} C_{vi} \cdot y_i. \tag{7.4}$$

R_v is the radius and U the wall heat-transfer coefficient of the vessel. Parameters h_i and C_{vi} are the enthalpy and the constant-volume specific heat capacity of the ith species, respectively, and their temperature dependence has been accounted for using the NASA polynomials (Gordon and McBride, 1976).

The local sensitivity of the ith variable y_i, with respect to the generic independent parameter ϕ, $s(y_i; \phi)$, is defined as

$$s(y_i; \phi) = \frac{\partial y_i}{\partial \phi} \tag{7.5}$$

Following the direct differential method discussed in Section 2.2.1, we differentiate Eqs. (7.1) and (7.2) with respect to ϕ and, using Eq. (7.5), obtain the local sensitivity equations:

$$\frac{ds(y; \phi)}{dt} = \underline{\underline{J}}(t) \cdot s(y; \phi) + \frac{\partial f}{\partial \phi} \tag{7.6}$$

with ICs

$$s(y_i; \phi)|_{t=0} = \delta(\phi - y_i^i) \qquad (i = 1, 2, \ldots N_s + 1) \tag{7.7}$$

where $s(y; \phi)$ is the sensitivity vector, $J(t)$ the Jacobian matrix, and δ the Kronecker delta function. By simultaneously solving Eqs. (7.1) and (7.6) with ICs (7.2) and (7.7), the values of the system variables y and of the corresponding local sensitivities s are obtained as functions of time.

As mentioned earlier, when applying the generalized criterion to ignition phenomena, we need the local sensitivities of selected objectives. Let us consider the concentration maximum of a chosen radical, y_r^*, or the temperature maximum, T^*, during the time evolution of the process. The corresponding objective sensitivities with respect to the generic independent parameter ϕ are defined as follows:

$$s\left(y_r^*; \phi\right) = \frac{\partial y_r^*}{\partial \phi}, \qquad s(T^*; \phi) = \frac{\partial T^*}{\partial \phi} \tag{7.8}$$

More appropriate quantities in sensitivity analysis are the normalized objective sensitivities

$$S\left(y_r^*; \phi\right) = \frac{\phi}{y_r^*} \cdot \frac{\partial y_r^*}{\partial \phi} = \frac{\phi}{y_r^*} \cdot s\left(y_r^*; \phi\right) \tag{7.9a}$$

$$S(T^*; \phi) = \frac{\phi}{T^*} \cdot \frac{\partial T^*}{\partial \phi} = \frac{\phi}{T^*} \cdot s(T^*; \phi) \tag{7.9b}$$

which have a clearer physical meaning since they normalize the magnitudes of the parameter ϕ and the variable y_r^* or T^*. Therefore, according to the generalized criterion, for a set of given operating conditions, when the normalized objective sensitivity $S(y_r^*; \phi)$ or $S(T^*; \phi)$ is plotted against a chosen input parameter ϕ' (which may or may not coincide with ϕ), the criticality for ignition to occur is defined as the ϕ' value at which the normalized objective sensitivity, $S(y_r^*; \phi)$ or $S(T^*; \phi)$, exhibits a maximum or minimum.

7.2 Explosion Limits in Hydrogen–Oxygen Mixtures

7.2.1 Application of the Sensitivity Criterion

Let us now illustrate the application of the above criterion through the description of the explosion limits in H_2–O_2 mixtures shown in Fig. 7.1. We consider a nonisothermal model with eight chemical species, H_2, O_2, H, O, OH, H_2O, HO_2, H_2O_2, and 53 elementary (reversible) reactions, as shown in Table 7.1. The adopted values of the kinetic parameters in Table 7.1 have been taken from the collections reported by Dougherty and Rabitz (1980) and Stahl and Warnatz (1991), except for the wall termination reactions 25, 26, and 27, whose kinetic parameters have been estimated according to Semenov (1959).

In this case, Eq. (7.1) contains nine equations, for the nine components in $y = [y_{H_2}\ y_{O_2}\ y_H\ y_O\ y_{OH}\ y_{H_2O}\ y_{HO_2}\ y_{H_2O_2}\ T]^T$, and so does the sensitivity equation (7.6). To apply the above criterion, we need to define objectives. As mentioned above, since we are dealing with a reaction system involving both chain branching (where the H radical concentration is known to be critical) and thermal phenomena, we consider both the concentration maximum of the H radical, y_H^*, and the temperature maximum, T^*, as objectives. Thus, among all the sensitivities given by Eq. (7.6), we are interested in only two, $s(y_H; \phi)$ and $s(T; \phi)$, calculated at $y_H = y_H^*$ and $T = T^*$, respectively.

For a given initial pressure P^i, let us consider the normalized sensitivity of the concentration maximum of the H radical, y_H^*, and of the temperature maximum, T^*, with respect to the initial temperature, i.e., $\phi = T^i$, $S(y_H^*; T^i)$ and $S(T^*; T^i)$, as a function of the initial temperature, T^i. The explosion limit at the given initial pressure is then defined as the initial temperature at which $S(y_H^*; T^i)$ or $S(T^*; T^i)$ exhibits a maximum or minimum. Following similar arguments, but now referring to a fixed initial temperature, we calculate the normalized objective sensitivity, $S(y_H^*; T^i)$ or $S(T^*; T^i)$, as a function of the initial pressure, and the explosion limit is defined as the initial pressure at which $S(y_H^*; T^i)$ or $S(T^*; T^i)$ exhibits a maximum or minimum. From Eq. (7.9), the two normalized objective sensitivities can be expressed as

$$S\left(y_H^*; T^i\right) = \frac{T^i}{y_H^*} \cdot \left(\frac{\partial y_H^*}{\partial T^i}\right) = \frac{T^i}{y_H^*} \cdot s\left(y_H^*; T^i\right) \qquad (7.10a)$$

$$S(T^*; T^i) = \frac{T^i}{T^*} \cdot \left(\frac{\partial T^*}{\partial T^i}\right) = \frac{T^i}{T^*} \cdot s(T^*; T^i) \qquad (7.10b)$$

It should be mentioned that although both $s(y_H^*; T^i)$ and $s(T^*; T^i)$ in Eq. (7.10) can be obtained by simultaneously solving Eqs. (7.1) and (7.6) with ICs (7.2) and (7.7), some numerical difficulties often arise due to the stiffness of the system. To reduce the numerical effort, one may integrate only the model equations (7.1), while the values of the local sensitivities are computed by using the finite difference method discussed in Section 2.2.1. This approximation, as indicated by Kramer *et al.* (1984), may require more computation time relative to the other methods when the sensitivities with respect to all the input parameters are computed. However, since we are interested in only two sensitivities, the finite difference method is more convenient and requires less computation time.

It is worth noting that, from Eqs. (7.10), as the initial temperature or the initial pressure increases in passing through the boundary from a nonignited to an ignited region, the objective sensitivities are both *positive* and exhibit a *maximum* value. On the other hand, as the initial temperature or the initial pressure increases through the boundary from an ignited to a nonignited region, the objective sensitivities are both *negative* and exhibit a *minimum* value. In both cases, the absolute value of the sensitivity exhibits a maximum when crossing the boundary between the ignited and nonignited regions.

Table 7.1. Detailed reaction scheme and values of the kinetic parameters for the H_2–O_2 system[a]

No.	Reaction	A_F	B_F	E_F/R_g	A_R	B_R	E_R/R_g	$\Delta H_{1000 K}$
1	$H + HO_2 = H_2 + O_2$	2.50E13[b]	0.00	3.50E02	3.10E13	0.00	2.87E04	−5.25E+04
2	$H + O_2 = OH + O$	2.20E14	0.00	8.50E03	3.00E12	0.28	0.00E00	1.62E+04
3	$O + H_2 = OH + H$	1.80E10	1.00	4.48E03	8.30E09	1.00	5.50E03	1.86E+03
4	$OH + H_2 = H_2O + H$	1.00E08	1.60	1.66E03	1.40E14	−0.03	1.02E04	−1.53E+04
5	$O + H_2O = OH + OH$	5.80E13	0.00	9.07E03	6.30E12	0.00	5.50E02	1.71E+04
6	$H + H + M^c = H_2 + M$	2.57E18	−1.00	0.00E00	6.29E14	0.00	4.83E04	−1.06E+05
7	$H + OH + M = H_2O + M$	2.40E22	−2.00	0.00E00	1.14E24	−2.20	5.90E04	−1.21E+05
8	$O + O + M = O_2 + M$	5.43E13	0.00	−9.00E2	5.14E18	−1.00	5.94E04	−1.21E+05
9	$H + O_2 + M = HO_2 + M$	2.30E18	−0.80	0.00E00	6.00E15	0.00	2.30E04	−5.37E+04
10	$H + HO_2 = OH + OH$	2.50E14	0.00	9.50E02	1.20E13	0.00	2.02E04	−3.45E+04
11	$H + HO_2 = H_2O + O$	5.00E13	0.00	5.00E02	4.80E14	0.45	2.87E04	−5.16E+04
12	$O + HO_2 = OH + O_2$	5.00E13	0.00	5.00E02	1.30E13	0.18	2.82E04	−5.07E+04
13	$OH + HO_2 = H_2O + O_2$	5.00E13	0.00	5.00E02	5.60E13	0.17	3.66E04	−6.78E+04
14	$HO_2 + HO_2 = H_2O_2 + O_2$	1.80E13	0.00	5.00E02	3.00E13	0.00	2.16E04	−3.27E+04
15	$H_2O_2 + M = OH + OH + M$	1.71E17	0.00	2.29E04	7.71E14	0.00	−2.65E03	5.19E+04
16	$H + H_2O_2 = H_2 + HO_2$	1.70E12	0.00	1.90E03	1.32E12	0.00	1.01E04	−1.98E+04

#	Reaction	A	B	E	A	B	E	
17	$H + H_2O_2 = H_2O + OH$	1.00E13	0.00	1.80E03	2.40E14	0.00	4.05E04	$-6.96E+04$
18	$O + H_2O_2 = OH + HO_2$	2.00E13	0.00	2.95E03	5.20E10	0.50	1.06E04	$-1.80E+04$
19	$OH + H_2O_2 = H_2O + HO_2$	1.00E13	0.00	9.10E02	1.80E13	0.00	1.51E04	$-3.51E+04$
20	$O + OH + M = HO_2 + M$	2.29E17	0.00	0.00E00	1.94E20	-0.43	3.22E04	$-6.99E+04$
21	$H_2 + O_2 = OH + OH$	1.70E15	0.00	2.42E04	1.70E13	0.00	2.41E04	1.80E+04
22	$H + O + M = OH + M$	2.26E16	0.00	0.00E00	1.74E16	0.00	5.11E04	$-1.04E+05$
23	$H_2 + HO_2 = H_2O + OH$	6.50E11	0.00	9.40E03	7.20E09	0.43	3.61E04	$-4.98E+04$
24	$O + H_2O_2 = H_2O + O_2$	8.40E11	0.00	2.13E03	3.40E10	0.52	4.48E04	$-8.57E+04$
25	$H \xrightarrow{\text{wall}} 0.5H_2$	1.90E01	0.00	0.00E00				$-5.31E+04$
26	$H \xrightarrow{\text{wall}} 0.5O_2$	9.20E01	0.00	0.00E00				$-6.03E+04$
27	$OH \xrightarrow{\text{wall}} 0.5H_2O + 0.25O_2$	9.20E01	0.00	0.00E00				$-3.87E+04$
28	$HO_2 \xrightarrow{\text{wall}} 0.5H_2 + O_2$	1.00E-1	0.00	0.00E00				5.90E+02
29	$H_2O_2 \xrightarrow{\text{wall}} H_2 + O_2$	1.00E-2	0.00	0.00E00				3.39E+04

From Dougherty and Rabitz (1980) and Stahl and Warnatz (1984).

[a] The reaction rate constant is given by $k = A \cdot T^B \cdot \exp(-E/R_g T)$, with units in mol $\cdot$ cm $\cdot$ s $\cdot$ K $\cdot$ cal. The subscript F denotes the forward reaction and R the reverse reaction.

[b] Read as 2.50×10^{13}.

[c] M indicates a third body. The values of the pre-exponential factor A for the reactions involving a third body, originally developed for M = Ar or N_2, have been corrected for M = H_2. In the calculations reported in this book the third-body concentration has been evaluated as $y_M = y_{H_2} + 0.4y_{O_2} + 6.5y_{H_2O}$, according to the efficiency of each component (Warnatz, 1984).

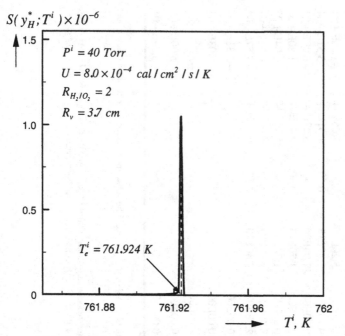

$$S(y_H^*;T^i)\times 10^{-6}$$

$P^i = 40 \; Torr$

$U = 8.0 \times 10^{-4} \; cal \, / \, cm^2 \, / \, s \, / \, K$

$R_{H_2/O_2} = 2$

$R_v = 3.7 \; cm$

$T_e^i = 761.924 \; K$

T^i, K

Figure 7.2. Normalized sensitivity of the concentration maximum of the H radical with respect to the initial temperature $S(y_H^*;T^i)$ as a function of the initial temperature T^i. The explosion limit is indicated by T_e^i.

An application of the ignition criterion defined above is illustrated in Fig. 7.2, where, for a fixed initial pressure value, $P^i = 40 \, Torr$, the normalized sensitivity $S(y_H^*;T^i)$ is shown as a function of T^i. It appears that $S(y_H^*;T^i)$ exhibits a sharp maximum at the value $T_e^i = 761.924$ K, which may be regarded as the explosion limit separating the nonignited ($T^i < T_e^i$) from the ignited ($T^i > T_e^i$) regions.

In order to verify the reliability of the adopted explosion criterion, the temperature and concentration profiles of all species as a function of time are shown in Fig. 7.3 for two initial temperature values considered in Fig. 7.2. In particular, Fig. 7.3a corresponds to an initial temperature value ($T^i = 761.92$ K) that is near but lower than the explosion limit ($T_e^i = 761.924$ K), while Fig. 7.3b corresponds to an initial temperature value ($T^i = 761.93$ K) near but greater than the explosion limit. It can be seen that, although the difference in the initial temperature between Fig. 7.3a and Fig. 7.3b is only 0.01 K, the system behavior is vastly different. In Fig. 7.3a, a typical slow reaction behavior is observed. In the time period explored, the hydrogen conversion is very small, the concentrations of radicals H, O, and OH, reach their shallow maxima between $t = 0.5$ and 1 s, and then decrease with time. Also, the temperature of the system does not increase significantly. The above features remain true during the entire reaction process. This can be shown by taking the H_2 conversion instead of time as the independent variable to integrate the model equations. In this way, it is seen that the system evolves from zero hydrogen conversion to the value corresponding to

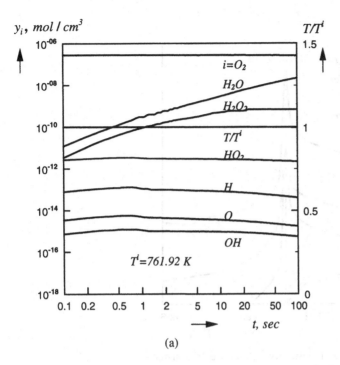

(a)

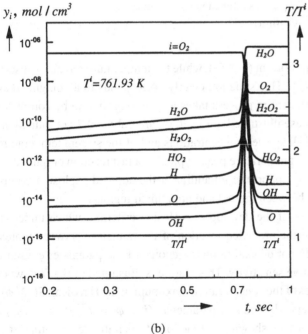

(b)

Figure 7.3. Profiles of chemical species concentrations and temperature as functions of time: (a) $T^i = 761.92\,\text{K} < T_e^i$ and (b) $T^i = 761.93\,\text{K} > T_e^i$. $P^i = 40$ Torr; $R_{H_2/O_2} = 2$; $U = 8.0 \times 10^{-4}$ cal/cm^2/s/K; $R_v = 3.7$ cm.

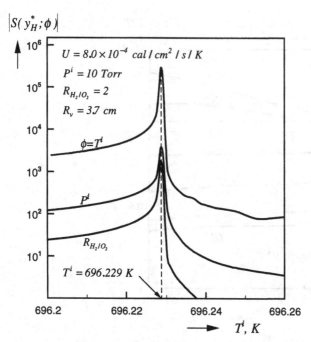

Figure 7.4. Normalized sensitivity of the concentration maximum of the H radical, $S(y_H^*; \phi)$, as a function of the initial temperature; $\phi = T^i$, P^i, or R_{H_2/O_2}.

the thermodynamic equilibrium (>99%), while the temperature remains substantially constant, *i.e.*, $T/T^i \approx 1$. This situation clearly corresponds to a nonignited system behavior, and in fact the entire process takes a very long time to be completed. On the other hand, in the case illustrated in Fig. 7.3b, after about 0.7 s of induction time, significant changes of all species concentrations and of the system temperature take place, leading to a sharp temperature peak, typical of a fast reaction or ignited-system behavior. These results support the reliability of the adopted explosion criterion in locating the boundary between ignited and nonignited regions.

An important feature of the developed explosion criterion, which underlines its intrinsic nature, is that the predicted location of the boundary between ignited and nonignited region does not depend upon the choice of the parameter ϕ used in the definition of the objective sensitivity. This feature is illustrated in Fig. 7.4, where the normalized sensitivities of the concentration maximum of the H radical, $S(y_H^*; \phi)$, with respect to various independent model parameters (*i.e.*, $\phi = T^i$, P^i, and R_{H_2/O_2}, the initial H_2/O_2 molar ratio) are shown as a function of the initial temperature. It can be seen that, although the absolute values of the objective sensitivities are different, they all exhibit their maxima at the same initial temperature value. Thus, according to the adopted explosion criterion, they indicate the same location of the boundary between ignited and nonignited regions, *i.e.*, the explosion limit. The simultaneous occurrence of such sensitivity peaks with respect to each of the independent parameters represents

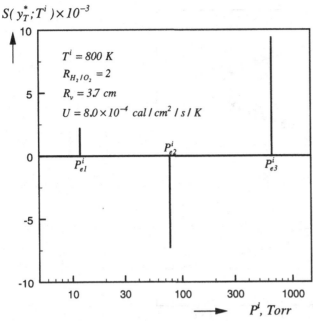

$S(y_T^*; T^i) \times 10^{-3}$

Figure 7.5. Normalized sensitivity of the temperature maximum with respect to the initial temperature, $S(T^*; T^i)$, as a function of the initial pressure P^i, where P_{e1}^i, P_{e2}^i, and P_{e3}^i indicate the three explosion limits.

an intrinsic feature of system behavior in the vicinity of the ignition boundary. This is the most qualifying aspect of the adopted explosion criterion, as evidenced in previous chapters with reference to simpler reacting systems involving only one or two reactions.

In Fig. 7.5 the normalized sensitivity of the temperature maximum with respect to the initial temperature, $S(T^*; T^i)$, is shown as a function of P^i. It appears that, for a fixed initial temperature value, three explosion limits, P_{e1}^i, P_{e2}^i, and P_{e3}^i, exist. Positive maxima at P_{e1}^i and P_{e3}^i and a negative minimum at P_{e2}^i suggest that explosion occurs for initial pressure values in the intervals $P_{e1}^i \leq P^i \leq P_{e2}^i$ and $P^i \geq P_{e3}^i$, while no explosion occurs in the intervals $P^i < P_{e1}^i$ and $P_{e2}^i < P^i < P_{e3}^i$. It is remarkable that the predicted occurrence of three critical transitions in the system behavior, which are encountered by increasing the initial pressure at fixed initial temperature, is in agreement with the typical inverse-S-shaped curve representing the explosion limits measured experimentally for the H_2–O_2 system (see Fig. 7.1). These findings, which are elaborated below, further support the reliability of the adopted explosion criterion.

7.2.2 Comparison between Experimental and Calculated Explosion Limits

Dougherty and Rabitz (1980) were the first to investigate the explosion behavior of the H_2–O_2 system, using a detailed kinetic model based on a number of elementary

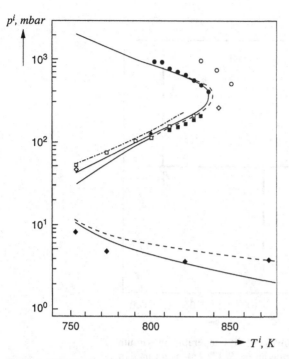

Figure 7.6. Explosion limits for the stoichiometric H_2-O_2 mixture in a spherical vessel of 7.4 cm in diameter, calculated by Maas and Warnatz (1988) with the rate constants for all the wall destruction rates: $---$ $k_{wall} = 10^{-2}$; $\underline{\quad\quad}$ $k_{wall} = 10^{-3}$; $-\cdot-$ $k_{wall} = 10^{-4}$ s^{-1}, and experimentally measured data: (1) in a spherical vessel of 7.4 cm in diameter, from Heiple and Lewis (1941), ■ thinly KCl-coated, ○ heavily KCl-coated, ● KCl-coated; from von Elbe and Lewis (1942), ◇ KCl-coated, ▲ clean Pyrex; from Egerton and Warren (1951), □ B_2O_3-coated. (2) In a cylindrical silica vessel of 1.8 cm in diameter, from Hinshelwood and Moelwyn-Hughes (1932), ♦. (From Maas and Warnatz (1988).)

reactions. However, they used an isothermal model, which does not predict the third explosion limit. A quantitative comparison with the experimental explosion limits was reported later by Maas and Warnatz (1988) using a slightly different kinetic scheme and a nonisothermal, two-dimensional reactor model. The obtained results together with the experimental data are shown in Fig. 7.6. It is seen that the model predicts the three explosion limits, and that the agreement between the measured and computed explosion boundaries is good. Note that, unlike those in Fig. 7.1, the experimental data reported in Fig. 7.6 come from different sources in the literature and were obtained under different experimental conditions. In both studies mentioned above, the explosion boundary was located empirically.

Let us now apply the explosion criterion described above to simulate the explosion boundary measured by Lewis and von Elbe (1961) using the kinetic scheme reported

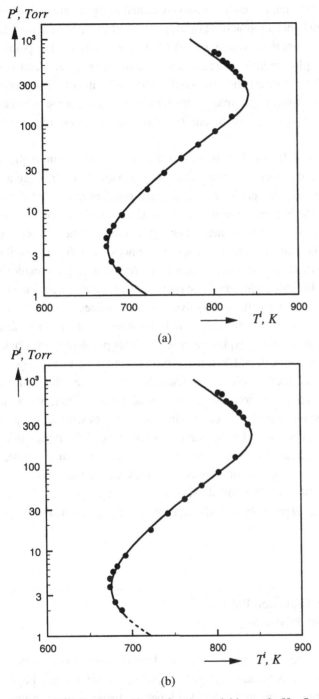

Figure 7.7. Explosion limits for the stoichiometric H_2–O_2 mixture in a KCl-coated vessel as measured experimentally by Lewis and von Elbe (1961) (●) and predicted (solid curves) using the generalized criterion based on (a) $S(y_H^*; T^i)$ and (b) $S(T^*; T^i)$. $U = 8.0 \times 10^{-4}$ cal/cm^2/s/K; $R_v = 3.7$ cm. From Morbidelli and Wu (1992).

in Table 7.1. Figures 7.7a and b show the results obtained using the normalized sensitivity of the maximum concentration of H radical, *i.e.*, $S(y_H^*; T^i)$, and the normalized sensitivity of the maximum temperature, *i.e.*, $S(T^*; T^i)$, respectively. The points represent the measured explosion limits. It can be seen that in both Figs. 7.7a and 7.7b the agreement between the computed and measured explosion boundaries is excellent. It should be noted that only one adjustable parameter has been used in the calculations, *i.e.*, the wall overall heat-transfer coefficient, U, whose value was not reported in the original work.

By comparing Figs. 7.7a and 7.7b it is seen that the explosion limits predicted based on the normalized objective sensitivities, $S(y_H^*; T^i)$ and $S(T^*; T^i)$, are always coincident, except for very low pressure values, corresponding to the first explosion limit represented by the broken curve in Fig. 7.7b. Under the latter conditions, the strength of the explosion in terms of thermal energy is low, and the explosion tends to be driven only by the radical branching reactions without a significant contribution by the thermal processes. Accordingly, the temperature rise during the explosion is small, thus making it difficult to identify the maximum temperature value. Under these conditions the temperature normalized objective sensitivity values, $S(T^*; T^i)$, are also small and their maximum as a function of the initial temperature tends to be shallow, making it impossible to apply the explosion criterion. This problem arises obviously because, under these extreme conditions, the system temperature is no longer a significant variable in describing the system behavior. A better choice, which allows one to overcome this problem, is to consider the H radical concentration, which undergoes a rapid change when an explosion occurs, independently of whether temperature changes occur. This is confirmed by the curves shown in Fig. 7.7a representing the explosion limits as predicted by the criterion based on $S(y_H^*; T^i)$. In this case, even at very low pressure values, the absolute sensitivity maxima are still large, so that no difficulty arises in establishing the explosion boundary. *Thus, it is preferable to use the objective sensitivity based on the H radical concentration, $S(y_H^*; \phi)$, in determining the explosion limits.*

7.3 Further Insight into Explosion Behavior in Hydrogen–Oxygen Mixtures

The sensitivity analysis reported above can be used not only to provide a criterion to identify the explosion limits but also as a tool to investigate the relative importance of the various elementary steps in the detailed kinetic scheme. This is an important aspect of the study of complex kinetic systems and will be discussed in more detail in Chapter 8. In the following, we describe an additional application of sensitivity analysis to the H_2–O_2 system, by discussing the characteristics of explosions at low and high pressures.

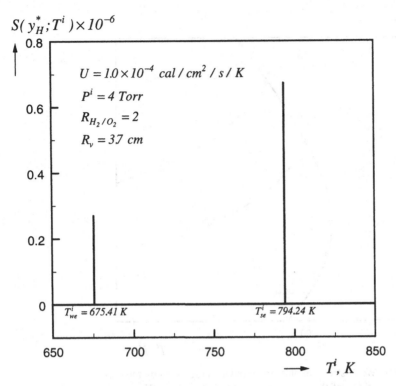

$$S(y_H^*;T^i)\times 10^{-6}$$

$U = 1.0 \times 10^{-4}\ cal\,/\,cm^2\,/\,s\,/\,K$

$P^i = 4\ Torr$

$R_{H_2/O_2} = 2$

$R_v = 3.7\ cm$

$T_{we}^i = 675.41\ K$

$T_{se}^i = 794.24\ K$

T^i, K

Figure 7.8. Normalized sensitivity of the concentration maximum of the H radical with respect to the initial temperature, $S(y_H^*; T^i)$, as a function of the initial temperature in the low-pressure region.

7.3.1 Explosion in the Low-Pressure Region

Some further insight into system behavior at low initial pressures can be obtained by investigating the sensitivity of the concentration maximum of the H radical with respect to, say, the initial temperature, *i.e.*, $S(y_H^*; T^i)$. The values of this quantity at low initial pressure, *i.e.*, at $P^i = 4$ Torr, are shown in Fig. 7.8 as a function of the initial temperature. It can be seen that the sensitivity $S(y_H^*; T^i)$ exhibits a peculiar shape, characterized by two sharp maxima occurring at two different values of the initial temperature. By repeating these calculations for sensitivities with different input parameters, such as the initial pressure $S(y_H^*; P^i)$ or the initial H_2/O_2 molar ratio $S(y_H^*; R_{H_2/O_2})$, the same behavior is found with the two maxima located at the same positions. This indicates that these maxima represent critical transitions between two different types of system behavior. Moreover, since in contrast to the situation shown in Fig. 7.5, in this case the sensitivity exhibits two sharp maxima and no sharp minimum, we can conclude that both of them correspond to transitions where, as the initial temperature increases, the system goes from a slower to a faster reaction regime. This is a unique feature of this reacting system at low-pressure values, which has not been reported before in kinetic studies.

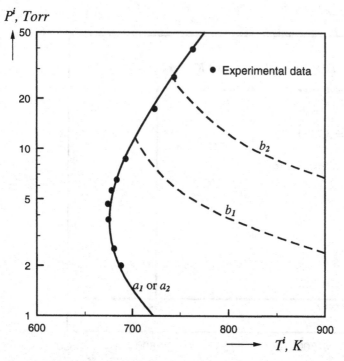

Figure 7.9. Weak-strong explosion boundaries predicted by the normalized sensitivity of the concentration maximum of the H radical, $S(y_H^*; T^i)$. $U = 1.0 \times 10^{-4}$ cal/cm^2/s/K in curves a_1 and b_1, and $U = 8.0 \times 10^{-4}$ cal/cm^2/s/K in curves a_2 and b_2.

The loci of the two maxima, as a function of the initial pressure, are shown in Fig. 7.9 by the two pairs of curves (a_1, b_1) and (a_2, b_2), corresponding to two different values of the overall heat-transfer coefficient, $U = 1.0 \times 10^{-4}$ and $U = 8.0 \times 10^{-4}$ cal/cm^2/s/K. In order to illustrate more in detail the system behavior in this region, let us consider Figs. 7.10, 7.11, and 7.12, where calculated values of the concentrations of the involved species and temperature are shown as a function of hydrogen conversion, x_{H_2}, for three different initial temperature values. These have been selected so as to illustrate the system behavior corresponding to initial conditions falling in the regions on the left-hand side of curve a_1, between curves a_1 and b_1, and on the right-hand side of curve b_1.

In Fig. 7.10 it is seen that the concentrations of all the radicals reach their maximum value at about $x_{H_2} = 1.0 \times 10^{-4}$ and then decrease continuously as x_{H_2} increases. The concentrations of the radicals are always very low and the system temperature remains substantially constant and very close to the initial value throughout the entire reaction, thus indicating a slow reaction or nonignited behavior.

In Fig. 7.11, the concentrations of all the species and the temperature are reported as functions of both (a) hydrogen conversion and (b) time. In Fig. 7.11a, the concentrations of H, O, and OH radicals reach their maximum value at about $x_{H_2} = 0.01$,

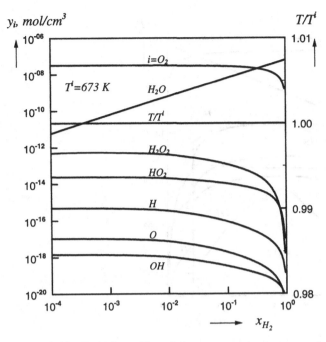

Figure 7.10. Typical profiles of the system temperature and chemical species concentrations as functions of the H_2 conversion, x_{H_2}, when the initial pressure and temperature are located on the left-hand side of curve a_1 in Fig. 7.9. $T^i = 673$ K; $P^i = 4.0$ Torr; $R_{H_2/O_2} = 2$; $U = 1.0 \times 10^{-4}$ cal/cm^2/s/K; $R_v = 3.7$ cm.

while the HO_2 radical exhibits its concentration maximum at about $x_{H_2} = 0.3$. It can be seen that the system behavior for hydrogen conversion values below $x_{H_2} = 0.3$ is different from that shown in Fig. 7.10. All the radical species exhibit much higher concentration values, thus leading to larger rates of reaction. This is confirmed by the significant, even though not very high, temperature rise in the system. In this case, as can be seen from Fig. 7.11b, it takes less than 50 s for the H_2 conversion to change from 0.01, where the H concentration and the temperature reach their maximum, to 0.3, indicating a fast reaction or ignited behavior when compared with that shown in Fig. 7.10. At a hydrogen conversion value close to 0.3, the system exhibits a peculiar behavior, which in Figs. 7.11a appears as a sharp, almost discontinuous, change in the radical concentration values. If the conversion scale is suitably expanded, as shown by Wu *et al.* (1993), it can be seen that this behavior corresponds to a very narrow multiplicity region; *i.e.*, at each given conversion value, three different values exist for each species concentration. This anomalous situation arises, because in this region, the hydrogen conversion reaches a local maximum, decreases to a local minimum, and then starts increasing again, as shown in Fig. 7.11b. Thus, when the species concentrations are plotted as a function of hydrogen conversion, we obtain the unusual multiplicity region in Fig. 7.11a. After this region, *i.e.*, for $x_{H_2} > 0.3$, the system behavior changes and becomes similar to the typical nonignited behavior illustrated in

237

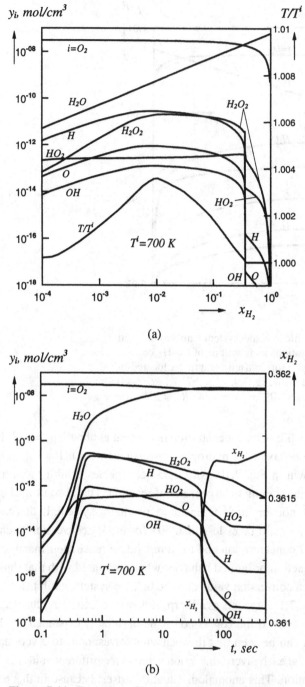

Figure 7.11. Typical profiles of the chemical species concentrations and system temperature as functions of (a) H_2 conversion, x_{H_2}, and (b) time, when the initial pressure and temperature are located in the region bounded by curves a_1 and b_1 in Fig. 7.9. $T^i = 700$ K; other parameter values as in Fig. 7.10.

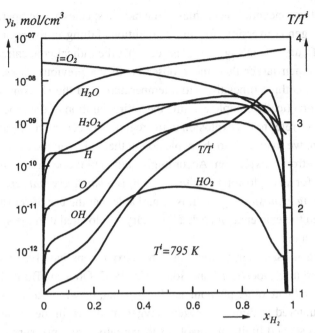

Figure 7.12. Typical profiles of the chemical species concentrations and system temperature as functions of the H_2 conversion, x_{H_2}, when the initial pressure and temperature are located on the right-hand side of curve b_1 in Fig. 7.9. $T^i = 795$ K; other parameter values as in Fig. 7.10.

Fig. 7.10. This is clearly indicated by the value of the ratio T/T^i, which, as shown in Fig. 7.11a, remains close to unity as the hydrogen conversion increases from 0.3 to 1.

The same behavior is exhibited by all systems whose initial temperature falls in the region bounded by curves a_1 and b_1 in Fig. 7.9. In general, in the case where $P^i = 4$ Torr and $U = 1.0 \times 10^{-4}$ cal/cm^2/s/K, the explosion starts at some relatively low hydrogen conversion (*i.e.*, $x_{H_2} < 0.15$) and reaches very rapidly a larger hydrogen conversion value, where it stops and then the reaction proceeds slowly, with a typical nonignited behavior, up to complete hydrogen conversion. For a given initial pressure, the maximum temperature rise and the hydrogen conversion value where the explosion ends increase for increasing values of the initial temperature.

The system behavior described above is in good agreement with the experimental findings reported for the stoichiometric mixture H_2–O_2 at low initial pressure (Semenov, 1959; Lewis and von Elbe, 1961; Dixon-Lewis and Williams, 1977). When the latter exceeds the first explosion limit by only a few Torr (*i.e.*, presumably in the region bounded by curves a_1 and b_1) it is reported that the system undergoes a mild explosion, generally referred to as a flash. This involves small changes of the concentrations of H_2 and O_2 with respect to their initial values and leads to a modest temperature rise.

Figure 7.12 shows the concentration profiles of the radical species and temperature as a function of hydrogen conversion, for initial conditions falling in the region on the right-hand side of curve b_1 in Fig. 7.9. It is seen that the radical concentrations and temperature are much larger than those observed in the previous cases. The temperature maximum is three times the initial temperature, and the H_2 conversion after the explosion is practically complete, while the entire duration of the process is less than 1 s. Thus, if we refer to the explosion in the region between curves a_1 and b_1 as a "weak" explosion, we may refer to the explosion in the region on the right-hand side of curve b_1 as a "strong" explosion. Accordingly, although it does not represent the critical conditions for an explosion to occur, *curve b_1 is a boundary that separates the weak from the strong explosion region*. It is remarkable that the use of sensitivity analysis has uncovered the existence of such a boundary and located it in the system operating parameter plane.

It is worth noting that the same terms, weak and strong explosions, have been used in prior studies in the literature (Voevodsky and Soloukhin, 1965; Oran and Boris, 1982; Yetter *et al.*, 1991) to indicate other phenomena. It particular, they refer to explosion regions that are encountered in dilute hydrogen–oxygen mixtures, in the vicinity of the so-called extended second limit, and involve substantially larger pressure values than those examined here.

Let us now further consider the behavior of the system during the weak explosion illustrated in Fig. 7.11a and use sensitivity analysis to investigate the mechanism of the explosion quenching, which is not due to the complete depletion of hydrogen (or oxygen) but involves a decrease of hydrogen conversion in time. This unusual behavior is responsible for the peculiar shape of the radical concentration versus conversion profiles shown in Fig. 7.11a. Through the sensitivity analysis of the hydrogen conversion ($\hat{x}_{H_2}$) in the region where the weak explosion quenches, with respect to the pre-exponential factor A of all the reactions in Table 7.1, it is found, as shown in Fig. 7.13, that the process is dominated by the following reactions:

$$H + O_2 \rightarrow OH + O \tag{2F}$$

$$H + O_2 + M \rightarrow HO_2 + M \tag{9F}$$

$$H \xrightarrow{\text{wall}} 0.5H_2 \tag{25F}$$

$$H + HO_2 \rightarrow OH + OH \tag{10F}$$

$$H + HO_2 \rightarrow H_2 + O_2 \tag{1F}$$

where the number in the parentheses indicates the reaction number in Table 7.1 and F denotes the forward reaction. Note that all the above reactions involve the H radical, whose concentration determines the rate of hydrogen oxidation under these conditions. The evolution of the system is conditioned by the competition between the branching reactions 2F and 10F and the termination reactions 9F, 25F, and 1F.

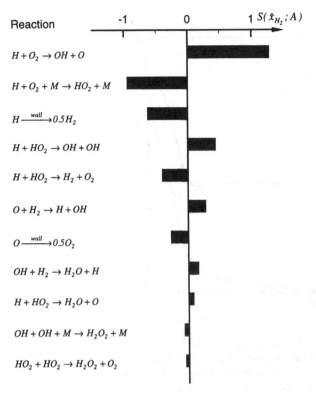

Figure 1.13. Normalized sensitivity of the H_2 conversion value where the weak explosion ends ($\hat{x}_{H_2}$) with respect to the pre-exponential factor A for the most important elementary reactions among those reported in Table 7.1. From Wu *et al.* (1993).

From Table 7.1, we see that the branching reactions have activation energy values larger than the termination reactions and therefore are favored at higher temperature. Thus, for relatively low initial temperature values, *i.e.*, in the region bounded by curves a_1 and b_1 in Fig. 7.9, the system undergoes a weak explosion up to a given hydrogen conversion value, where termination reactions take over branching reactions and the explosion extinguishes. Since termination reactions involve the production of hydrogen, this leads to the decrease of the hydrogen conversion in time. As noted above, the hydrogen conversion value, where the weak explosion extinguishes, increases as the initial temperature increases. This is due to the increased importance of branching reactions over termination reactions. When the initial temperature value crosses the boundary given by curve b_1 in Fig. 7.9, then the weak explosion is no longer extinguished and it continues up to complete hydrogen consumption. The increased rate of the branching reactions also justifies the higher strength of the explosion under these conditions, as shown in Fig. 7.12. In this figure one can still detect the traces of the weak explosion mechanism observed at lower initial temperature values. It can be seen that the hydrogen radical concentration first goes through a maximum (corresponding

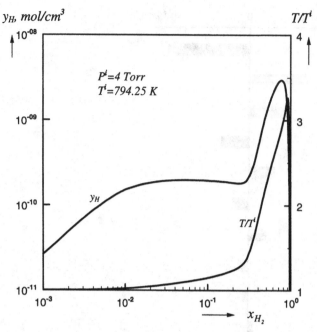

Figure 1.14. Profiles of the concentration of the H radical and the system temperature as functions of the H_2 conversion when the initial pressure and temperature are located in the region on the right-hand side of, but close to, curve b_1 in Fig. 7.9. $T^i = $ 794.25 K; other parameter values as in Fig. 7.10. The boundary b_1 is located at $T^i = 794.2$ K.

to the weak explosion) and then decreases slightly. However, since in this case the temperature value is too large for termination reactions to prevail over branching reactions, such a decrease does not continue and the system does not extinguish. The hydrogen radical concentration goes in fact through a minimum and starts increasing again, leading to the true explosion (which leads to much larger values of T/T^i) as indicated by the occurrence of a second maximum. This behavior is more evident in the case shown in Fig. 7.14, where an initial temperature value in the region on the right-hand side of, but close to, curve b_1 in Fig. 7.9 is considered.

A further support of this interpretation is the strong dependence of the location of the boundary between weak and strong explosions on the thermal behavior of the system. This is illustrated in Fig. 7.9, where it is seen that this boundary moves to higher initial temperature values, i.e., from curve b_1 to curve b_2, as the wall heat-transfer coefficient increases. On the other hand, the location of the weak explosion limit is not affected by changes in the heat-transfer coefficient, as illustrated by curves a_1 and a_2, which are essentially coincident.

Finally, it is worth comparing the results obtained above with those of Maas and Warnatz (1988), shown in Fig. 7.6. Let us first note that their rate constants for the wall destruction reactions of radicals are much lower than those in Table 7.1, lying in the

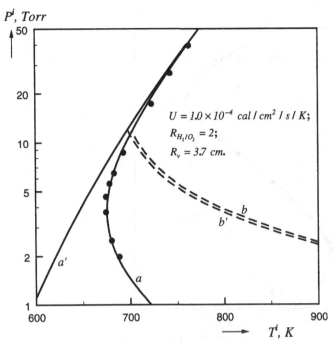

Figure 1.15. Effect of the wall termination rate constants of H, O, and OH radicals on the first explosion limit and on the weak–strong explosion boundary. Curves a and b: $k_{25} = 19$, $k_{26} = k_{27} = 92$. Curves a' and b': $k_{25} = k_{26} = k_{27} = 10^{-2}$.

range between 10^{-4} and $10^{-2}\,\mathrm{s}^{-1}$. Figure 7.15 compares the first explosion limit and the boundary between weak and strong explosions calculated using the wall destruction rate constants in Table 7.1 (curves a and b, respectively) with those adopted by Maas and Warnatz (1988) (curves a' and b', respectively). It can be seen that, using the data of Maas and Warnatz, the second explosion limit decreases continuously (curve a') as the initial pressure decreases in the range under examination, without exhibiting the turning point that characterizes the occurrence of the first explosion limit. This is instead correctly exhibited by curve a. This may explain why this ignition limit was not identified in the work of Maas and Warnatz. The only critical transition that they identified, referred to as the first explosion limit (in Fig. 7.6), is most likely that given by curve b' (in Fig. 7.15), which as shown above represents the transition between the weak and strong explosions.

1.3.2 Explosion in the High-Pressure Region

Explosion occurring in the high-pressure region is bounded by the so-called third explosion limit. Previous findings reported in the literature (Oran and Boris, 1982; Foo and Yang, 1971; Griffiths *et al.*, 1981) indicate that this limit is controlled by thermal processes. This can be verified by determining whether the position of the third limit is sensitive to changes in the value of the overall heat-transfer coefficient U

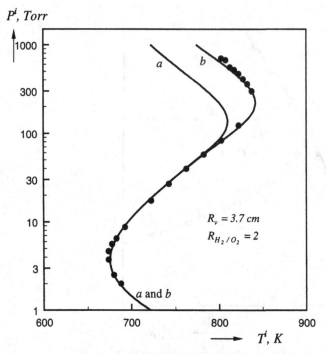

Figure 1.16. Effect of the overall wall heat-transfer coefficient, U, on the calculated explosion limits. Curve a: $U = 1.0 \times 10^{-4}$ cal/cm^2/s/K. Curve b: $U = 8.0 \times 10^{-4}$ cal/cm^2/s/K.

in Eq. (7.3). Curves a and b in Fig. 7.16 were computed using two different values of the overall heat-transfer coefficient, $i.e.$, $U = 1.0 \times 10^{-4}$ and 8.0×10^{-4} cal/cm^2/s/K, respectively. It is seen that the third explosion limit moves to higher initial temperature values as the wall heat-transfer coefficient increases, whereas the first and the second limits remain unchanged. Thus, we can confirm that unlike the first and the second limits, the third limit is dominated by the thermal behavior of the system.

Note: In this chapter we have analyzed the parametric sensitivity behavior in an example system involving a large number of reactions, through applications of the generalized sensitivity criterion. Similar investigations can also be carried out for other complex reaction systems. The reader may refer to papers by Tjahjadi *et al.* (1987) and Kapoor *et al.* (1989) for parametric sensitivity analyses in the context of polymerization processes.

References

Dixon-Lewis, G., and Williams, D. J. 1977. The oxidation of hydrogen and carbon monoxide. In *Comprehensive Chemical Kinetics*, C. H. Bamford and C. F. H. Tipper, eds. Vol. 17, p. 1. Amsterdam: Elsevier.

Dougherty, E. P., and Rabitz, H. 1980. Computational kinetics and sensitivity analysis of hydrogen-oxygen combustion. *J. Chem. Phys.* **72**, 6571.

Egerton, A. C., and Warren, D. R. 1951. Kinetics of the hydrogen/oxygen reaction: I. The explosion region in boric acid-coated vessels. *Proc. R. Soc. London A* **204**, 465.

Foo, K. K., and Yang, C. H. 1971. On the surface and thermal effects on hydrogen oxidation. *Combust. Flame* **17**, 223.

Gray, B. F., and Yang, C. H. 1965. On the unification of the thermal and chain theories of explosion limits. *J. Phys. Chem.* **69**, 2747.

Gray, B. F., and Yang, C. H. 1967. The present theoretical position in explosion theory. In *11th Symposium (International) on Combustion*, p. 1057. Pittsburgh: The Combustion Institute.

Griffiths, J. F., Scott, S. K., and Vandamme, R. 1981. Self-heating in the $H_2 + O_2$ reaction in the vicinity of the second explosion limit. *J. Chem. Soc. Faraday Trans. I* **77**, 2265.

Heiple, H. R., and Lewis, B. 1941. The reaction between hydrogen and oxygen: kinetics of the third explosion limit. *J. Chem. Phys.* **9**, 584.

Hinshelwood, C. N., and Moelwyn-Hughes, E. A. 1932. *Proc. R. Soc. London A* **138**, 311.

JANAF. 1986. *JANAF Thermochemical Tables*. Washington, DC: American Chemical Society; New York: American Institute of Physics for the National Bureau of Standards.

Kapoor, B. I. R., Gupta, S. K., and Varma, A. 1989. Parametric sensitivity of chain polymerization reactors exhibiting the Trommsdorff effect. *Polym. Eng. Sci.* **29**, 1246.

Kordylewski, W., and Scott, S. K. 1984. The influence of self-heating on the second and third explosion limits in the $O_2 + H_2$ reaction. *Combust. Flame* **57**, 127.

Kramer, M. A., Rabitz, H., Calo, J. M., and Kee, R. J. 1984. Sensitivity analysis in chemical kinetics: recent developments and computational comparisons. *Int. J. Chem. Kinet.* **16**, 559.

Lewis, B., and von Elbe, G. 1961. *Combustion, Flames and Explosions of Gases*. New York: Academic.

Maas, U., and Warnatz, J. 1988. Ignition processes in hydrogen-oxygen mixtures. *Combust. Flame* **53**, 74.

Morbidelli, M., and Wu, H. 1992. Critical transitions in reacting systems through parametric sensitivity. In *From Molecular Dynamics to Combustion Chemistry*, S. Carrà and N. Rahman, eds., p. 117. Singapore: World Scientific.

Morbidelli, M., and Varma, A. 1988. A generalized criterion for parametric sensitivity: application to thermal explosion theory. *Chem. Eng. Sci.* **43**, 91.

Oran, E. S., and Boris, J. P. 1982. Weak and strong ignition: II. Sensitivity of the hydrogen-oxygen system. *Combust. Flame* **48**, 149.

Semenov, N. N. 1928. Zur theorie des verbrennungsprozesses. *Z. Phys.* **48**, 571.

Semenov, N. N. 1959. *Some Problems of Chemical Kinetics and Reactivity*. London: Pergamon.

Stahl, G., and Warnatz, J. 1991. Numerical investigation of time-dependent properties and extinction of strained methane- and propane-air flamelets. *Combust. Flame* **85**, 285.

Tjahjadi, M., Gupta, S. K., Morbidelli, M., and Varma, A. 1987. Parametric sensitivity in tubular polymerization reactors. *Chem. Eng. Sci.* **42**, 2385.

Voevodsky, V. V., and Soloukhin, R. I. 1965. On the mechanism and explosion limits of hydrogen-oxygen chain self-ignition in shock waves. In *10th Symposium (International) on Combustion*, p. 279. Baltimore: Williams and Wilkins.

von Elbe, G., and Lewis, B. 1942. Mechanism of the thermal reaction between hydrogen and oxygen. *J. Chem. Phys.* **10**, 366.

Warnatz, J. 1984. Rate coefficients in the C/H/O system. In *Combustion Chemistry*, W. C. Gardiner, Jr., ed., p. 197. New York: Springer-Verlag.

Wu, H., Cao, G., and Morbidelli, M. 1993. Parametric sensitivity and ignition phenomena in hydrogen-oxygen mixtures. *J. Phys. Chem.* **97**, 8422.

Yang, C. H., and Gray, B. F. 1967. The determination of explosion limits from a unified thermal and chain theory. In *11th Symposium (International) on Combustion*, p. 1099. Pittsburgh: The Combustion Institute.

Yetter, R. A., Rabitz, H., and Hedges, R. M. 1991. A combined stability-sensitivity analysis of weak and strong reactions of hydrogen/oxygen mixtures. *Int. J. Chem. Kinet.* **23**, 251.

Zeldovich, Ya. B., Barenblatt, G. I., Librovich, V. B., and Makhviladze, G. M. 1985. *The Mathematical Theory of Combustion and Explosions*. New York: Consultants Bureau.

8

Sensitivity Analysis in Mechanistic Studies and Model Reduction

D ETAILED OR RIGOROUS KINETIC MODELS, consisting of a large number of elementary reactions, are used increasingly to simulate complex reacting processes. An example is given in Chapter 7, where, using detailed kinetic models, we predicted the explosion limits of hydrogen–oxygen mixtures. The main advantage of a detailed versus a *simplified or empirical kinetic model* is its wider operating window. In other words, detailed models generally describe the kinetics of complex processes for a larger range of operating conditions, while simplified models can be used only for specific conditions. Moreover, detailed models are able to provide proper estimation of the radical concentrations involved in complex processes. Thus, detailed kinetic modeling is an important tool for the analysis and design of complex reacting systems.

A related aspect of detailed kinetic models is that, although they may provide satisfactory simulations of experimental results, their complexity often prevents the understanding of the key features of a process. For example, when using detailed kinetic models, it is often difficult to identify the main reaction paths in a complex reacting system.

Simplified kinetic models, on the other hand, offer several advantages in practical applications. A complex reacting (*e.g.*, combustion) process typically involves a few hundreds of elementary reactions, and hence includes several hundred kinetic parameters. This is true not only for the combustion of complex fuels but also for simpler ones, such as hydrogen or methane. The computational effort associated with the application of complex kinetic models forces the introduction of very simple models to describe the transport processes in reactors where the reaction takes place (*e.g.*, plug-flow or perfectly mixed). In the case of complex reactors (*e.g.*, geometrically complicated combustion chambers) where three-dimensional fluid dynamic models are required, the use of detailed kinetic models is still beyond our current computational capabilities. In this case, a simplified but reliable kinetic model would be attractive.

For both understanding the main reaction paths and extracting a simplified or reduced kinetic model from a detailed one, an effective tool is sensitivity analysis. This implies that one should calculate the sensitivity of the specific objective of interest with respect to each of the reactions comprising the detailed kinetic model. From the values of these sensitivities, it is possible to formulate a simplified mechanism of the reacting process, which accounts for only the most relevant reaction pathway, leading to a simplified kinetic model. In this chapter, we first discuss the application of sensitivity analysis to identify the process mechanism in several complex reacting systems, such as oxidation of wet carbon monoxide, Belousov-Zhabotinsky oscillations, and hydrogen–oxygen explosions. Then we illustrate, in Section 8.2, the procedure of kinetic model reduction using sensitivity analysis. In particular, we derive reduced kinetic models to describe some experimental observations, such as the three explosion limits of hydrogen–oxygen systems in a batch reactor and the chemical species concentration in the outlet of a well-stirred continuous flow reactor fed with methane–ethane–air mixtures.

8.1   Sensitivity Analysis in Mechanistic Studies

For any given objective, it is possible to classify the various reactions constituting a detailed model in terms of their importance with respect to that objective, by simply examining the corresponding sensitivity values. Thus, through a local sensitivity analysis of the dynamics of a process with respect to each of the elementary reactions in the detailed kinetic scheme, we can classify the reactions, determine their importance, and finally understand the mechanism for the process dynamics.

Let us consider a generic reacting system involving n chemical species and m elementary reactions. The changes of the species concentrations in a closed, uniform vessel can be described by the following system of ordinary differential equations in vector form:

$$\frac{dy}{dt} = f(t, y, k, \psi, \phi) \tag{8.1}$$

with initial conditions (ICs)

$$y = y^i \quad \text{at} \quad t = 0 \tag{8.2}$$

where $y = [y_1 y_2 \cdots y_n]^T$ is the vector of the n chemical species concentrations and f is the vector of functions representing the formation rate of each species. In previous chapters, ϕ was used to denote the model input parameters. Here, since we are investigating the reaction mechanism, it is convenient to distinguish between the kinetic parameters, denoted by k, and the remaining physicochemical parameters, denoted by ϕ.

Based on the definition in Chapter 2, the sensitivity of the given process dynamics is an *objective sensitivity*. According to Eq. (2.13), the sensitivity of an objective I

with respect to the rate constant of the jth reaction, k_j, is given by

$$s(I;k_j) = \frac{\partial I}{\partial k_j} = \lim_{\Delta k_j \to 0} \frac{I(k_j + \Delta k_j) - I(k_j)}{\Delta k_j} \tag{8.3}$$

For comparative analysis, it is often convenient to compute the *normalized objective sensitivity*, defined as

$$S(I;k_j) = \frac{k_j}{I} \cdot \frac{\partial I}{\partial k_j} = \frac{\partial \ln I}{\partial \ln k_j} = \frac{k_j}{I} \cdot s(I;k_j) \tag{8.4}$$

which removes any artificial variation due to the magnitudes of I and k_j.

The objective sensitivity, $s(I;k_j)$ for $j = 1, 2, \ldots m$, can be determined by using the different numerical techniques, such as those based on direct differentiation, finite differences, and Green's functions, discussed in Chapter 2. In practical applications, since the number of elementary reactions in a detailed kinetic scheme is usually larger than the number of dependent variables in the system, i.e., $m \gg n$, the Green's function method is the most convenient. However, the finite difference method is also widely used, particularly in two cases. The first case arises when the objective is given implicitly in complex functional form or is not even given by a mathematical formula, so that the objective sensitivity cannot be evaluated from the local sensitivity values obtained by solving the sensitivity equations. The second case is when numerical difficulties occur in solving the model and sensitivity equations simultaneously, for example, due to stiffness of the resulting system. In the following, we illustrate the application of sensitivity analysis in mechanistic studies using both the Green's function and the finite difference methods to compute objective sensitivities.

8.1.1 Applications of the Green's Function Method

Equation (8.1) may be differentiated with respect to the rate constant of the jth reaction, k_j, to yield the following sensitivity equations:

$$\frac{\partial s(y;k_j)}{\partial t} = J(t)s(y;k_j) + \frac{\partial f}{\partial k_j} \tag{8.5}$$

where $J(t)$ is the $n \times n$ Jacobian matrix with elements $\partial f_i/\partial y_j$. This set of equations can be solved simultaneously with the system equations (8.1) to obtain the local sensitivities, $s(y;k_j)$. This is the so-called direct differential method. Since the sensitivity analysis involves m rate constants, this leads to the problem of solving a system of $(m+1) \times n$ differential equations. In order to reduce the computational effort, the Green's function method described in Chapter 2 can be used. Accordingly, we first solve the following Green's function problem:

$$\frac{dG(t,\tau)}{dt} = J(t)G(t,\tau), \qquad t > \tau \tag{8.6a}$$

$$G(\tau,\tau) = 1 \tag{8.6b}$$

and then compute the local sensitivities, $s(y; k_j)$, from the integral equations

$$s(y; \phi_j) = \frac{\partial y(t)}{\partial \phi_j} = G(t, 0) \cdot \delta + \int_0^t G(t, \tau) \frac{\partial f(\tau)}{\partial \phi_j} \, d\tau \qquad (8.7)$$

where each element δ_k in vector δ is a Kronecker delta function, defined as

$$\delta_k = \delta\left(\phi_j - y_k^i\right) \qquad (8.8')$$

With this method to obtain the local sensitivities with respect to m rate constants, we need to solve only $n \times n$ differential equations (8.6) plus n integrals (8.7). Thus, when $m \gg n$, the Green's function method leads to a significant reduction of the required numerical effort, as compared to the direct differential method.

Example 8.1 Oxidation of wet carbon monoxide. Oxidation of wet carbon monoxide, *i.e.*, the chemical reaction occurring in the $CO-H_2O-O_2$ system, has been studied extensively in the literature. A detailed kinetic scheme for this process, based on several elementary reactions, is reported in Table 8.1 (Westbrook *et al.*, 1977; Yetter *et al.*, 1985). In order to better understand the reaction mechanism, Yetter *et al.* have performed sensitivity analysis of the time evolution of CO concentration with respect to rate constants of the elementary reactions, using the Green's function method to compute the local sensitivities. The adopted initial conditions are $y_{CO}^i = 2000$ ppm, $y_{O_2}^i = 2.8\%$, $y_{H_2O}^i = 1.0\%$, and $P = 1$ atm. The system is assumed to be isothermal at $T = 1100$ K. The calculated concentrations of the involved species are shown as functions of time in Fig. 8.1, where it may be seen that ignition occurs just before 0.01 s.

The normalized sensitivities of CO concentration with respect to various reaction rate constants are shown as functions of time in Figs. 8.2a, b, and c, depending on whether the sensitivity value is larger than 1, between 0.1 and 1, and between 0.01 and 0.1, respectively. Reactions leading to normalized sensitivities lower than 0.01 have not been considered.

Figure 8.2a shows the most important reactions for the wet CO oxidation process, which, in decreasing order of importance, are

$$CO + OH \rightarrow CO_2 + H \qquad (6F)$$

$$H + O_2 + M \rightarrow HO_2 + M \qquad (24R)$$

$$H + O_2 \rightarrow OH + O \qquad (8F)$$

$$O + OH \rightarrow H + O_2 \qquad (8R)$$

$$O + H_2O \rightarrow OH + OH \qquad (10F)$$

$$OH + OH \rightarrow O + H_2O \qquad (10R)$$

Table 8.1. Detailed kinetic scheme for the oxidation of wet carbon monoxide $(CO-H_2O-O_2)^a$

No.	Reaction	k_f	k_r
1	$HCO + H = CO + H_2$	$3.32E-10^b$	$2.85E-27$
2	$HCO + OH = CO + H_2O$	$1.66E-10$	$6.07E-30$
3	$O + HCO = CO + OH$	$5.00E-11$	$4.76E-28$
4	$HCO + O_2 = CO + HO_2$	$5.00E-12$	$4.30E-18$
5	$CO + HO_2 = CO_2 + OH$	$5.12E-15$	$2.95E-26$
6	$CO + OH = CO_2 + H$	$3.19E-13$	$1.31E-15$
7	$CO_2 + O = CO + O_2$	$7.36E-22$	$1.41E-21$
8	$H + O_2 = OH + O$	$1.87E-13$	$2.30E-11$
9	$H_2 + O = H + OH$	$5.62E-13$	$6.36E-13$
10	$O + H_2O = OH + OH$	$2.74E-14$	$6.82E-12$
11	$H + H_2O = OH + H_2$	$1.28E-14$	$3.05E-12$
12	$H_2O_2 + OH = H_2O + HO_2$	$6.03E-12$	$1.18E-17$
13	$HO_2 + O = O_2 + OH$	$3.60E-11$	$4.11E-22$
14	$H + HO_2 = OH + OH$	$1.87E-10$	$2.30E-19$
15	$H + HO_2 = H_2 + O_2$	$4.16E-11$	$4.13E-22$
16	$OH + HO_2 = H_2O + O_2$	$2.18E-11$	$9.18E-25$
17	$H_2O_2 + O_2 = HO_2 + HO_2$	$2.24E-19$	$1.05E-11$
18	$HO_2 + H_2 = H_2O_2 + H$	$2.33E-16$	$5.07E-13$
19	$O_2 + M = O + O + M^c$	$3.37E-32$	$5.13E-34$
20	$H_2 + M = H + H + M$	$3.07E-29$	$8.27E-33$
21	$OH + M = O + H + M$	$3.00E-28$	$2.76E-32$
22	$H_2O_2 + M = OH + OH + M$	$1.81E-16$	$2.55E-32$
23	$H_2O + M = H + OH + M$	$5.00E-29$	$3.19E-31$
24	$HO_2 + M = H + O_2 + M$	$2.91E-18$	$7.18E-33$
25	$CO_2 + M = CO + O + M$	$5.37E-32$	$2.50E-33$
26	$HCO + M = H + CO + M$	$4.61E-14$	$8.85E-34$

From Yetter *et al.* (1985).
$^a k_f$ and k_r are rate constants of forward and reverse reactions, respectively, evaluated at 1100 K, with units in molecule $\cdot$ cm $\cdot$ s.
^{b} Read as 3.32×10^{-10}.
$^c M$ refers to third body.

where the number in parentheses corresponds to the reaction number in Table 8.1, while the letter F (R) denotes the forward (reverse) reaction. It is seen that reactions 6F, 8F, and 10F lead to negative sensitivity values. Thus, the CO consumption rate increases as the rates of these reactions increase, *i.e.*, these reactions promote CO oxidation. This implies that CO is consumed mainly through reaction 6F, which also produces an H radical. This leads to the chain branching reaction, 8F, which is followed by another chain branching reaction, 10F. The OH radicals formed in these two branching reactions promote further the consumption of CO through reaction 6F. Thus, reactions 6F, 8F, and 10F constitute a reaction cycle, which is expected to be the dominant reaction path in the wet CO oxidation process.

Reactions 24R, 8R, and 10R in Fig. 8.2a exhibit positive sensitivity values, which implies that they have an inhibition effect on CO oxidation. They compete for the H,

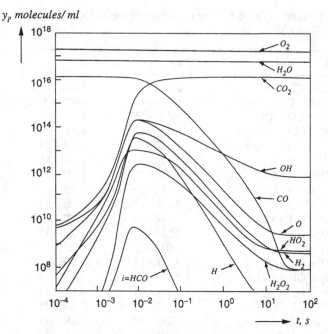

Figure 8.1. Species concentrations as functions of time for wet CO oxidation with the kinetic scheme in Table 8.1. Initial conditions: $T = 1100$ K; $P = 1013$ kPa; $y_{CO} = 1.337 \times 10^{16}$ molecule/cm^3; $y_{O_2} = 1.867 \times 10^{17}$ molecule/cm^3; $y_{H_2O} = 6.686 \times 10^{16}$ molecule/cm^3. From Yetter *et al.* (1985).

O and OH radicals with the reaction cycle noted above. Reactions 8R and 10R are the reverse of reactions 8F and 10F, respectively. The reverse of reaction 6F, *i.e.*, reaction 6R, has an effect only when the system nears equilibrium, where the sensitivity values of the two reactions become equal in magnitude but opposite in sign as shown.

Figure 8.2b illustrates the sensitivity values of the next six most important reactions for CO concentration. Note that they all involve the intermediate species HO$_2$ and H$_2$O$_2$:

$$OH + HO_2 \rightarrow H_2O + O_2 \tag{16F}$$

$$HO_2 + M \rightarrow H + O_2 + M \tag{24F}$$

$$H_2O_2 + M \rightarrow OH + OH + M \tag{22F}$$

$$OH + OH + M \rightarrow H_2O_2 + M \tag{22R}$$

$$H_2O_2 + OH \rightarrow H_2O + HO_2 \tag{12F}$$

$$H_2O + O_2 \rightarrow OH + HO_2 \tag{16R}$$

In agreement with the sign of the corresponding sensitivities, it is found that reactions 16F, 22R, and 12F inhibit CO oxidation since they consume the intermediate radical

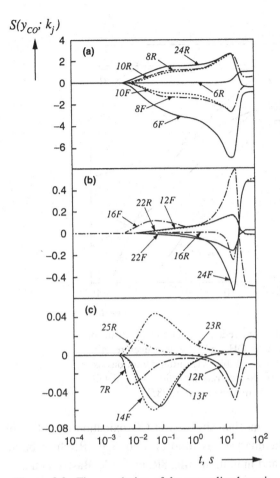

Figure 8.2. Time evolution of the normalized sensitivities of the CO concentration with respect to rate constants of the most important reactions in the detailed kinetic scheme shown in Table 8.1. Initial conditions as in Fig. 8.1. From Yetter *et al.* (1985).

OH, which, as discussed above, has a key role in the oxidation process. On the other hand, reactions 24F, 22F, and 16R promote CO oxidation.

Even though reactions shown in Fig. 8.2c exhibit very low sensitivity values, and are therefore not significant with respect to CO concentration, they still provide some useful information about the mechanism of wet CO oxidation. Let us consider in particular the sensitivity values corresponding to reaction 7R:

$$CO + O_2 \rightarrow CO_2 + O \tag{7R}$$

Besides the negative sign, which implies that reaction 7R promotes CO oxidation, it is seen that the sensitivity of this reaction exhibits a maximum during the induction period of the process, which clearly precedes that of all the other chemical species. This is a strong indication that reaction 7R is the chain initiator of the entire process.

y_i, molecules/ml

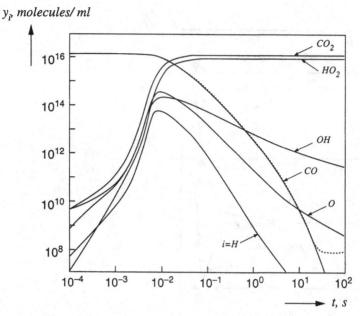

Figure 8.3. Concentration profiles for a reduced kinetic mechanism consisting of species CO, O_2, H_2O, H, O, OH, and HO_2 and seven reactions (6F, 7R, 8F, 8R, 10F, 10R, and 24R) in Table 8.1. Initial conditions as in Fig. 8.1. The dotted curve is the CO concentration profile for the detailed kinetic scheme shown in Fig. 8.1. From Yetter *et al.* (1985).

In conclusion, the dominant mechanism for wet CO oxidation identified by the sensitivity analysis can be summarized as follows: initiation 7R, chain branching 10F and 8F, propagation 6F, and inhibition 24R, 8R, and 10R. Based on this result, Yetter *et al.* (1985) constructed a simplified kinetic model including only these seven elementary reactions and computed, for the same system and initial conditions as in Fig. 8.1, the time evolution of the concentrations of the involved chemical species shown in Fig. 8.3. As expected, the obtained CO concentration profile is substantially coincident with that in Fig. 8.1. The most significant differences occur after about 20 s, when CO oxidation has reached its thermodynamic equilibrium with CO conversion values above 99.99%.

It is worth noting that in Example 8.1, above, the objective for the sensitivity analysis is CO concentration, which is one of the system dependent variables. In this case, the objective sensitivity coincides with a local sensitivity and can be computed directly from Eq. (8.7) using the Green's function method. In the following, we discuss a second example, where the objective is not a dependent variable described by the model equations. In this case, the evaluation of the objective sensitivity is more complicated and requires proper combination of the local sensitivity values.

Example 8.2 Sensitivity analysis of the Belousov-Zhabotinsky oscillating reaction. The Belousov-Zhabotinsky (BZ) reaction is a typical example of chemically oscillating

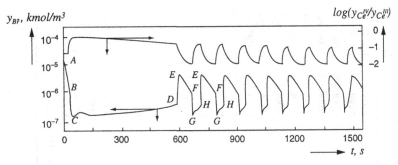

Figure 8.4. Oscillation behavior of bromide and cerium ions observed experimentally for the Belousov–Zhabotinsky system. Initial conditions: $y_{CH_2(COOH)_2} = 0.032$ kmol/m^3; $y_{KBrO_3} = 0.063$ kmol/m^3; $y_{KBr} = 1.5 \times 10^{-5}$ kmol/m^3; $y_{Ce(NH_4)_2(NO_3)_5} = 0.001$ kmol/m^3; $y_{H_2SO_4} = 0.8$ kmol/m^3. From Field et al. (1972).

system. The classical BZ oscillation, based on the Ce(III)-catalyzed oxidation and the bromination of malonic acid by acidic bromate, has been studied extensively. Oscillations can be either spatial or temporal, depending on the degree of mixing in the reacting solution. Figure 8.4 shows typical temporal oscillations as observed experimentally in a batch reactor by Field et al. (1972). They investigated this system using the detailed kinetic scheme proposed by Edelson et al. (1979), shown in Table 8.2. In particular, a comprehensive sensitivity analysis was performed in order to understand the oscillation mechanism and the effect of each elementary reaction on the oscillation period and shape.

Let us consider the BZ oscillation with perfectly repetitive waves occurring in a well-stirred system. The model differential equations derived from the mass balances of the involved chemical species must satisfy the following condition:

$$y_i(t, \phi) = y_i(t + \tau, \phi) \tag{E8.1}$$

where y_i is the concentration of the ith component that exhibits oscillatory behavior, τ is its oscillation period, and ϕ is the rate-constant vector.

In order to illustrate the oscillation mechanism, let us consider the sensitivities of the oscillation period and the shape characteristics of the cerium concentration wave, i.e., crest, trough, and height (crest-trough), with respect to the rate constants of the involved elementary reactions. Since the shape characteristics of the cerium wave are directly related to the cerium concentration, their sensitivities can be obtained by solving simultaneously the model and sensitivity equations. For the oscillation period, τ, however, since it is not given by the direct solution of the model equations, the sensitivity values cannot be obtained in the same way. In this case, the above equation (E8.1) needs to be used to derive the sensitivity values (Edelson and Thomas, 1981). In particular, differentiating Eq. (E8.1) with respect to the jth rate constant, k_j, leads to

$$\frac{dy_i(t, \phi)}{dk_j} = \frac{dy_i(t + \tau, \phi)}{dk_j} \tag{E8.2}$$

255

Table 8.2. Detailed kinetic scheme for the Belousov-Zhabotinsky reaction[a]

No.	Reaction	k_f	k_r
1	$2H^+ + Br^- + BrO_3^- \longleftrightarrow HOBr + HBrO_2$	2.10	1.00E+4[b]
2	$H^+ + HBrO_2 + Br^- \longleftrightarrow 2HOBr$	2.00E+9	5.00E-5
3	$HOBr + Br^- + H^+ \longleftrightarrow Br_2 + H_2O$	8.00E+9	2.00
4	$CH_2(COOH)_2 \longleftrightarrow (OH)_2C = CHCOOH$	3.00E-3	2.00E+2
5	$Br_2 + (OH)_2C = CHCOOH \longrightarrow H^+ + Br^- + BrCH(COOH)_2$	6.00E+4	0.00E+0
6	$HOBr + (OH)_2C = CHCOOH \longrightarrow H_2O + BrCH(COOH)_2$	0.00E+0	0.00E+0
7	$HBrO_2 + BrO_3^- + H^+ \longrightarrow 2BrO_2 \bullet + H_2O$	1.00E+4	3.64E+5
8	$BrO_2 \bullet + Ce^{(III)} + H^+ \longrightarrow Ce^{(IV)} + HBrO_2$	6.50E+5	2.40E+7
9	$Ce^{(IV)} + BrO_2 \bullet + H_2O \longrightarrow BrO_3^- + 2H^+ + Ce^{(III)}$	1.70E-1	1.30E-4
10	$2HBrO_2 \longleftrightarrow HOBr + BrO_3^- + H^+$	4.00E+7	2.00E-10
11	$Ce^{(IV)} + CH_2(COOH)_2 \longrightarrow \bullet CH(COOH)_2 + Ce^{(III)} + H^+$	1.00E+0	
12	$\bullet CH(COOH)_2 + BrCH(COOH)_2 + H_2O \longrightarrow Br^- + CH_2(COOH)_2 + HO\dot{C}(COOH)_2 + H^+$	1.00E+4	
13	$Ce^{(IV)} + BrCH(COOH)_2 + H_2O \longrightarrow Br^- + HO\dot{C}(COOH)_2 + Ce^{(III)} + 2H^+$	7.30E-3	
14	$2HO\dot{C}(COOH)_2 \longrightarrow HOCH(COOH)_2 + O = CHCOOH + CO_2$	1.00E+9	
15	$Ce^{(IV)} + HOCH(COOH)_2 \longrightarrow HO\dot{C}(COOH)_2 + Ce^{(III)} + H^+$	2.13E+0	
16	$Ce^{(IV)} + O = CHCOOH \longrightarrow O = \dot{C}COOH + Ce^{(III)} + H^+$	5.00E+1	
17	$2O = \dot{C}COOH + H_2O \longrightarrow O = CHCOOH + HCOOH + CO_2$	1.80E+7	
18	$Br_2 + HCOOH \longrightarrow 2Br^- + CO_2 + 2H^+$	6.00E-3	
19	$HOBr + HCOOH \longrightarrow Br^- + H^+ + CO_2 + H_2O$	0.00	
20	$2 \bullet CH(COOH)_2 + H_2O \longrightarrow CH_2(COOH)_2 + HOCH(COOH)_2$	1.80E+7	
21	$Br_2 + BrCH(COOH)_2 \longrightarrow Br_2CHCOOH + Br^- + H^+ + CO_2$	0.00	
22	$HO\dot{C}(COOH)_2 + BrCH(COOH)_2 + H_2O \longrightarrow Br^- + HOCH(COOH)_2 + HO\dot{C}(COOH)_2 + H^+$	0.00	

From Edelson et al. (1979).

[a] k_f and k_r are rate constants of forward and reverse reactions, respectively, with units in mol · l · s.

[b] Read as 1.00×10^4.

which corresponds to

$$\left[\frac{\partial y_i}{\partial t} \cdot \frac{\partial t}{\partial k_j} + \frac{\partial y_i}{\partial k_j}\right]_t = \left[\frac{\partial y_i}{\partial (t+\tau)} \cdot \frac{\partial (t+\tau)}{\partial k_j} + \frac{\partial y_i}{\partial k_j}\right]_{t+\tau} \tag{E8.3}$$

Since

$$\frac{\partial y_i}{\partial t} = \frac{\partial y_i}{\partial (t+\tau)} \tag{E8.4}$$

Equation (E8.3) becomes

$$\left(\frac{\partial y_i}{\partial k_j}\right)_t = \left(\frac{\partial y_i}{\partial t} \cdot \frac{\partial \tau}{\partial k_j}\right)_t + \left(\frac{\partial y_i}{\partial k_j}\right)_{t+\tau}, \tag{E8.5}$$

from which the expression for the sensitivity of τ with respect to k_j is obtained

$$s(\tau;k_j) = \frac{\partial \tau}{\partial k_j} = \frac{(\partial y_i/\partial k_j)_t - (\partial y_i/\partial k_j)_{t+\tau}}{(\partial y_i/\partial t)_t} = \frac{(s(y_i;k_j))_t - (s(y_i;k_j))_{t+\tau}}{(\partial y_i/\partial t)_t} \tag{E8.6}$$

where the species local sensitivities, $s(y_i;k_j)$, can be computed from the solution of the model and sensitivity equations. The corresponding normalized sensitivity is given by

$$S(\tau;k_j) = \frac{k_j}{\tau} \cdot s(\tau;k_j) \tag{E8.7}$$

Edelson (1981) evaluated the normalized sensitivities of the period and the characteristics of the cerium concentration wave with respect to the rate constants of the elementary reactions, using the Green's function method for computing the species local sensitivities. In Fig. 8.5 the normalized sensitivities of the period are listed in decreasing order, while in Fig. 8.6 are given the sensitivities of the characteristics of the cerium concentration wave, for some of the elementary reactions reported in Table 8.2. Similarly to Example 8.1, from these results it is possible to elucidate the mechanism of the BZ oscillation. It should be noted, however, that the results in Figs. 8.5 and 8.6 indicate that, when considering different objectives in the sensitivity analysis, the importance of each elementary reaction can be substantially different. For instance, reactions 4F, 4R, and 5F are the most important ones for determining the oscillation period, while they become secondary when considering the characteristics of the cerium concentration wave. Moreover, even for the latter, the trough is most sensitive to reactions 1F, 2F, 7F, and 11, while the crest and height are most sensitive to reactions 7F, 7R, 8F, and 8R. Therefore, to fully understand the system dynamics, it is generally required to perform the sensitivity analysis for each model output of

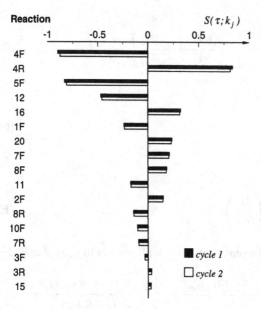

Figure 8.5. Normalized sensitivity of the BZ oscillation period with respect to the reaction rate constants in Table 8.2, in two adjacent cycles of oscillation (cycle 1, $t = 128.7-242.7$ s; cycle 2, $t = 242.7-356.7$ s).

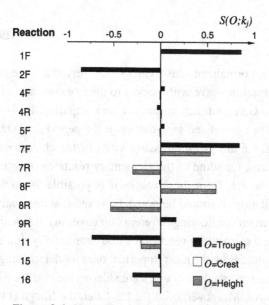

Figure 8.6. Normalized sensitivity of the characteristics of the cerium concentration wave with respect to the rate constants of the elementary reactions in Table 8.2.

Table 8.3. Values of the sensitivity of the oscillation period with respect to the reactions with zero rate constants in Table 8.2, computed at two consecutive cycles

Reaction	$s(\tau; k)$	
	Cycle 1	Cycle 2
5R	8.331×10^{-2}	8.122×10^{-2}
6F	4.300×10^{-6}	3.982×10^{-6}
6R	3.203×10^{4}	2.910×10^{4}
19	-9.438×10^{-2}	-9.201×10^{-2}
21	-4.924×10^{-1}	-4.799×10^{-1}
22	1.141×10^{-2}	1.114×10^{-2}

From Edelson (1981).

interest. This will be illustrated in the next several examples (8.3 to 8.6) with reference to $H_2 - O_2$ explosion.

It is worth mentioning another technique, developed by Edelson (1981), for the investigation of complex kinetic mechanisms. This refers to the problem of establishing the importance of reactions that are left out of the detailed kinetic scheme considered or whose rate constants are not known. For example, reactions 5R, 6F, 6R, 19, 21, and 22 in Table 8.2 have zero rate constants, indicating that they have been omitted in the model simulations. However, the sensitivity analysis with respect to the rate constants of these reactions can still be performed around their nominal values (*i.e.*, zero). Of course, in this case, we calculate the absolute rather than the normalized sensitivities, since from Eq. (E8.7), the normalized sensitivity is always zero ($k_j = 0$). Table 8.3 shows the sensitivities of the oscillation period with respect to all the elementary reactions with zero rate constants in Table 8.2. It is seen that reaction 6F with a rate of the order of 10^5 would be sufficient to compete with reaction 5F in the bromination of malonic acid enol, and reaction 6R with a very small rate could also be significant. Note that these conclusions involve an extrapolation of the sensitivities computed for a zero value of the specific rate constants. Therefore, the sensitivity evaluation should be repeated at the estimated nonzero value to confirm the importance of the considered elementary reaction. Moreover, it was shown (Edelson, 1981) that if a value of 10^4 is given to the constant of reaction 22, the oscillation disappears in either an open or a closed system for the kinetic scheme in Table 8.2. This information is important for retrofitting the kinetic scheme and for organizing specific experiments to test the reaction rate constants that have been initially omitted.

8.1.2 Applications of the Finite Difference Method

In the above two examples, the objective for the sensitivity analysis was explicit. However, as discussed in Chapter 2, there are cases in which either the objective is not available explicitly, or it cannot even be represented by a mathematical formula. In such cases, the finite difference method can be used.

From definition (2.13), when the variation, $\Delta\phi_j$, is small, the objective sensitivity can be approximated as

$$s(I;\phi_j) = \frac{\partial I}{\partial\phi_j} \approx \frac{\Delta I}{\Delta\phi_j} = \frac{I(t,\phi_j+\Delta\phi_j) - I(t,\phi_j)}{\Delta\phi_j} \tag{8.8}$$

where $I(t,\phi_j)$ and $I(t,\phi_j+\Delta\phi_j)$ are evaluated from the solution of the system model equations with $\phi_j = \phi_j$ and $\phi_j = \phi_j + \Delta\phi_j$, respectively. Thus, if $I(t,\phi_j+\Delta\phi_j)$ is computed for each $\Delta\phi_j(j = 1, 2, \ldots, m)$ through Eq. (8.8), we can obtain the values of the objective sensitivity, $s(I;\phi_j)$, with respect to all the elementary reactions in the kinetic scheme. This is the so-called finite difference method.

Note that in general the value to be given to the variation $\Delta\phi_j$ depends on j. As discussed in Chapter 2, there exists a minimum variation $\Delta\phi_{j,\min}$ that allows the proper estimation of the sensitivity value for a given objective, I. It satisfies the condition

$$\frac{\varepsilon_I \cdot I}{\Delta\phi_{j,\min}} = \varepsilon_s \cdot s(I;\phi_j) \tag{8.9}$$

where ε_I and ε_s are the tolerated fractional errors in I and $s(I;\phi_j)$, respectively. This implies that $\Delta\phi_{j,\min}$ increases as $s(I;\phi_j)$ decreases. Since the sensitivity $s(I;\phi_j)$ can vary significantly for different parameters ϕ_j, in general the variation $\Delta\phi_j$ is different for different j values. Moreover, even for the same ϕ_j, since the objective sensitivity is a function of time, it may be required to vary $\Delta\phi_j$ with time. This can complicate the use of the finite difference method from a numerical point of view.

Example 8.3 Explosion mechanism in hydrogen–oxygen systems: the first limit. In Chapter 7 we discussed the mathematical modeling of the explosion of hydrogen–oxygen systems in a closed vessel. In particular, we have shown that, by using a detailed kinetic scheme and a proper definition of the explosion limits, it is possible to predict the explosion boundaries measured experimentally. Figure 8.7 compares the measured (dark points) and predicted (solid curve) explosion limits in the pressure-temperature plane, reproduced from Figs. 7.7 and 7.9. In the model predictions, the detailed kinetic scheme given in Table 7.1 has been used, which includes eight species, H_2, O_2, H, O, OH, H_2O, HO_2, and H_2O_2, and 53 elementary reactions, including forward and reverse reactions. The critical initial temperature for explosion to occur at a given value of the initial pressure was defined, based on the Morbidelli and Varma (1988) generalized criterion, as the initial temperature where the normalized sensitivity of the concentration maximum of the H radical with respect to any of the system parameters reaches its maximum or minimum. The inverse-S-shaped explosion boundaries in Fig. 8.7 are divided by two bifurcation points b_1 and b_2 into three branches, which from low to high initial pressure are usually referred to as the first, second, and third explosion

P^i, Torr

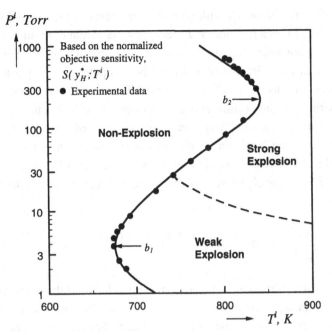

Figure 8.7. Explosion limits of the stoichiometric H_2-O_2 mixture in the pressure-temperature plane. Values measured experimentally (•) in a KCl-coated vessel by Lewis and von Elbe (1961) and predicted (solid curve) using the detailed kinetic scheme in Table 7.1. The broken curve is the boundary separating the weak from the strong explosion regions; b_1 and b_2 are bifurcation points representing the transition from the first to second and from the second to third explosion limits, respectively.

limits. In this example, we investigate in detail the mechanism for the occurrence of the first explosion limit through objective sensitivity analysis, while the mechanisms for the second and third limits will be discussed in Examples 8.4 and 8.5, respectively.

To analyze the mechanism of the first explosion limit, it is sufficient to choose a representative point on the explosion boundary, *i.e.*, a critical initial temperature at a typical initial pressure in the region of the first limit, as objective for sensitivity analysis. Then, the sensitivity of this critical initial temperature with respect to variation in the pre-exponential factor of each elementary reaction in Table 7.1 is computed.

From Eq. (8.4), the normalized sensitivity of the critical initial temperature or explosion temperature, T_{e1}^i, with respect to the pre-exponential factor of the kth reaction in Table 7.1, A_k, can be defined as

$$S(T_{e1}^i; A_k) = \frac{A_k}{T_{e1}^i} \cdot s(T_{e1}^i; A_k) = \frac{A_k}{T_{e1}^i} \cdot \frac{\partial T_{e1}^i}{\partial A_k} \tag{E8.8}$$

where subscript 1 denotes the first explosion limit. The objective T_{e1}^i is defined by the generalized criterion as the initial temperature, where the normalized sensitivity of the H concentration maximum y_H^* with respect to any choice of the model input parameter ϕ, i.e., $S(y_H^*; \phi)$, as a function of T^i reaches a maximum or minimum.

It should be noted, however, that since the explosion limit T_{e1}^i, defined by the generalized criterion, is given implicitly and must be computed using appropriate numerical techniques, its sensitivity $s(T_{e1}^i; A_k)$ in Eq. (E8.8) is also given implicitly. For this reason, neither the direct differential nor Green's function method can be applied, since neither a relevant sensitivity equation can be derived nor can $s(T_{e1}^i; A_k)$ be expressed in terms of the local sensitivities of the system dependent variables. Thus, the finite difference method should be used for evaluating $s(T_{e1}^i; A_k)$, i.e., the above expression (E8.8) for the objective sensitivity is replaced by the finite difference equation,

$$S\left(T_{ei}^i; A_k\right) \approx \frac{A_k}{T_{e1}^i} \cdot \frac{\Delta T_{e1}^i}{\Delta A_k} = \frac{A_k}{T_{e1}^i} \cdot \frac{T_{e1}^i(A_k + \Delta A_k) - T_{e1}^i(A_k)}{\Delta A_k} = \frac{\Delta T_{e1}^i / T_{e1}^i}{\delta_k} \quad \text{(E8.9)}$$

where δ_k is a small relative variation in the pre-exponential factor of the kth reaction, and its value needs to be adjusted based on Eq. (8.9) during the computation. Both $T_{e1}^i(A_k)$ and $T_{e1}^i(A_k + \Delta A_k)$ are computed as described in Section 7.1.1. Then, the general procedure to obtain the sensitivity value can be described as follows: (1) give a small perturbation δ_k to the pre-exponential factor of the kth reaction, and find the variation in the explosion limit, ΔT_{e1}^i; (2) evaluate the sensitivity value using Eq. (E8.9); (3) varify if δ_k satisfies Eq. (8.9). If $\delta_k < \delta_{k,\min}$, we need to increase δ_k and repeat steps (1) and (2). The same procedure is used for all the reactions in Table 7.1.

When sensitivity analysis is used to identify the less important reactions as well as the rate-limiting steps in a large kinetic scheme, it is necessary to investigate the effect of each elementary reaction on all the involved chemical species. This can be done using a Euclidean norm of the concentration sensitivity matrix, such as

$$[a_j] = \left[\left(\sum_{i=1}^{N_S} \left(\frac{\partial \ln y_i}{\partial \ln k_j} \right)^2 \right)^{1/2} \right]$$

or the rate sensitivity matrix

$$[B_j] = \left[\left(\sum_{i=1}^{N_S} \left(\frac{\partial \ln R_i}{\partial \ln k_j} \right)^2 \right)^{1/2} \right]$$

where R_i is the overall formation rate for the ith species (Hwang, 1982; Turanyi et al., 1989). A similar result can be obtained using the objective sensitivity analysis described above. The objective of the analysis is the explosion temperature, which in turn is computed through a sensitivity analysis carried out on the entire nonisothermal reaction path, from the reactants to the final products. Thus, the explosion limit contains, in implicit form, information related to the entire time evolution of temperature and concentrations of all the involved chemical species. In addition, the two sets of reactions that are responsible for the system behavior under ignited and nonignited conditions somehow balance each other at the explosion boundary. Small changes of the initial temperature value are in fact sufficient to make one or the other of the two sets prevail. In other words, the explosion limit is a characteristic of the system behavior that is rather sensitive to the various elementary reactions constituting the kinetic mechanism. Therefore, it seems reasonable to expect that the explosion limit can be used as an indicator to determine whether or not a given elementary reaction is important relative to all the others comprising the kinetic mechanism.

We choose the explosion temperature T_{e1}^i at $P^i = 2$ Torr as representative of the first explosion limit to perform the sensitivity analysis. Both the forward and reverse reactions in Table 7.1 are considered as independent elementary reactions. The sensitivity values with respect to the pre-exponential factor of each reaction in Table 7.1 are shown in Fig. 8.8 in decreasing order. It is worth noting the meaning of the algebraic sign of sensitivity in the present case. For a given $\delta_k > 0$, if the computed variation in the explosion temperature is negative, $i.e.$, $\Delta T_{e1}^i < 0$, then the explosion limit occurs at a lower initial temperature, which implies that this reaction promotes the occurrence of explosion. In this case, from Eq. (E8.9) we obtain a negative sensitivity value. On the contrary, when a reaction inhibits the occurrence of explosion, it leads to a positive sensitivity value. Thus, in general, branching (termination) reactions exhibit negative (positive) sensitivity values.

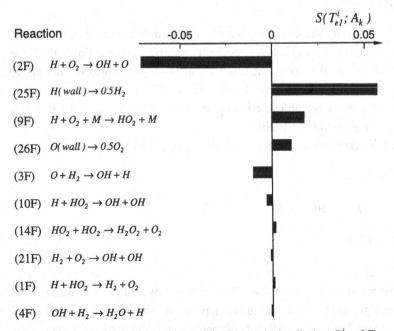

Figure 8.8. Normalized sensitivities of the first explosion limit at $P^i = 2$ Torr with respect to the reaction rate constants in Table 7.1, listed in decreasing order. From Wu $et\ al.$ (1998).

Among all reactions for chain initiation in Table 7.1,

$$H_2 + O_2 \rightarrow HO_2 + H \tag{2R}$$

$$H_2 + M \rightarrow H + H + M \tag{6R}$$

$$H_2 + O_2 \rightarrow OH + OH, \tag{21F}$$

only 21F is present in Fig. 8.8, while the effect of reactions 2R and 6R on the first limit is negligible. This means that reactions 2R and 6R can be eliminated without changing the position of the explosion limit. Thus, the chain combustion process is initiated substantially only by reaction 21F. This supports the hypothesis of Willbourn and Hinshelwood (1946), who were the first to propose reaction 21F as the principal initiator in the H_2–O_2 combustion.

The chain initiator together with the chain propagation

$$OH + H_2 \rightarrow H_2O + H \tag{4F}$$

produces the H radical. This leads to the chain branching

$$H + O_2 \rightarrow OH + O \tag{2F}$$

followed by

$$O + H_2 \rightarrow OH + H. \tag{3F}$$

Very high sensitivity values for reactions 2F and 3F indicate that they are rate-limiting branching steps for the explosion to occur. The above three reactions, 4F, 2F, and 3F, are well known as the dominant reactions in H_2–O_2 combustion, and are usually referred to as the *reaction cycle* (Lewis and von Elbe, 1961). We will see in later examples that this reaction cycle is important not only for the first but also for the second and third explosion limits.

The chain termination reactions included in Fig. 8.8 are

$$H_2 \xrightarrow{\text{wall}} 0.5H_2 \tag{25F}$$

$$H + O_2 + M \rightarrow HO_2 + M \tag{9F}$$

$$O \xrightarrow{\text{wall}} 0.5O_2, \tag{26F}$$

where 25F and 26F are wall termination reactions. It is worth noting that these reactions, like reactions 27–29 in Table 7.1, represent overall rates of wall terminations that lump together the two consecutive steps of diffusion toward the wall and surface reaction. In the framework of the one-dimensional model used in this work, these two contributions cannot be properly distinguished. Thus, the finding that wall terminations play an important role for the occurrence of the first explosion limit is in good agreement with the observation based on experimental results (Lewis and von Elbe, 1961; Dixon-Lewis and Williams, 1977), that at fixed initial partial pressures of H_2 and O_2, increasing inert gas pressure causes a lowering of the first explosion-limit temperature since it hinders the diffusion of radicals to the wall. On the other hand, reaction 9F produces the less reactive component, HO_2, and hence can be regarded as a termination reaction. This is confirmed by the large positive value of its sensitivity.

In Fig. 8.8, the following three reactions are also included:

$$H + HO_2 \rightarrow OH + OH \qquad\qquad\qquad (10F)$$

$$H + HO_2 \rightarrow H_2 + O_2 \qquad\qquad\qquad (1F)$$

$$HO_2 + HO_2 \rightarrow H_2O_2 + O_2 \qquad\qquad\qquad (14F)$$

which may either promote or inhibit the occurrence of the first explosion limit. All these reactions involve the species HO_2, and can be regarded as consecutive with respect to reaction 9F.

Thus, the first explosion limit can be interpreted in terms of the following kinetic mechanism: initiation, 21F; propagation, 4F; chain branching, 2F and 3F; termination, 25F, 9F, and 26F. Such a mechanism has been identified previously in the literature by various researchers based upon considerations about the chemistry of the process and various experimental evidences (see Hinshelwood and Williamson, 1934; Semenov, 1959; Lewis and von Elbe, 1961; Minkoff and Tipper, 1962; Dixon-Lewis and Williams, 1977). Here, the importance of these reactions has been determined based solely on the objective sensitivity analysis of a rather general detailed kinetic scheme, using the explosion limit as the objective.

Example 8.4 Explosion mechanism in hydrogen–oxygen systems: the second limit. From Fig. 8.7 it is seen that the explosion temperature at $P^i = 40$ Torr can be taken as a representative of the second explosion limit for the sensitivity analysis. The normalized sensitivities of this critical initial temperature with respect to the pre-exponential factor of each of the elementary reactions in Table 7.1, $S(T_{e2}^i; A_K)$, have been computed based on the procedure described in Example 8.3, and are shown in Fig. 8.9 in decreasing order.

It appears that except for reaction 21F, all the other reactions in Fig. 8.9 involve radicals H and HO_2 or both. Thus, it is clear that these species play the dominant role in the occurrence of the second explosion limit. Moreover, we see that among all the reactions that cause explosion to occur (i.e., negative sensitivity values), reaction 2F exhibits the highest sensitivity. Thus, also in the second limit region, similarly to the first limit, the explosion process is dominated by chain branching. Reaction 21F is evidently the chain initiator, and the reaction cycle, 4F, 2F, and 3F, is again dominating. Although 3F (11th rank) and 4F (15th rank) are not included in Fig. 8.9, they are required for consumption of O and OH species, respectively.

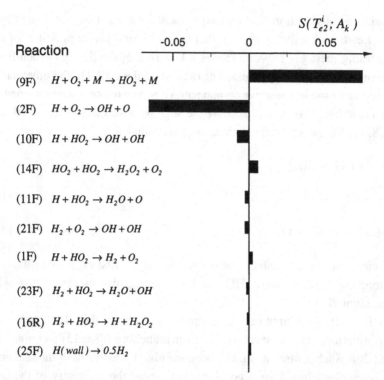

Figure 8.9. Normalized sensitivities of the second explosion limit at $P^i = 40$ Torr with respect to the reaction rate constants in Table 7.1, listed in decreasing order. From Wu *et al.* (1998).

The most significant difference between the kinetic mechanisms involved in first and second explosion limits is in the termination reactions. In the latter case, we see from Fig. 8.9 that reaction 9F as a termination reaction leads to the highest sensitivity, competing with reaction 2F for the H radical. This agrees with the conclusion of earlier studies (Lewis and von Elbe, 1961), which related the occurrence of the second explosion limit to the presence of a third-order termination reaction. Reaction 25F, which was the dominant termination step for the first explosion limit, now has only the 10th rank in Fig. 8.9. This indicates that surface reaction or wall diffusion is much less important in determining the second limit. This coincides with the experimental evidence (cf. Dixon-Lewis and Williams, 1977) that the second explosion limit is independent of the vessel size, provided that the volume is sufficiently large. Note that this difference in the termination reactions for the two explosion limits is in fact the reason for the occurrence of the bifurcation point, b_1.

All the remaining reactions in Fig. 8.9 involve the species HO_2, and therefore are in series to reaction 9F which is the only one producing HO_2. The reactions identified in the figure have been included in the reduced kinetic mechanism proposed by Baldwin *et al.* (1974) to interpret the second explosion limit. They also included other reactions involving the chemical species H_2O_2, such as 15F, 16F, 17F, 18F, and

Table 8.4. Reduced mechanism for the second explosion limit of H_2–O_2 system derived from sensitivity analysis

No.	Reaction	I[a]	II[b]	Note
21F	$H_2 + O_2 \rightarrow OH + OH$	yes	yes	initiation
6R	$H_2 + M \rightarrow H + H + M$	yes		initiation
4F	$OH + H_2 \rightarrow H_2O + H$	yes	yes	propagation
2F	$H + O_2 \rightarrow OH + O$	yes	yes	branching
3F	$O + H_2 \rightarrow OH + H$	yes	yes	branching
4R	$H + H_2O \rightarrow OH + H_2$	yes		4F reverse
10F	$H + HO_2 \rightarrow OH + OH$	yes	yes	H–HO_2
11F	$H + HO_2 \rightarrow H_2O + O$	yes	yes	H–HO_2
1F	$H + HO_2 \rightarrow H_2 + O_2$	yes	yes	H–HO_2
2R	$OH + O \rightarrow H + O_2$	yes		2F reverse
22F	$H + O + M \rightarrow OH + M$	yes		inhibition
16R	$HO_2 + H_2 \rightarrow H + H_2O_2$		yes	HO_2
14F	$HO_2 + HO_2 \rightarrow H_2O_2 + O_2$		yes	HO_2
23F	$HO_2 + H_2 \rightarrow H_2O + OH$		yes	HO_2
9F	$H + O_2 + M \rightarrow HO_2 + M$	yes	yes	termination
25F	$H \xrightarrow{\text{wall}} 0.5H_2$		yes	termination
7F	$H + OH + M \rightarrow H_2O + M$	yes		termination

From Wu *et al.* (1998).
[a] By Dougherty and Rabitz (1980).
[b] By Wu *et al.* (1998).

24F. However, simulations indicate that these reactions can be eliminated without changing the position of the explosion limit. Thus, they are not essential to interpret the second explosion limit.

The sensitivity analysis of the second explosion limit has also been performed by Dougherty and Rabitz (1980) and Maas and Warnatz (1988). Dougherty and Rabitz did not take the explosion limit as the objective, but instead studied the sensitivities of the H radical concentration with respect to each elementary reaction around the explosion limits. They also used a more limited model, which assumed isothermal explosion conditions and it was then unable to predict the occurrence of the third explosion limit. The reduced kinetic mechanism for the second explosion limit identified by Dougherty and Rabitz is compared with the one derived above in Table 8.4. It appears that the reactions included in the two mechanisms are somewhat different. In particular, it can be observed that all the reactions in the Dougherty and Rabitz mechanism involve the H radical, except for the initiation reaction 21F. This is because they used the H radical concentration rather than the explosion limit as the objective for the sensitivity analysis. Accordingly, the derived mechanism is most suited to describe the generation of the H radicals around the second explosion limit, rather than the explosion limit itself. In addition, the use of a different model, due to the isothermality assumption mentioned above, is also responsible for differences in the derived mechanisms.

On the other hand, Maas and Warnatz (1988) took the explosion limit as the objective in the sensitivity analysis, and used a more sophisticated, nonisothermal two-dimensional reactor model. Their sensitivity analysis was based on the variation of the explosion temperature with respect to *simultaneous* variations in the pre-exponential factors of both forward and reverse reactions. In other words, they considered the sensitivity of the explosion limit with respect to the net rate of each elementary reaction. Obviously, this approach does not allow discrimination between forward and reverse reactions. As an example, consider the objective sensitivity analysis based on the net rate of each elementary reaction for $P^i = 40$ Torr. The normalized sensitivity $S(T_{e2}^i; A_{k,FR})$ of the critical initial temperature T_{e2}^i with respect to simultaneous variations in the pre-exponential factors of both the forward and the reverse kth reactions $A_{k,FR}$ is defined as

$$S\left(T_{e2}^i; A_{k,FR}\right) = \frac{\Delta T_{e2}^i / T_{e2}^i}{\delta_{k,FR}},$$
(E8.10)

where $\delta_{k,FR}$ represents the small relative variation given to the pre-exponential factors of the kth forward and reverse reactions. The sensitivity values computed in this manner are shown in Fig. 8.10 in decreasing order. Comparing Figs. 8.9 and 8.10, it appears that the mechanisms deduced are similar. The only difference is that using the sensitivities (E8.10), it is not possible to discriminate between the forward and the reverse reactions, so that both have to be included in the mechanism. For example, in

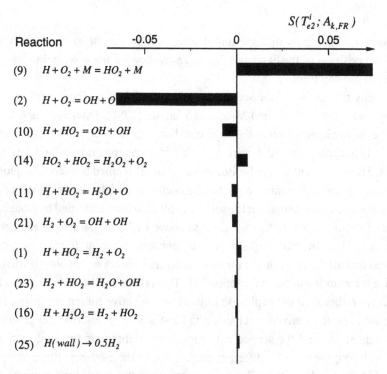

Figure 8.10. Normalized sensitivities of the second explosion limit at $P^i = 40$ Torr with respect to the pair of forward and reverse reaction rate constants in Table 7.1, listed in decreasing order.

Fig. 8.10, the pair of reactions

$$H + H_2O_2 = H_2 + HO_2 \tag{16F\&16R}$$

is identified among the top 10 ones, while Fig. 8.9 indicates that only the reverse reaction,

$$H_2 + HO_2 \rightarrow H + H_2O_2 \tag{16R}$$

is rate-limiting. Thus, when using sensitivity analysis to identify the rate-limiting steps in kinetic schemes involving reversible reactions, it is better to consider separately the forward and the reverse reactions.

Example 8.5 Explosion mechanism in hydrogen–oxygen systems: the third limit. It is generally agreed in the combustion literature that the occurrence of the third explosion limit is strictly related to thermal processes (see Griffiths *et al.*, 1981; Kordylewski and Scott, 1984; Foo and Yang, 1971). This is confirmed for example by the observation that an isothermal model does not predict the third explosion limit (Dougherty and Rabitz, 1980), or that the explosion temperature increases as the wall overall heat-transfer coefficient increases (see Fig. 7.16). To explain how thermal processes affect this limit, Kordylewski and Scott (1984) introduced the endothermic decomposition of hydrogen peroxide (reaction 15F in Table 7.1):

$$H_2O_2 + M \rightarrow OH + OH + M \tag{15F}$$

which provides a new route for the production of the OH radical. Sensitivity analysis of the third limit performed by Maas and Warnatz (1988) supports this proposal, by indicating the reversible reaction 15 (hence both 15F and 15R) exhibiting the largest sensitivity value. Let us now revisit these conclusions and derive the mechanism for the third explosion limit through the objective sensitivity analysis described above.

The sensitivity analysis of the explosion temperature at $P^i = 500$ Torr with respect to the variation in the pre-exponential factor of each elementary reaction in Table 7.1 leads to the sensitivity values shown in Fig. 8.11. Comparing with the corresponding values for the first and second explosion limits, *i.e.*, Figs. 8.8 and 8.9 respectively, it appears that in this case a larger number of elementary reactions are simultaneously important. Thus we should expect the mechanism for the third limit to be more complex than those for the first and second limits.

Reaction 21F is also in this case the only chain-initiator of the process. However, its rank in each explosion limit varies significantly, from the 8th position in Fig. 8.8 to the 6th in Fig. 8.9 and finally to the 4th in Fig. 8.11. In this last figure it is also seen that reaction 15F leads to the highest sensitivity. A proper reaction flux analysis indicates that this reaction also exhibits the largest flux when explosion occurs. This is in agreement with Kordylewski and Scott (1984) who suggested that under these conditions the decomposition of hydrogen peroxide represents a major path for

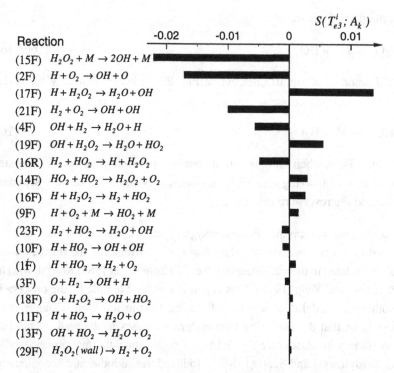

Figure 8.11. Normalized sensitivities of the third explosion limit at $P^i = 500$ Torr with respect to the reaction rate constants in Table 7.1, listed in decreasing order. From Wu *et al.* (1998).

producing the OH radicals. The reaction cycle (4F, 2F, and 3F) is again included in Fig. 8.11, together with another reaction,

$$H_2 + HO_2 \rightarrow H + H_2O_2 \qquad (16R)$$

which also exhibits high sensitivity. This is so because this reaction produces H radicals, which promote reaction 15F.

Reactions 25F and 9F, which are the terminations exhibiting the highest sensitivity for the first and second limit, respectively, have much lower or even negligible influence on the third limit. The wall termination reaction 25F, which is unfavored at high pressure since it is controlled by diffusion, is not included in Fig. 8.11. The small effect of reaction 9F in terminating the radical species is related to the importance of H_2O_2 decomposition 15F characteristic of the third limit as discussed above. The species HO_2 produced by 9F is mostly transformed into H_2O_2 through reactions 14F and 16R, whose decomposition actually produces new radical species. With respect to the radical termination process in the third limit, the largest sensitivity is exhibited by reaction 17F, which is *per se* a chain propagation reaction. However, this propagation occurs at the expense of H_2O_2, which is potentially a strong radical generator 15F; hence 17F has the overall effect of decreasing the radical concentration, just like a

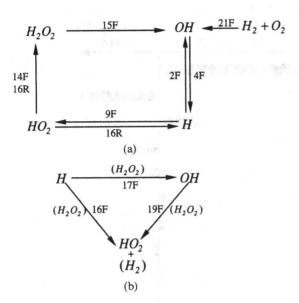

Figure 8.12. Mechanism for the occurrence of the third explosion limit: (a) evolution of the explosion; (b) termination. From Wu *et al.* (1998).

termination reaction. Other important terminations shown in Fig. 8.11 are reactions 19F and 16F, which both imply consumption of species H_2O_2.

Thus, the mechanism that interprets the occurrence of the third limit can be summarized as shown in Fig. 8.12, where the explosion evolution and termination mechanisms are illustrated separately. The evolution mechanism is similar to that proposed by Kordylewski and Scott (1984), except for the formation of species H_2O_2. They considered only the reaction

$$HO_2 + HO_2 \rightarrow H_2O_2 + O_2 \tag{14F}$$

as the main source for H_2O_2, while the above sensitivity as well as reaction flux analyses indicate that both reactions 14F and 16R are important. Moreover, in the mechanism of Kordylewski and Scott, reaction 9F is considered as the dominant termination reaction (similarly to the first and second limits), but from Fig. 8.11 it is seen that it plays only a secondary role.

As a final comment, let us explain the reason why the H_2O_2 decomposition reaction 15F becomes so important in the region of the third explosion limit. This is essentially due to the increase of both reactant concentrations: the third body M and H_2O_2 itself (Wu *et al.*, 1998). The first is a direct consequence of the increased pressure, while the second is due to the accumulation of H_2O_2 through the sequence of reactions $4F \Rightarrow 9F \Rightarrow 14F + 16R$.

Example 8.6 Explosion mechanism in hydrogen–oxygen systems: the weak–strong explosion boundary (WSEB). As discussed in Section 7.2, for the H_2–O_2 system there exists a boundary in the P^i–T^i plane, given by the broken curve in Fig. 8.7, that divides the

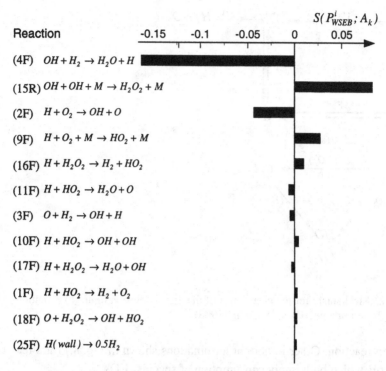

Figure 8.13. Normalized sensitivities of the critical initial pressure for the transition between weak and strong explosions (broken curve in Fig. 8.7) at $T^i = 800$ K, with respect to the reaction rate constants in Table 7.1, listed in decreasing order. From Wu *et al.* (1998).

explosion region into two portions: the weak explosion region at lower pressure and the strong one at higher pressure. We now investigate the mechanism that leads to the transition from weak to strong explosions, by analyzing the sensitivity behavior of the WSEB.

For the sensitivity analysis of WSEB, we consider the initial temperature $T^i = 800$ K, where the transition between weak and strong explosions occurs at the initial pressure $P^i = 11.8$ Torr. The sensitivities of this critical initial pressure with respect to the pre-exponential factors of all reactions in Table 7.1 have been computed, and the largest ones are listed in Fig. 8.13.

As discussed in Section 7.2, the weak explosion is one that remains incomplete because after a relatively low hydrogen conversion, termination reactions (9F, 25F, and 1F) take over branching reactions (2F and 10F). At larger temperature values, branching reactions are favored, since they have higher activation energies. Thus, if the heat produced in the system leads to a sufficient temperature rise, the termination reactions may not be able to take over the branching ones, so that explosion goes to completion, *i.e.*, the strong explosion regime. This picture matches nicely with the results of the sensitivity analysis as shown in Fig. 8.13. In particular, the reaction exhibiting the largest sensitivity (4F) is an exothermic reaction, which indicates that the heat required for the transition from weak to strong explosion comes mainly from this reaction. Moreover it is expected that, since reaction 4F requires an OH radical, reactions 2F and 3F, which produce OH, also are important. These in turn depend strongly on reactions involving radicals H and O, which thus also exhibit high sensitivities. These conclusions agree with the sensitivity values reported in Fig. 8.13.

About the termination process, reaction 15R exhibits the highest sensitivity. This can be readily understood considering that it competes with reaction 4F for the OH radical. Two other important (high-sensitivity value) termination reactions are 9F and 16F, which compete with reaction 2F for the H radical.

Thus summarizing, the occurrence of the transition from the weak to strong explosion is mainly a thermal phenomenon, since it is the temperature that changes the balance between the various elementary reactions in the explosion mechanism. However, the explosion mechanism is the same on both sides of this boundary and it is based primarily on the chain branching. Below the boundary, the explosion extinguishes, resulting in a weak explosion, because the temperature rise caused by the exothermicity of the process is low, and terminations can overtake branching reactions after a relatively low H_2 conversion. Above the boundary, on the other hand, the temperature rise (caused mainly by reaction 4F) is high enough so that branching reactions *always* dominate over terminations, leading to a strong (*i.e.*, complete) explosion.

8.2   Reduction of Detailed Kinetic Models

Another relevant topic in application of sensitivity analysis is to extract important (or to eliminate unimportant) elementary reactions from a detailed kinetic model so as to obtain a *reduced model, i.e.*, a simpler model that provides essentially equivalent predictions. The decision whether or not an elementary reaction is included in a reduced model can be taken based on whether or not the sensitivity value of the specific objective of interest with respect to this reaction is sufficiently large. In order to apply this concept in practice, we need a quantitative criterion. For this, Rota *et al.* (1994a) proposed to define an important elementary reaction as one that leads to a sensitivity value higher than a prescribed fraction, ε, of the highest sensitivity value among all the elementary reactions in the detailed kinetic model. In other words, the following condition should be satisfied:

$$|S(I; A_k)| \geq \varepsilon \cdot \max[|S(I; A_k)|, \quad \text{for } k = 1, 2, 3 \ldots] \tag{8.10}$$

where I is the chosen objective for the sensitivity analysis. It is clear that the number of important elementary reactions included in the reduced model depends upon the value chosen for ε. In particular, for $\varepsilon = 0$, all the elementary reactions in the detailed kinetic model are included, while for $\varepsilon = 1$ the reduced model contains only one reaction, *i.e.*, the one leading to the highest sensitivity. Thus, ε is an adjustable parameter that can be used to tune the accuracy of the reduced kinetic model. We refer to it in the following as the *reduction factor*.

It should be noted that the necessary condition for a reaction to be omitted in a reduced mechanism is that it not be a main reaction path. The sensitivity analysis can identify only the rate-limiting steps, but not all the main reaction paths. For example, β decompositions in hydrocarbon combustion are not rate limiting (thus leading to low sensitivity values), but they cannot be omitted in the reduced mechanism since they are main paths. Thus, it is generally recommended that kinetic model reduction through sensitivity analysis be accompanied by the reaction flux analysis (cf. Yetter *et al.*, 1991). A reaction can be omitted when it results in small values of both reaction flux and objective sensitivity. In the following, we focus on the role of sensitivity analysis in the determination of reduced models. Actually, in some cases, such as the two examples discussed below, objective sensitivity analysis alone is sufficient to obtain a reduced mechanism. The first of these involves use of the explosion temperature as the objective of the sensitivity analysis, which is a particularly convenient choice for model reduction as explained in detail in the context of Example 8.3.

Example 8.7 Minimum reduced kinetic model for the explosion limits of hydrogen–oxygen systems. The three explosion limits and the WSEB for the hydrogen–oxygen system are shown in Fig. 8.7. Their sensitivity analysis with respect to each elementary reaction of the detailed kinetic model in Table 7.1 was performed in Examples 8.3 to 8.6. Recall that the important reactions vary with the explosion limit examined. Thus, in order to derive a reduced kinetic model that can describe all three explosion limits as well as the WSEB, we need to include all the important reactions identified by the individual sensitivity analyses.

Let us consider the reduction factor value, $\varepsilon = 0.01$. With this, we first identify the important reactions for each explosion limit, as well as for the WSEB, and then combine all of them to form the reduced model shown in Table 8.5, which contains 22 elementary reactions out the 53 listed in Table 7.1. For comparison, the three explosion limits and the WSEB have been recomputed using this reduced kinetic model, as shown in Fig. 8.14. The results are very close to those obtained with the original detailed model, to the extent that the two curves plotted in Fig. 8.14 are superimposed. Further, a comparison between the explosion temperatures predicted using the reduced and detailed kinetic models is shown in Table 8.6 for various pressure values. It appears that in all cases the difference between the predictions of the two models involves at most the fourth significant digit. Therefore, we can conclude that the reduced kinetic model including 22 reactions (Table 8.5) can well replace the detailed model (Table 7.1) for describing the H_2–O_2 explosion limits.

Table 8.7 compares the critical initial pressures for the weak-strong explosion transition (WSEB) computed using the reduced and detailed kinetic models, for various initial temperature values. It appears that the difference between the two models is

Table 8.5. Reduced kinetic model for the explosion limits in the H_2–O_2 system with $\varepsilon = 0.01$

No.	Reaction	Note[a]	Action
21F	$H_2 + O_2 \rightarrow OH + OH$	1, 2, 3	initiation
4F	$OH + H_2 \rightarrow H_2O + H$	1, 3, 4	propagation
2F	$H + O_2 \rightarrow OH + O$	1, 2, 3, 4	branching
3F	$O + H_2 \rightarrow OH + H$	1, 3, 4	branching
10F	$H + HO_2 \rightarrow OH + OH$	1, 2, 3, 4	H
11F	$H + HO_2 \rightarrow H_2O + O$	2, 3, 4	H
1F	$H + HO_2 \rightarrow H_2 + O_2$	1, 2, 3, 4	H
17F	$H + H_2O_2 \rightarrow H_2O + OH$	3, 4	H
16F	$H + H_2O_2 \rightarrow H_2 + HO_2$	3, 4	H
16R	$HO_2 + H_2 \rightarrow H + H_2O_2$	2, 3	HO_2
14F	$HO_2 + HO_2 \rightarrow H_2O_2 + O_2$	1, 2, 3	HO_2
23F	$HO_2 + H_2 \rightarrow H_2O + OH$	2, 3	HO_2
13F	$HO_2 + OH \rightarrow H_2O + O_2$	3	HO_2
19F	$H_2O_2 + OH \rightarrow H_2O + HO_2$	3	H_2O_2
18F	$H_2O_2 + O \rightarrow OH + HO_2$	3, 4	H_2O_2
15F	$H_2O_2 + M \rightarrow OH + OH + M$	3	H_2O_2
15R	$OH + OH + M \rightarrow H_2O_2 + M$	4	termination
9F	$H + O_2 + M \rightarrow HO_2 + M$	1, 2, 3, 4	termination
25F	$H \xrightarrow{wall} 0.5H_2$	1, 2, 4	termination
26F	$O \xrightarrow{wall} 0.5O_2$	1	termination
27F	$OH \xrightarrow{wall} 0.5H_2O + 0.25O_2$	1	termination
29F	$H_2O_2 \xrightarrow{wall} H_2 + O_2$	3	termination

[a] 1, 2, 3, and 4 indicate the important reactions for the first, second, and third explosion limits and the weak-strong explosion boundary, respectively.

somewhat larger than that observed in Table 8.6 for the explosion limits. Nevertheless, this is still a very small difference, involving at most the third significant digit.

Further reduced models may be constructed using larger ε values. Tables 8.8, 8.9, 8.10, and 8.11 give the reduced kinetic models containing 16, 15, 13, and 10 elementary reactions, obtained using values of the reduction factor, $\varepsilon = 0.05, 0.1, 0.13$, and 0.15, respectively. For the first three cases, i.e., $\varepsilon = 0.05, 0.1$, and 0.13, the explosion limits and WSEB predicted using the reduced kinetic models are shown in Fig. 8.14. As may be seen, for $\varepsilon < 0.13$, all the reduced kinetic models give practically the same results. For $\varepsilon = 0.13$, some significant error with respect to the detailed model arises, but the reduced model can still predict the inverse-S shape of the explosion boundaries, indicating the existence of the three explosion limits. Thus the model behavior remains correct at least in qualitative terms. For reduction factors, $\varepsilon \geq 0.15$, the reduced model becomes too inaccurate to be useful. For example, the reduced model derived for $\varepsilon = 0.15$, summarized in Table 8.11, does not even predict the existence of three explosion limits. Moreover, as seen from Table 8.11, as a consequence of the reaction elimination procedure, reaction 2F in this model produces the O radical, but there is

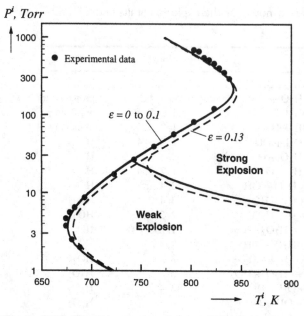

Figure 8.14. Explosion limits of the stoichiometric H_2–O_2 mixture in the pressure-temperature parameter plane, predicted by the Morbidelli-Varma generalized criterion using reduced models obtained with various values of the reduction factor, ε. (•) experimental data as in Fig. 8.7. From Wu *et al.* (1998).

no reaction to consume it! Thus, this model leads to the unrealistic result that the O radicals accumulate continuously in the system.

Thus, summarizing, we can conclude that *the minimum reduced kinetic model that can predict the occurrence of three explosion limits of the hydrogen–oxygen system is comprised of the 13 elementary reactions listed in Table 8.10, while the minimum reduced kinetic model that can reproduce the three explosion limits measured experimentally with satisfactory accuracy consists of the 15 elementary reactions, listed in Table 8.9.*

It is worth comparing the minimum reduced kinetic model in Table 8.9 with that proposed by Baldwin *et al.* (1974), which was used by Kordylewski and Scott (1984) to study the second and third limits as discussed in the examples above. Let us first note that, since this mechanism was devised to predict the second and third limits, it does not include the two termination reactions 25F and 26F, which are important only for the first explosion limit (see Example 8.3), and reaction 15R, which as seen in Example 8.6 is important only for the WSEB. Thus, if we do not consider these three reactions, the minimum model includes 12 reactions, as opposed to the model proposed by Baldwin *et al.*, which includes 16 elementary reactions. On the other hand, all 12 reactions are also included in the Baldwin kinetic model. This indicates that the latter is indeed valid for describing the three explosion limits, although it includes four redundant reactions.

Table 8.6. Comparison between the explosion temperatures for the H_2–O_2 system at various pressure values, predicted using the reduced model in Table 8.5 and the detailed model in Table 7.1

P^i (Torr)	T_e^i, K	
	Reduced model	Detailed model
1.000	721.34	721.50
1.259	706.97	707.25
1.585	694.95	695.29
1.995	685.55	685.91
2.512	679.04	679.39
3.162	675.60	675.91
3.981	675.13	675.39
5.012	677.36	677.57
6.310	681.88	682.05
7.943	688.26	688.40
10.00	696.12	696.23
15.85	715.10	715.20
25.12	737.19	737.29
39.81	761.54	761.67
63.10	787.68	787.86
100.0	814.50	814.78
125.9	826.94	827.27
158.5	836.87	837.23
199.5	842.25	842.61
251.2	842.04	842.40
316.2	836.71	836.99
398.1	827.47	827.70
501.2	815.67	815.84
631.0	802.41	802.53
794.3	788.46	788.54
1000	774.37	774.41

Table 8.7. For the H_2–O_2 system, comparison between the critical initial pressure for the transition between weak and strong explosion regions at various temperature values, predicted using the reduced model in Table 8.5 and the detailed model in Table 7.1

T^i (K)	P_{WSEB}^i (Torr)	
	Reduced model	Detailed model
900	6.687	6.814
850	8.451	8.584
800	11.64	11.79
750	21.80	22.24
743	25.82	26.61

Table 8.8. Reduced kinetic model for the ignition phenomena in the H_2-O_2 system with $\varepsilon = 0.05$

No.	Reaction	Note[a]	Action
21F	$H_2 + O_2 \rightarrow OH + OH$	3	initiation
4F	$OH + H_2 \rightarrow H_2O + H$	3, 4	propagation
2F	$H + O_2 \rightarrow OH + O$	1, 2, 3, 4	branching
3F	$O + H_2 \rightarrow OH + H$	1	branching
10F	$H + HO_2 \rightarrow OH + OH$	2, 3	H
17F	$H + H_2O_2 \rightarrow H_2O + OH$	3	H
16F	$H + H_2O_2 \rightarrow H_2 + HO_2$	3, 4	H
16R	$HO_2 + H_2 \rightarrow H + H_2O_2$	3	HO_2
14F	$HO_2 + HO_2 \rightarrow H_2O_2 + O_2$	2, 3	HO_2
23F	$HO_2 + H_2 \rightarrow H_2O + OH$	3	HO_2
19F	$H_2O_2 + OH \rightarrow H_2O + HO_2$	3	H_2O_2
15F	$H_2O_2 + M \rightarrow OH + OH + M$	3	H_2O_2
15R	$OH + OH + M \rightarrow H_2O_2 + M$	4	termination
9F	$H + O_2 + M \rightarrow HO_2 + M$	1, 2, 3, 4	termination
25F	$H \xrightarrow{\text{wall}} 0.5H_2$	1	termination
26F	$O \xrightarrow{\text{wall}} 0.5O_2$	1	termination

[a] 1, 2, 3, and 4 indicate the important reactions for the first, second, and third explosion limits and the weak-strong explosion boundary, respectively.

Table 8.9. Reduced kinetic model for the ignition phenomena in the H_2-O_2 system with $\varepsilon = 0.1$

No.	Reaction	Note[a]	Action
21F	$H_2 + O_2 \rightarrow OH + OH$	3	initiation
4F	$OH + H_2 \rightarrow H_2O + H$	3, 4	propagation
2F	$H + O_2 \rightarrow OH + O$	1, 2, 3, 4	branching
3F	$O + H_2 \rightarrow OH + H$	1	branching
10F	$H + HO_2 \rightarrow OH + OH$	2	H
17F	$H + H_2O_2 \rightarrow H_2O + OH$	3	H
16F	$H + H_2O_2 \rightarrow H_2 + HO_2$	3	H
16R	$HO_2 + H_2 \rightarrow H + H_2O_2$	3	HO_2
14F	$HO_2 + HO_2 \rightarrow H_2O_2 + O_2$	3	HO_2
19F	$H_2O_2 + OH \rightarrow H_2O + HO_2$	3	H_2O_2
15F	$H_2O_2 + M \rightarrow OH + OH + M$	3	H_2O_2
15R	$OH + OH + M \rightarrow H_2O_2 + M$	4	termination
9F	$H + O_2 + M \rightarrow HO_2 + M$	1, 2, 4	termination
25F	$H \xrightarrow{\text{wall}} 0.5H_2$	1	termination
26F	$O \xrightarrow{\text{wall}} 0.5O_2$	1	termination

[a] 1, 2, 3, and 4 indicate the important reactions for the first, second, and third explosion limits and the weak-strong explosion boundary, respectively.

Table 8.10. Reduced kinetic model for the ignition phenomena in the H_2–O_2 system with $\varepsilon = 0.13$

No.	Reaction	Note[a]	Action
21F	$H_2 + O_2 \rightarrow OH + OH$	3	initiation
4F	$OH + H_2 \rightarrow H_2O + H$	3, 4	propagation
2F	$H + O_2 \rightarrow OH + O$	1, 2, 3, 4	branching
3F	$O + H_2 \rightarrow OH + H$	1	branching
17F	$H + H_2O_2 \rightarrow H_2O + OH$	3	H
16R	$HO_2 + H_2 \rightarrow H + H_2O_2$	3	HO_2
14F	$HO_2 + HO_2 \rightarrow H_2O_2 + O_2$	3	HO_2
19F	$H_2O_2 + OH \rightarrow H_2O + HO_2$	3	H_2O_2
15F	$H_2O_2 + M \rightarrow OH + OH + M$	3	H_2O_2
15R	$OH + OH + M \rightarrow H_2O_2 + M$	4	termination
9F	$H + O_2 + M \rightarrow HO_2 + M$	1, 2, 4	termination
25F	$H \xrightarrow{\text{wall}} 0.5H_2$	1	termination
26F	$O \xrightarrow{\text{wall}} 0.5O_2$	1	termination

[a] 1, 2, 3, and 4 indicate the important reactions for the first, second, and third explosion limits and the weak-strong explosion boundary, respectively.

Table 8.11. Reduced kinetic model for the ignition phenomena in the H_2–O_2 system with $\varepsilon = 0.15$

No.	Reaction	Note[a]	Action
21F	$H_2 + O_2 \rightarrow OH + OH$	3	initiation
4F	$OH + H_2 \rightarrow H_2O + H$	3, 4	propagation
2F	$H + O_2 \rightarrow OH + O$	1, 2, 3, 4	branching
17F	$H + H_2O_2 \rightarrow H_2O + OH$	3	H
16R	$HO_2 + H_2 \rightarrow H + H_2O_2$	3	HO_2
19F	$H_2O_2 + OH \rightarrow H_2O + HO_2$	3	H_2O_2
15F	$H_2O_2 + M \rightarrow OH + OH + M$	3	H_2O_2
15R	$OH + OH + M \rightarrow H_2O_2 + M$	4	termination
9F	$H + O_2 + M \rightarrow HO_2 + M$	1, 2, 4	termination
25F	$H \xrightarrow{\text{wall}} 0.5H_2$	1	termination

[a] 1, 2, 3, and 4 indicate the important reactions for the first, second, and third explosion limits and the weak-strong explosion boundary, respectively.

An important aspect that should be stressed explicitly is that since the minimum model in Table 8.9 has been obtained through the sensitivity analysis of the three explosion limits and WSEB, it should be used only for describing these; *i.e.*, reduced models constructed with this procedure depend on the chosen objective, and therefore they can be used to replace the detailed model only when simulating that specific objective. Caution should be applied when the reduced model is used for a different purpose. An example is shown in Fig. 8.15, where, for given values of the initial temperature and pressure located within the strong explosion region, we compare the

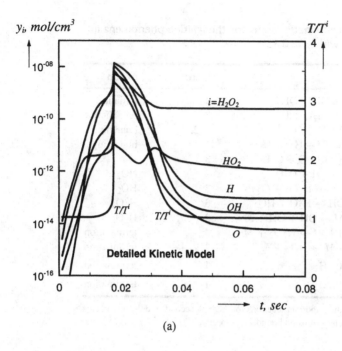

(a)

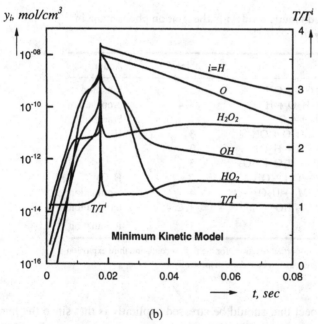

(b)

Figure 8.15. Species concentrations and temperature as functions of time computed with (a) the detailed model in Table 7.1 and (b) the minimum reduced model in Table 8.9. $T^i = 800$ K; $P^i = 20$ Torr; $R_v = 3.7$ cm; $U = 8.0 \times 10^{-4}$ cal/cm^2/s/K. From Wu *et al.* (1998).

evolution of the concentrations of some species and temperature with time computed with the detailed (a) and the minimum (b) kinetic models. It is seen that before reaching the temperature maximum, both models give similar predictions, but after the concentration and temperature values predicted by the two models are substantially different.

Example 8.8 Reduced kinetic model for the combustion of methane–ethane systems. In order to describe the combustion process in $CH_4-C_2H_6-O_2$ systems, several detailed kinetic models have been proposed in the literature (c.f. Miller and Bowman, 1989; Dagaut et al., 1991; Kilpinen et al., 1992). Some of these were compared with experimental data taken in a well-mixed flow reactor (Rota et al., 1994b), and it was found that the Kilpinen-Glarborg-Hupa (KGH) model, including 225 elementary reactions and 50 chemical species, exhibits the best agreement. In the same work, a new reaction accounting for the direct oxidation of the radical C_2H_3,

$$C_2H_3 + O_2 \rightarrow CH_2O + HCO$$

was introduced in order to better reproduce the acetylene experimental data. This leads to the so-called modified Kilpinen-Glarborg-Hupa (MKGH) model, including 226 elementary reactions and 50 chemical species.

   This detailed kinetic model can be used in the simulation of ideal reactors, such as well-mixed or plug-flow reactors, but it becomes cumbersome when simulating real reactors, where the interaction between fluid dynamics and chemical reactions has to be taken into account. Thus, a general procedure has been developed (Rota et al., 1994a) to derive a reduced kinetic model from the MKGH model, which allows correct prediction of the concentrations of the eight important chemical species (C_2H_6, C_2H_4, C_2H_2, CH_4, O_2, H_2, CO, and CO_2) for the range of operating conditions ($T = 990$ to 1100 K, $R_{fO_2} = 0.24$ to 1.5), where

$$R_{fO_2} = \frac{\text{fuel/O}_2}{(\text{fuel/O}_2)_{\text{stoichiometric}}} \tag{E8.11}$$

is the normalized fuel/O_2 ratio. The procedure involves two main steps. In the first, a reduced kinetic model for predicting the concentrations of the selected eight species is derived using objective sensitivity analysis and criterion (8.10), similar to the previous example. In the second step, a new reduced model is derived by including all elementary reactions that are responsible for the formation of each species. For this, the following criterion is adopted:

$$|r_{ij}| \geq \omega \cdot \max(|r_{ij}|, \text{ for } j = 1, 2, 3, \ldots) \tag{E8.12}$$

where r_{ij} is the production (or consumption) rate of the ith species due to the jth reaction and ω is another reduction factor, having a meaning similar to that of ε in Eq. (8.10). It may be noted that the first step accounts for the normalized sensitivity of the different chemical species to the various pre-exponential factors, while the second step accounts for the absolute production (or consumption) rate of each species.

In general, before carrying out the sensitivity analysis, some preliminary observations about the operating conditions can be made to reduce the computational effort. In this case, it is clear that for $R_{fO_2} < 1$, we have an oxidizing system, while for $R_{fO_2} > 1$, it is reducing. Thus, $R_{fO_2} = 1$ can be considered as a boundary, which contains the relevant information about both situations. This allows us to limit the sensitivity analysis only to the case $R_{fO_2} = 1$.

On the other hand, in most cases it is difficult to select only one temperature to represent a large range of temperature values for the sensitivity analysis, since different reactions often prevail at different temperatures. Thus, in the example considered here the temperature interval was divided and the sensitivity analysis was performed for 12 different temperature values.

The procedure for deriving the reduced kinetic model, proposed by Rota *et al.*, can be illustrated as follows:

(1) *For each temperature value, derive the reduced kinetic model for predicting the concentrations of the selected chemical species.* This is done following the procedure illustrated in the previous example. First, we extract from the detailed model eight subsets of important reactions through the sensitivity analysis of the concentration of each selected species with respect to all the elementary reaction rates in the detailed kinetic model. This uses criterion (8.10), where I is the concentration of one of a species. The value of the reduction factor, ε in Eq. (8.10), is chosen as the best compromise between a substantial reduction in the size of the kinetic model and a good agreement between the predictions of the reduced and detailed models. The combination of the eight subsets of reactions thus obtained forms the first reduced kinetic model.

(2) *For each temperature value, derive the reduced kinetic model that includes important reactions responsible for formation of the selected eight species, based on the criterion (E8.12).* This yields eight reaction subsets, and their union gives the second reduced kinetic model.

(3) *The union of the two reduced models obtained in steps (1) and (2) above constitutes the reduced model for the considered temperature value. By repeating this procedure for all the 12 temperature values and combining all the obtained reduced models, we get the final reduced model valid in the desired range of operating conditions.*

Following this procedure, with the reduction factors $\varepsilon = 0.01$ and $\omega = 0.1$, Rota *et al.* (1994b) derived a reduced model that includes 61 elementary reactions and 25 chemical species. A comparison between the concentrations of the considered eight species in the outlet stream of a well-mixed reactor, predicted by the detailed and reduced kinetic models, is shown in Fig. 8.16. It is apparent that the two models are in good agreement.

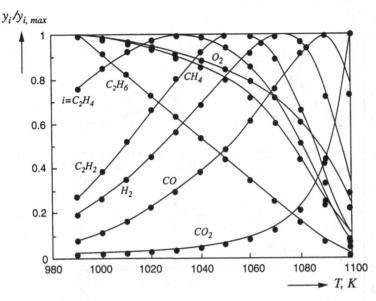

Figure 8.16. Concentrations of various chemical species in the outlet stream of a perfectly stirred reactor as functions of temperature, computed by the detailed ($\bullet$) and the reduced (solid curves) models. Combustion of the system $CH_4 - C_2H_6 - O_2$ with the normalized fuel/O_2 ratio, $R_{fO_2} = 1$. From Rota *et al.* (1994b).

The reduced model obtained above was derived for the normalized fuel/O_2 ratio, $R_{fO_2} = 1$. Let us now verify if this model can be applied to the entire range of R_{fO_2} values. For this, we consider the outlet concentrations of CO corresponding to the smallest and the greatest R_{fO_2} values studied, *i.e.*, $R_{fO_2} = 0.24$ and 1.5. The results obtained using the two models are in Fig. 8.17. From this, it can be concluded that the reduced model provides reliable predictions of the chemical species concentrations in the entire range of operating conditions.

Finally, it should be mentioned that in addition to the sensitivity-based methods described in Examples 8.7 and 8.8, other methodologies have also been proposed in the literature for kinetic model reduction. Most of these have been reviewed by Griffiths (1995) and Tomlin *et al.* (1996). Differences among the methods can be either their reduction strategy or applied technique. Some investigators (Frenklach *et al.*, 1986) aim at reducing directly the number of elementary reactions and indirectly the number of chemical species, while others aim at obtaining a reduced mechanism with a minimum number of species to describe the chemistry. Techniques used by different authors may involve different system properties, such as quasi steady state (Peters, 1991), the eigenvalues of the system Jacobian (Lam and Goussis, 1994), and the global reaction behavior (Jiang *et al.*, 1995). Details about these methods are not given here, since in the present book we discuss mainly the various applications of sensitivity concepts, and interested readers may refer directly to the original papers.

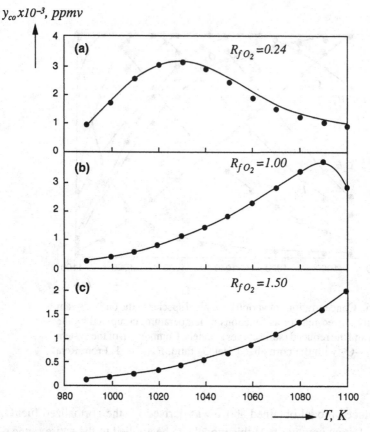

Figure 8.17. Concentration of CO in the outlet steam of a perfectly stirred reactor as a function of temperature, calculated by the detailed (●) and the reduced (solid curve) models. Combustion of the system $CH_4-C_2H_6-O_2$ for three values of the normalized fuel/O_2 ratio, R_{fO_2}. From Rota *et al.* (1994b).

References

Baldwin, R. R., Fuller, M. E., Hillman, J. S., Jackson, D., and Walker, R. W. 1974. Second limit of hydrogen-oxygen mixture: the reaction $H + HO_2$. *J. Chem. Soc. Faraday Trans.* **14**, 635.

Dagaut, P., Cathonnet, M., and Boettner, J. C. 1986. Kinetics of ethane oxidation. *Int. J. Chem. Kinet.* **23**, 437.

Dixon-Lewis, G., and Williams, D. J. 1977. The oxidation of hydrogen and carbon monoxide. In *Comprehensive Chemical Kinetics*, C. H. Bamford and C. F. H. Tipper, eds., Vol. 17, p. 1. Amsterdam: Elsevier.

Dougherty, E. P., and Rabitz, H. 1980. Computational kinetics and sensitivity analysis of hydrogen-oxygen combustion. *J. Chem. Phys.* **72**, 6571.

Edelson, D. 1981. Mechanistic details of the Belousov-Zhabotinsky oscillations. IV. Sensitivity analysis. *Int. J. Chem. Kinet.* **13**, 1175.

Edelson, D., Noyes, R. M., and Field, R. J. 1979. Mechanistic details of the Belousov-Zhabotinsky oscillations. II. The organic reaction subset. *Int. J. Chem. Kinet.* **11**, 155.

Edelson, D., and Thomas, V. M. 1981. Sensitivity analysis of oscillating reactions. 1. The period of the Oregonator. *J. Phys. Chem.* **85**, 1555.

Field, R. J., Koros, E., and Noyes, R. M. 1972. Oscillations in chemical systems: II. Thorough analysis of temporal oscillations in the bromate-cerium-malonic acid system. *J. Am. Chem. Soc.* **94**, 8649.

Foo, K. K., and Yang, C. H. 1971. On the surface and thermal effects on hydrogen oxidation. *Combust. Flame* **17**, 223.

Frenklach, M., Kailasanath, K., and Oran, E. S. 1986. Systematic development of reduced reaction mechanisms for dynamic modeling. *Prog. Astronaut. Aeronaut.* **105**, 365.

Griffiths, J. F. 1995. Reduced kinetic models and their application to practical combustion systems. *Prog. Energy Combust. Sci.* **21**, 25.

Griffiths, J. F., Scott, S. K., and Vandamme, R. 1981. Self-heating in the $H_2 + O_2$ reaction in the vicinity of the second explosion limit. *J. Chem. Soc. Faraday Trans. I* **77**, 2265.

Hinshelwood, C. N., and Williamson, A. T. 1934. *The Reaction Between Hydrogen and Oxygen*. Oxford: Oxford University Press.

Hwang, J. T. 1982. On the proper usage of sensitivities of chemical kinetics models to the uncertainties in rate coefficients. *Proc. Nat. Sci. Counc. B. ROC* **6**, 270.

Jiang, B., Ingram, D., Causon, D., and Saunders, R. 1995. A global simulation method for obtaining reduced reaction mechanisms for use in reactive blast wave flows. *Shock Waves* **5**, 81.

Jost, W. 1946. *Explosion and Combustion Processes in Gases*. New York: McGraw-Hill.

Kilpinen, P., Glarborg, P., and Hupa, M. 1992. Reburning chemistry: a kinetic modeling study. *Ind. Eng. Chem. Res.* **31**, 1477.

Kordylewski, W., and Scott, S. K. 1984. The influence of self-heating on the second and third explosion limits in the $O_2 + H_2$ reaction. *Combust. Flame* **57**, 127.

Lam, S. H., and Goussis, D. 1994. The CSP method for simplifying kinetics. *Int. J. Chem. Kinet.* **26**, 461.

Lewis, B., and von Elbe, G. 1961. *Combustion, Flames and Explosions of Gases*. New York: Academic.

Maas, U., and Warnatz, J. 1988. Ignition processes in hydrogen-oxygen mixtures. *Combust. Flame* **53**, 74.

Miller, J. A., and Bowman, C. T. 1989. Mechanism and modeling of nitrogen chemistry in combustion. *Prog. Energy Combust. Sci.* **15**, 287.

Minkoff, G. J., and Tipper, C. F. H. 1962. *Chemistry of Combustion Reactions*. London: Butterworths.

Morbidelli, M., and Wu, H. 1992. Critical transitions in reacting systems through parametric sensitivity. In *From Molecular Dynamics to Combustion Chemistry*, S. Carrà and N. Rahman, eds., p. 117. Singapore: World Scientific.

Morbidelli, M., and Varma, A. 1988. A generalized criterion for parametric sensitivity: application to thermal explosion theory. *Chem. Eng. Sci.* **43**, 91.

Peters, N. 1991. *Systematic Reduction of Flame Kinetics: Principles and Details* (Pitman Research Notes in Mathematics Series, Vol. 223). London: Longman Scientific & Technical.

Rota, R., Bonini, F., Servida, A., Morbidelli, M., and Carrà, S. 1994a. Analysis of detailed kinetic models for combustion processes: application to a methane-ethane mixture. *Chem. Eng. Sci.* **49**, 4211.

Rota, R., Bonini, F., Servida, A., Morbidelli, M., and Carrà, S. 1994b. Validation and updating of detailed kinetic mechanisms: the case of ethane oxidation. *Ind. Eng. Chem. Res.* **33**, 2540.

Semenov, N. N. 1959. *Some Problems of Chemical Kinetics and Reactivity*. London: Pergamon.

Tomlin, A. S., Turanyi, T., and Pilling, M. J. 1996. Mathematical tools for the construction, investigation and reduction of combustion mechanisms, In *Oxidation Kinetics and Autoignition of Hydrocarbons*, M. J. Pilling, ed. Amsterdam: Elsevier.

Turanyi, T., Berces, T., Vajda S., 1989. Reaction rate analysis of complex kinetic systems. *Int. J. Chem. Kinet.* **21**, 83.

Warnatz, J. 1984. Rate coefficients in the C/H/O system. In *Combustion Chemistry*, W. C. Gardiner, Jr., ed., p. 197. New York: Springer-Verlag.

Westbrook, C. K., Creighton, J., Lund, C., and Dryer, F. L. 1977. A numerical model of chemical kinetics of combustion in a turbulent flow reactor. *J. Phys. Chem.* **81**, 2542.

Willbourn, A. H., and Hinshelwood, C. N. 1946. The mechanism of the hydrogen-oxygen reaction: III. The influence of salts. *Proc. R. Soc. London A* **185**, 376.

Wu, H., Cao, G., and Morbidelli, M. 1993. Parametric sensitivity and ignition phenomena in hydrogen-oxygen mixtures. *J. Phys. Chem.* **97**, 8422.

Wu, H., Rota, R., Morbidelli, M., and Varma, A. 1998. Hydrogen-oxygen explosion mechanism and model reduction through sensitivity analysis. *Int. J. Chem. Kinet.* Submitted.

Yetter, R. A., Dryer, F. L., and Rabitz, H. 1985. Some interpretive aspects of elementary sensitivity gradients in combustion kinetics modeling. *Combust. Flame* **59**, 107.

9

Sensitivity Analysis in Air Pollution

THE PREDICTION OF POLLUTANT DISTRIBUTION in the atmosphere from emission sources and meteorological fields is a primary objective of air pollution studies. In general, this requires solving the so-called Eulerian atmospheric species diffusion-reaction equations, which describe the time evolution of the pollutant concentration in the atmosphere in the presence of wind, diffusion processes, reaction, and source terms. These models tend to be complex, involving many physicochemical parameters and complex reaction mechanisms, whose quantitative evaluation is frequently not straightforward. Thus, in order to assess the reliability of a model, it is important to evaluate the influence of uncertainties in physicochemical, kinetic, and meteorological parameters on model predictions. This can be done conveniently using sensitivity analysis, which often accompanies the solution of atmospheric diffusion-reaction models. An important problem, particularly for air quality control, is to determine the influence of a specific source on a specific target location (usually referred to as a receptor). This type of information is not given directly by the solution of the Eulerian model, but can be obtained from sensitivity analysis.

The sensitivity analysis of model predictions with respect to uncertainties in input parameters has been described in previous chapters, where we examined local sensitivities that account for small one-at-a-time parameter variations. In order to investigate air pollution problems noted above, we first need to introduce some new concepts of sensitivity analysis. In particular, in Section 9.2, we illustrate a specific technique that is well suited to evaluate the relations between receptor and emission sources, where the latter may vary in space and time. This is still a type of local sensitivity analysis, but with the sensitivity definition based on functional rather than partial derivatives. As an example, the receptor-to-source sensitivities for the emission sources in the eastern United States during specific meteorological conditions are examined. Next, in Section 9.3, applications of global sensitivity analysis described earlier in Chapter 2 are discussed. In this context, we report a full-scale sensitivity analysis of a model for photochemical air pollution, with respect to simultaneous, large variations in meteorological and emission parameters.

9.1 Basic Equations

Air pollution processes involving transport of chemically reactive species can be described by the Eulerian model, *i.e.*, the three-dimensional advection-diffusion equation (Seinfeld, 1986):

$$\frac{\partial C_i}{\partial t} + \nabla \cdot (v C_i) = \nabla \cdot (K \cdot \nabla C_i) + R_i + E_i; \qquad i = 1, 2, \ldots, q \tag{9.1}$$

where C_i is the gas-phase concentration of the ith species among a total number q, v is the wind velocity vector, K is the eddy diffusivity tensor, R_i is the rate of formation of the ith species by chemical reactions, and E_i is the rate of emission of the ith species from sources. This equation applies to all the chemical species involved. However, short-lived species, such as radicals, are most conveniently handled by using the quasi-steady-state approximation, where Eq. (9.1) is replaced by the algebraic equation $R_i = 0$.

When Eq. (9.1) is applied to a region of space enclosed by the earth surface, $z = h(x, y)$, and some prescribed height, $z = H(x, y)$, the boundary conditions (BCs) can be given as follows (Carmichael *et al.*, 1986):

$$n \cdot (v C_{i,b} - K \cdot \nabla C_i) = F_{i,e}, \qquad \text{for inflow (i.e., } n \cdot v < 0) \tag{9.2}$$

$$[n \cdot v C_i - n \cdot (K \cdot \nabla C_i)]_{t+\Delta t} = [n \cdot v C_i - n \cdot (K \cdot \nabla C_i)]_t,$$
$$\text{for outflow (i.e., } n \cdot v \geq 0) \tag{9.3}$$

$$n_h (K \cdot \nabla C_i) = v_{i,d} C_i - Q_i, \qquad \text{at the earth's surface} \tag{9.4}$$

where n is an outward unit vector normal to the top boundary, n_h is the inward vector normal to the earth's surface, $C_{i,b}$ is the concentration of species i outside the region, $F_{i,e}$ is a prescribed flux of the same species entering the region, Q_i is the surface emission rate, and $v_{i,d}$ is the deposition velocity of species i. For given values of the emission and meteorological fields, Eqs. (9.1) to (9.4) can be solved to yield the distribution of species concentrations in the entire region of interest. For this a numerical method has been proposed by Carmichael *et al.* (1986), the so-called locally one-dimensional finite element method, which combines the concepts of fractional time steps and one-dimensional finite elements. Specifically, this procedure (Mitchell, 1969) reduces the multidimensional partial differential equations into time-dependent, one-dimensional transport equations. These are then solved using a Crank-Nicolson-Galerkin finite element technique.

By introducing a new coordinate system moving with the mean horizontal wind velocities (*i.e.*, $\bar{v}_x$ and $\bar{v}_y$), the above Eulerian atmospheric diffusion equation is transformed into the Lagrangian *trajectory model*. In particular, by defining the two new variables: $\xi = x - \bar{v}_x t$ and $\eta = y - \bar{v}_y t$, the Lagrangian version of Eq. (9.1) is given by (Tilden and Seinfeld, 1982)

$$\frac{\partial C_i}{\partial t} + \left(\bar{v}_x - v_x \frac{\partial \xi}{\partial x}\right)\frac{\partial C_i}{\partial \xi} + \left(\bar{v}_y - v_y \frac{\partial \eta}{\partial y}\right)\frac{\partial C_i}{\partial y} + v_z \frac{\partial C_i}{\partial z}$$

$$= \frac{\partial}{\partial \xi}\left(K_{xx}\frac{\partial C_i}{\partial \xi}\right) + \frac{\partial}{\partial \eta}\left(K_{yy}\frac{\partial C_i}{\partial \eta}\right) + \frac{\partial}{\partial z}\left(K_{zz}\frac{\partial C_i}{\partial z}\right) + R_i + E_i \qquad (9.5)$$

If the last three terms on the left-hand side and the first two on the right-hand side of Eq. (9.5) are neglected (see Liu and Seinfeld, 1975), the so-called standard trajectory model formulation is obtained:

$$\frac{\partial C_i}{\partial t} = \frac{\partial}{\partial z}\left(K_{zz}\frac{\partial C_i}{\partial z}\right) + R_i + E_i \qquad (9.6)$$

which is widely used in applications as discussed in detail by Liu and Seinfeld (1975). Moreover, if a new vertical coordinate is introduced in Eq. (9.6): $\rho = z - h(\xi, \eta)$, where $h(\xi, \eta)$ is the surface elevation, it follows that

$$\frac{\partial C_i}{\partial t} = \frac{\partial}{\partial \rho}\left(K_{zz}\frac{\partial C_i}{\partial \rho}\right) + R_i + E_i \qquad (9.7)$$

with BCs,

$$v_{i,d} \cdot C_i - K_{zz}\frac{\partial C_i}{\partial \rho} = F_{i,g} \quad \text{at } \rho = 0 \qquad (9.8a)$$

$$-K_{zz}\frac{\partial C_i}{\partial \rho} = F_{i,e} \qquad \text{at } \rho = H(\xi, \eta) - h(\xi, \eta) \qquad (9.8b)$$

where $H(\xi, \eta)$ is the elevation at the top of the region under examination and $F_{i,g}$ and $F_{i,e}$ are the fluxes of species i through the lower and upper boundaries, respectively.

Equation (9.7) with the BCs (9.8) can be solved by dividing the region $\rho \in (0, H-h)$ into N cells, i.e., converting Eqs. (9.7) and (9.8) into N coupled ordinary differential equations for the average cell concentrations $C_i^1, C_i^2, \ldots, C_i^N$.

It should be noted that the validity of the above trajectory model is restricted by the approximations introduced in its derivation. For example, in cases where high wind shear, horizontal inhomogeneity in wind fields, and source distributions are present, significant deviations between the trajectory model predictions and pollutant concentrations observed in the atmosphere should be expected. In addition, the use of the trajectory model for calculations with travel times of more than a few hours is generally not recommended.

The principal advantages of the trajectory model versus the two- or three-dimensional models are its relatively small requirements for data and computational efforts. However, these advantages are often overcome by the fact that the trajectory model needs to be solved several times in order to properly cover the region of interest with a sufficient number of trajectories.

In the following sections, we illustrate two examples of sensitivity analysis of air quality. In the first one, the Eulerian model is used to investigate the receptor-to-source sensitivities in the eastern United States, while in the second we compute the sensitivity of the calculated trajectory occurring in the South Coast Air Basin of California with respect to simultaneous variations in the emission and meteorological parameters.

9.2 Sensitivity Analysis of Regional Air Quality with Respect to Emission Sources

There are several methods to obtain information about the influence of a given source on a specific target location or region (*i.e.*, receptor). Hsu and Chang (1987) developed a method for determining *source-receptor relationships* in Eulerian models by including a distinct time-dependent carrier signal on individual sources and decomposing the signal of the pollutant concentrations at specific receptor sites. This method requires accurate numerical techniques such that the computed concentrations can reflect small changes in the emission. Kleinman (1988) proposed to label sources by adding additional conservation equations for species emitted from different sources. In this way, the influence of individual sources on the secondary pollutant at specific receptor sites can be investigated. Cho *et al.* (1988) investigated the relationships between sources and regional air quality through sensitivity analysis. They also developed general techniques that allow one to calculate efficiently the sensitivities of individual species at specific target locations with respect to spatially distributed emissions.

In this section we describe the techniques developed by Cho *et al.* (1988) and their application to calculate emission sensitivities for sources in the eastern United States during specific meteorological conditions.

9.2.1 Definition of Sensitivities

Receptor-to-source point sensitivity

Let us consider a particular emission of the jth chemical species, $E_j(x', y', z', t')$. Its influence on the concentration of the ith chemical species, $C_i(x, y, z, t)$, generated by the jth species through direct or indirect reactions, can be expressed as follows:

$$s(C_i; E_j) = \frac{\delta C_i(x, y, z, t)}{\delta E_j(x', y', z', t')} \tag{9.9}$$

This quantity may be referred to as the *receptor-to-source point sensitivity* and represents the response of the concentration of species i at position (x, y, z) at time t to an *infinitesimal variation* of the emissions of species j at position (x', y', z') at time t'. In this definition, we use functional derivatives, instead of partial derivatives, because the emissions are distributed spatially and temporally within the model domain.

The receptor-to-source point sensitivity contains important information about the relation between source and receptor. A large value of $s(C_i; E_j)$ indicates that the emission of species j at (x', y', z', t') is important in determining the concentration of species i at (x, y, z, t). Moreover, if $s(C_i; E_j)$ is positive, the concentration of species i increases as the emission of the jth species increases.

Equations for describing the receptor-to-source point sensitivities may be derived by taking the first variation of Eq. (9.1), leading to

$$\frac{\partial s(C_i; E_j)}{\partial t} + \nabla \cdot (vs(C_i; E_j)) = \nabla \cdot K \cdot \nabla s(C_i; E_j) + \sum_n \frac{\partial R_i}{\partial C_n} s(C_n; E_j)$$
$$+ s(C_i; E_j)\delta(x - x')\delta(y - y')\delta(z - z')\delta(t - t')$$

$$(9.10)$$

Efficient procedures to compute these sensitivities can be found in the literature (see Cho *et al.* 1987).

Receptor-to-source region sensitivity

The receptor-to-source point sensitivity defined by Eq. (9.9) indicates the effect of an individual source at *point* (x', y', z', t') on the concentration at a specific receptor located at (x, y, z, t). In some instances, we are interested in determining how sources distributed *in a specific region* affect the concentration at a specific receptor. The variation of the concentration of species i due to the variation of the emission of the jth species distributed in a particular region can be determined by integrating the receptor-to-source point sensitivity in the entire region, *i.e.*,

$$\delta C_i(x, y, z, t; V', t') = \int_{V'} s(C_i; E_j)\delta E_j(x', y', z', t')\, dV' \qquad (9.11)$$

where $V' = (a_1 < x' < b_1, a_2 < y' < b_2, a_3 < z' < b_3)$ is the volume of the source region of interest. When the variation $\delta E_j(x', y', z', t')$ is a constant, the sensitivity of the concentration of species i to the emission of species j in a chosen region, referred to as the *receptor-to-source region sensitivity*, can be defined, based on Eq. (9.11) as follows:

$$s(C_i; E_j, V') = \delta C_i(x, y, z, t; V', t')/\delta E_j(x', y', z', t')$$
$$= \int_{V'} s(C_i; E_j)\, dV' \qquad (9.12)$$

The corresponding *normalized receptor-to-source region sensitivity* is defined by

$$S(C_i; E_j, V') = \int_{V'} \frac{E_j(x', y', z', t')}{C_i(x, y, z, t)} \cdot s(C_i; E_j)\, dV' \qquad (9.13a)$$
$$= D_j(V', t') \cdot C(x, y, z, t) \qquad (9.13b)$$

where

$$D_j(V', t') = \int_{V'} \frac{\delta}{\delta \ln[E_j(x', y', z', t')]} \, dV' \tag{9.14}$$

is an integral operator.

The general procedure for calculating the receptor-to-source region sensitivity implies first the solution of Eqs. (9.1) and (9.10) to obtain the receptor-to-source point sensitivity and then the integration of Eq. (9.12) or Eq. (9.13a). An alternative technique is to apply the operator D_j defined in Eq. (9.13b) to Eq. (9.1). In this way, one can obtain the value of the normalized sensitivity $S(C_i; E_j, V')$ directly, without first calculating the point sensitivities.

In the case where emissions change in a given time interval, $t \in [t_1, t_2]$, the receptor-to-source region sensitivity has to be integrated over time as well, i.e.,

$$s(C_i; E_j, V', t_1 < t' < t_2) = \int_{t_1}^{t_2} \int_{V'} s(C_i; E_j) \, dV' \, dt' \tag{9.15}$$

and the normalized sensitivity becomes

$$S(C_i; E_j, V', t_1 < t' < t_2) = \int_{t_1}^{t_2} \int_{V'} S(C_i; E_j, V', t') \, dV' \, dt'$$

$$= \int_{t_1}^{t_2} D_j(V', t') C_i(x, y, z, t) \, dt' \tag{9.16}$$

In the most general case, where we are interested in the variation of the concentration of species i at a specific receptor with respect to the variations of the emissions of *several chemical species in several regions*, Eq. (9.11) becomes

$$\delta C_i\left(x, y, z, t; V'_1, \dots, V'_p, t'\right) = \sum_{l=1}^{p} \sum_{j=1}^{q} \int_{V'_l} s(C_i; E_j) \delta E_j(x', y', z', t') \, dV'_l \tag{9.17}$$

where p and q are the number of regions and species being considered, respectively, and V'_l is the volume of source region l.

Finally, it should be pointed out that although the receptor-to-source region sensitivity given above represents the sensitivity of the species concentration with respect to emission sources, this approach can be extended to sensitivities of different objectives with respect to various parameters. For each objective, the general procedure is that the individual objective-to-point sensitivities are calculated first, and then used to obtain the objective-to-region sensitivities through proper integration.

9.2.2 A Case Study: Emissions of NO_x and SO_2 in the Eastern United States

To illustrate the sensitivity analysis discussed above, we report the results of the study of Cho *et al.* (1988), who considered NO_x and SO_2 emissions in the eastern United

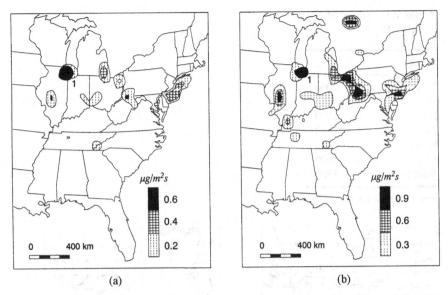

Figure 9.1. Ground-level emissions of (a) NO_x and (b) SO_2, for July 4, 1974. The number 1 indicates the emission source located in the vicinity of Gary. From Cho *et al.* (1988).

States during the meteorological conditions on July 4, 1974. These were obtained from the Sulfate Regional Experiment (SURE) and are shown in Fig. 9.1 (Hidy and Mueller, 1976).

Sensitivities of ground-level concentrations to a specific source point

We now investigate the sensitivities of ground-level pollutant concentrations to the SO_2 and NO_x sources in the vicinity of Gary, Indiana (*i.e.*, source point 1 in Fig. 9.1), assuming that the variations of the emission are constant relative to emission strength during the entire simulation. Obviously, these are receptor-to-source point sensitivities.

Let us first evaluate the sensitivities of ground-level SO_2 and sulfate concentrations with respect to the SO_2 source in the vicinity of Gary. To do this, we need to solve simultaneously the sensitivity [Eqs. (9.10)] and model equations [Eq. (9.1)]. For this, a reaction kinetic model, the model input parameters, and the initial and boundary conditions need to be given.

The reaction kinetic model was taken from Carmichael *et al.* (1986). The initial conditions of SO_2 and sulfate were generated through an inverse r-squared interpolation of observed surface data (Goodin *et al.*, 1979). The initial vertical profiles of all species were estimated by the relation

$$C(z) = C(z = 0) \cdot \exp(-z/H_s) \tag{9.18}$$

where H_s is a function of pollutant properties (*e.g.*, dry deposition velocity and solubility). The H_s values used in this study are given in Table 9.1.

Table 9.1. Values of H_s in Eq. (9.18)

Species	H_s (m)
NO_x	2000
O_x	∞
HNO_3	3000
NH_3	2000
H_2O_2	3500
SO_2	2000
Sulfate	3500

From Cho *et al.* (1988).

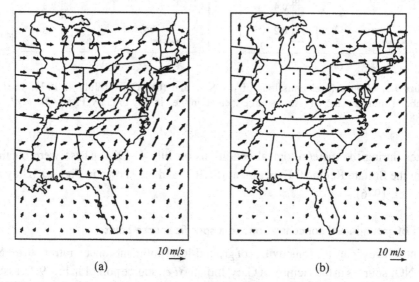

$10\ m/s$ $10\ m/s$

(a)   (b)

Figure 9.2. The ground-level wind fields for July 4, 1974, for the hours (a) 00:00–12:00, and (b) 12:00–14:00. From Cho *et al.* (1988).

The upper bound of the modeling region was set at $H = 3\,\text{km}$, which is well above the maximum mixing-layer height. Simulations were performed from midnight to midnight, over a 48-h period, using the meteorology of July 4, 1974. The wind field was derived from the meteorological data and is shown in Fig. 9.2, indicating a high-pressure area located off the coast of North Carolina and a low-pressure area located over Lake Superior. The predicted winds were generally from the west and southwest and increased in magnitude with elevation.

The time evolution of the sensitivity of the ground-level SO_2 concentration to the variation of the SO_2 source in the vicinity of Gary is shown in Fig. 9.3 in terms of percentage changes in concentration (*i.e.*, 100% indicates that a unit magnitude change in the SO_2 emission results in a unit magnitude change in the ground-level concentration). It is seen that the high-sensitivity region (shaded in the figure) at noon has an elongated shape in the direction of the wind field (see Fig. 9.2 for wind

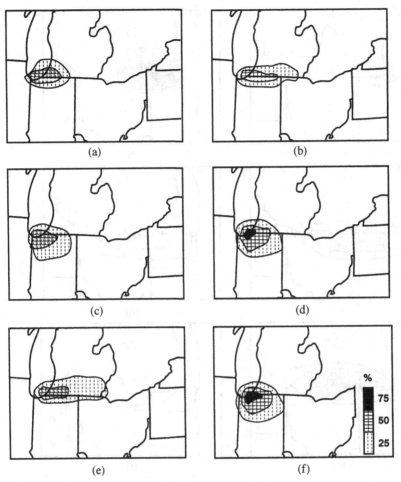

Figure 9.3. Sensitivities of the ground-level SO$_2$ concentration to the variation of the SO$_2$ emission source in the vicinity of Gary, Indiana, on July 4 at (a) 6:00, (b) 12:00, (c) 18:00, and (d) 24:00, and on July 5 at (e) 12:00 and (f) 24:00. From Cho *et al.* (1988).

field). The sensitivity of the ground-level sulfate concentration to variation of the SO$_2$ source in the vicinity of Gary is shown in Fig. 9.4. The variation of the SO$_2$ source affects the sulfate concentration mostly through photochemical reactions. A 3% or more sensitivity of the ground-level sulfate concentration first appears 6 h after the simulation starts. This is so because during the first 6 h of the simulation was nighttime, when no photochemical processes are feasible and the only reaction mechanism for producing sulfate is the heterogeneous process, due to the interaction between gas and aerosol phases. The highest sensitivity of the ground-level sulfate concentration occurs at noon, as it clearly appears in Fig. 9.4b. In both Figs. 9.3 and 9.4, the sensitivity profiles for the next 24-h period have patterns similar to those during the first 24 h.

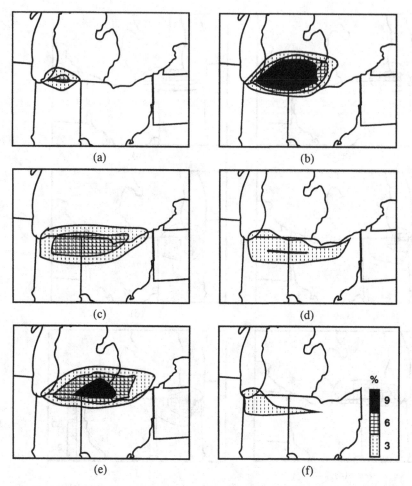

Figure 9.4. Sensitivities of the ground-level sulfate concentration to the variation of the SO_2 emission source in the vicinity of Gary on July 4 at (a) 6:00, (b) 12:00, (c) 18:00, and (d) 24:00, and on July 5 at (e) 12:00 and (f) 24:00. From Cho *et al.* (1988).

A comparison between Figs. 9.3 and 9.4 indicates that the sensitivity profiles for SO_2 are substantially different from those for sulfate. The sensitivity maximum of SO_2 at the ground level with respect to variations of the SO_2 source increases as time goes from noon to 18:00 to midnight, while that of sulfate decreases during this period. The increase of the sensitivity maximum of SO_2 during nighttime is due to the reduction in the photochemical SO_2 conversion and to the limited mixing implied by the low mixing-layer height. The reduction in the photochemical SO_2 conversion to sulfate in this time period also results in a decrease of the sensitivity maximum of sulfate.

Some further insight into the sensitivity behavior of SO_2 and sulfate can be obtained by considering the percentage variations of the ground-level SO_2 and sulfate concentrations along the same latitude as the SO_2 source in the vicinity of Gary. These variations are shown in Fig. 9.5, both at noon and at midnight. It is seen that

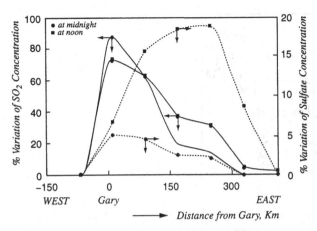

Figure 9.5. Sensitivity (%) profiles of the ground-level SO_2 and sulfate concentrations along the straight line parallel to the abscissa and passing through the SO_2 source in the vicinity of Gary, Indiana. From Cho *et al.* (1988).

the maximum variations of the ground-level SO_2 are observed in the vicinity of Gary at both noon and midnight. Similar behavior is shown also in Fig. 9.3, where a 25% or more sensitivity of SO_2 occurs only in the vicinity and downwind of the SO_2 source. This indicates that *the influence of the SO_2 source on the ground-level SO_2 concentration is restricted mostly to the vicinity of the source.* In the case of sulfate, the maximum sensitivity is found close to the source region at midnight, while about 250 km away from the source region at noon. This can be seen also in Fig. 9.4b, corresponding to noon, where the source lies only at one end of the 3% or more sensitivity region, while the maximum sensitivity region occurs significantly downwind. As noted earlier, this is due to the diurnal variations of the chemical conversion rate of SO_2 to sulfate. Thus, *the influence of the SO_2 source on the ground-level sulfate concentration can be both in the vicinity of and away from the source.*

In addition, it has been found that the sensitivity behavior of the ground-level NO_x to the variation of the NO_x source in the vicinity of Gary is similar to that of the ground-level SO_2 to the SO_2 source variation, *i.e.*, the high sensitivity region of NO_x is restricted to the vicinity and downwind of the NO_x source region, and the sensitivity maximum at nighttime is larger than that during daytime.

Influence of several emission sources on one receptor location

In the previous section, we examined the sensitivities of the ground-level pollutant concentrations to only one emission source. Let us now consider the influence of several emission sources on one receptor location. Two cases are investigated: (1) influence of several *independent emission sources* on one receptor and (2) influence of several *related emission sources* (*i.e.*, a source region) on one receptor. The former is connected to the receptor-to-source point sensitivity problem, while the latter is related to the receptor-to-source region sensitivity.

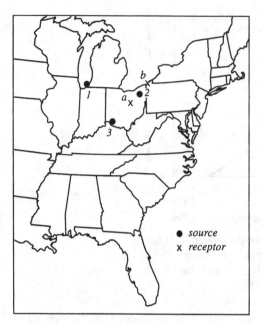

Figure 9.6. Locations of the three emission sources, 1, 2, and 3 in the vicinity of Gary, Pittsburgh, and Cincinnati, respectively, and the two receptors, a and b.

For the first case, let us select SO_2 sources at three locations in the vicinities of Gary, Pittsburgh, and Cincinnati, as indicated by 1, 2, and 3, respectively in Fig. 9.6. The influence of these sources on two receptors, a and b in Fig. 9.6, is estimated by a direct evaluation of the receptor-to-source point sensitivities with respect to each of these emissions. In Fig. 9.7 are shown the sensitivities of the ground-level sulfate concentration at the two receptors to the emissions of SO_2 at the three different locations. The diurnal variation of the sensitivities reflects temporal and spatial changes in the chemical reaction rate, wind velocity, eddy diffusivity, and dry deposition velocity fields. It is seen that the ground-level sulfate concentration is most sensitive to the nearby SO_2 sources (*i.e.*, the source in the vicinity of Cincinnati for receptor a and the source in the vicinity of Pittsburgh for receptor b) during daytime and the far-away source (*i.e.*, the source in the vicinity of Gary for both receptors a and b) during nighttime. This occurs because of changes in chemical reaction rates with time, as discussed earlier.

To study the influence of an emission source region on a specific receptor, let us consider the case shown in Fig. 9.8, where the source region in Ohio includes nine grid points, and two receptors are indicated by A and B. The sensitivity of pollutant concentrations at a specific receptor with respect to an emission source area can be evaluated by integrating the receptor-to-source point sensitivity equations over the given source area, *i.e.*, using Eq. (9.11). If the variation δE in the source area is assumed to be constant, we may use Eq. (9.12) or Eq. (9.13) to obtain the normalized sensitivities. The diurnal normalized sensitivities of the ground-level sulfate

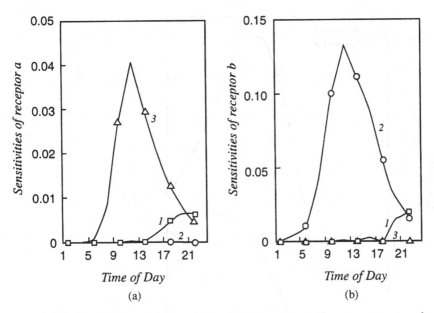

Figure 9.7. Sensitivities of the ground-level sulfate concentration at receptors a and b with respect to variations of SO_2 sources in the vicinity of Gary (1), Pittsburgh (2), and Cincinnati (3) as functions of time. From Cho *et al.* (1988).

Figure 9.8. Locations of the receptor regions, A and B, and the emission source region in Ohio, indicated by the shaded area.

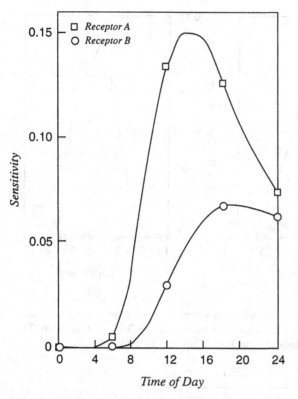

Figure 9.9. Sensitivities of the ground-level sulfate concentration at the two receptor regions, A and B, with respect to variations of SO$_2$ emissions in Ohio. From Cho *et al.* (1988).

concentration at receptors A and B with respect to variations of the SO$_2$ emissions in Ohio are shown in Fig. 9.9, where a constant δE has been assumed. The maximum sensitivity of receptor A occurs at about 14:00, while the response of receptor B first becomes noticeable around 6:00.

Relations between primary emissions and regional air quality

In both cases of receptor-to-source point and receptor-to-source region sensitivity analyses, we compute sensitivities of the pollutant concentration at each point in the region of interest with respect to source emissions, from which sensitivity distributions in the specific region are obtained. For example, Figs. 9.3 and 9.4 show the sensitivity distributions in the eastern United States at the ground level. When the sensitivity distributions are computed at different vertical positions, they form a sensitivity space. However, in regional air quality control, it is often required to have an average quantity in the region of interest to represent such spatially distributed sensitivities, referred to as the *lumped sensitivity*. This value indicates the sensitivity of the regional air quality to source emissions, and can be obtained by integrating the receptor-to-source point or the receptor-to-source region sensitivities over the domain of interest. For example, in the latter case, we have

$$\bar{S} = \frac{1}{V} \int_V [S(C_i; E_j, V', t_1 < t' < t_2)] \, dV \qquad (9.19)$$

where V is the volume of the domain and $S(C_i; E_j, V', t_1 < t' < t_2)$ is the receptor-to-source region sensitivity defined by Eq. (9.16). If we are interested in the air quality only in a surface at a specific height (e.g., the ground level), then Eq. (9.19) reduces to

$$\bar{S} = \frac{1}{A} \int_A [S(C_i; E_j, V', t_1 < t' < t_2)] \, dx \, dy \qquad (9.20)$$

where A is the area of the region and x and y are the surface directions.

Two examples are given in Fig. 9.10, which shows the lumped sensitivities of the ground-level sulfate concentration integrated over the entire domain (i.e., the eastern United States) with respect to variations of the SO_2 emissions in Ohio (curve 1) or in the entire domain (curve 2) as functions of time. For both cases, the sensitivity maximum occurs at about noon, but the sensitivities for the case where only the source in Ohio is varied are 6 to 7 times smaller than when all the sources in the

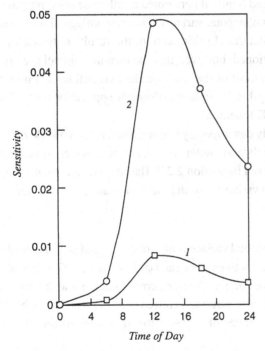

Figure 9.10. Sensitivities of the averaged ground-level sulfate concentration in the entire eastern United States region with respect to variations of the SO_2 source in Ohio only (curve 1) and all the SO_2 sources in the entire region (curve 2). From Cho *et al.* (1988).

domain are varied. Note that even in the latter case, the sensitivity of the ground-level sulfate concentration is still relatively small; *i.e.*, its maximum is less than 5% of the source variation. Note that the ratio between the sensitivity maxima with respect to the emissions in Ohio and in the entire eastern United States region is approximately the same as the ratio between the SO_2 emissions in Ohio and in the entire region.

9.3 Global Sensitivity Analysis of Trajectory Model for Photochemical Air Pollution

9.3.1 Global Sensitivities and the FAST Method

Sensitivity analysis of atmospheric diffusion-reaction models to uncertainties in model input parameters has been reported in several studies (*e.g.*, Gelinaś and Vajk, 1978; Koda *et al.*, 1979b; Falls *et al.*, 1979; Dunker, 1980, 1981; Tilden and Seinfeld, 1982; Cho *et al.*, 1987). Most of these were based on local sensitivity analysis (*i.e.*, one-at-a-time parameter variations) and considered only variations of small magnitude. However, Falls *et al.* (1979) and Tilden and Seinfeld (1982) used global sensitivity analysis. In particular, Tilden and Seinfeld performed a full-scale sensitivity analysis to understand the effect of simultaneous variations in meteorological and emission parameters on the model predictions. In this section, the results obtained by them are reported. It should be mentioned, however, that the chemical model used in their studies is out of date, although most of the final conclusions still remain true today. Thus, our main purpose of using this example is to illustrate applications of the Fourier amplitude sensitivity test (FAST) method.

The FAST method, originally developed by Cukier, Shuler, and coworkers (Cukier *et al.*, 1973, 1975, 1978; Schaibly and Shuler, 1973), has been widely used for many chemical systems and is discussed in Section 2.2.2. Based on this method, the global sensitivity of y_i with respect to variations of the jth parameter ϕ_j in ϕ is defined as

$$S_g(y_i; \phi_j) = \sigma_{i,\omega_j}^2(t)/\sigma_i^2(t) \tag{9.21}$$

where σ_{i,ω_j}^2 and σ_i^2 are partial and total variances of y_i due to ϕ_j and ϕ, respectively. The central idea of the FAST method is to express each ϕ_j in ϕ as $\phi_j = G_j(\sin \omega_j \xi)$, $j = 1, 2, \ldots$, where G_j is a set of known functions, referred to as the search curves, ω_j is a set of frequencies, and ξ is a scalar variable, referred to as the search variable. Then, by properly selecting G and the range of ξ, the following equations for σ_{i,ω_j}^2 and σ_i^2 can be derived:

$$\sigma_i^2(t) = \langle y_i^2(t) \rangle - \langle y_i(t) \rangle^2 = 2 \sum_{k=1}^{\infty} \left[A_{i,k}^2(t) + B_{i,k}^2(t) \right] \tag{9.22}$$

$$\sigma_{i,\omega_j}^2(t) = 2 \sum_{r=1}^{\infty} \left[A_{i,r\omega_j}^2(t) + B_{i,r\omega_j}^2(t) \right] \tag{9.23}$$

where

$$A_{i,k}(t) = \frac{1}{2\pi} \int_{-\pi}^{\pi} y_i(t, \phi) \cos(k\xi) \, d\xi \tag{9.24}$$

$$B_{i,k}(t) = \frac{1}{2\pi} \int_{-\pi}^{\pi} y_i(t, \phi) \sin(k\xi) \, d\xi \tag{9.25}$$

are the Fourier coefficients. The coefficients $A_{i,r\omega_j}$ and $B_{i,r\omega_j}$ are also given by Eqs. (9.24) and (9.25), but replacing k with $r\omega_j$.

Thus, evaluation of the global sensitivity requires the calculation of the y_i values in the chosen range of the search variable ξ and the construction of the Fourier coefficients for each of the assigned frequencies ω_j and its harmonics. A computational implementation of the FAST method, based on the work of Seinfeld and co-workers (Koda *et al.*, 1979; McRae *et al.*, 1982b), is described in Section 2.2.2.

9.3.2 A Case Study: Emissions of NO, NO₂, Reactive Hydrocarbons, and O₃

Following the work by Tilden and Seinfeld (1982), we consider the evaluation of the concentrations of various chemical species in the air parcel that travels along the trajectory shown in Fig. 9.11. These conditions occurred on June 26, 1974, the first of the three-day oxidant episode in the South Coast Air Basin of California. The trajectory starts near downtown Los Angeles at 5:00 local time and ends at 20:00 in

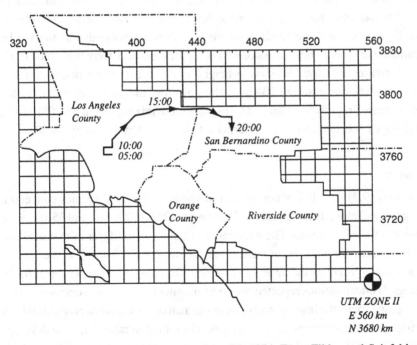

Figure 9.11. Air parcel trajectory on June 26, 1974. From Tilden and Seinfeld (1982).

Table 9.2. Trajectory conditions for the model simulation

Time	Wind Speed (m/s)	Mixing Height (m)	T (°C)	$K_{zz,max}$ (m²/s)	Solar Intensity (W/m²)	Emission Rates (ppm/m/min)	
						NO_x	RHC
5:30	0.7	154	17	36.3	136.1	0.0323	0.0403
6:30	0.5	104	19	9.9	334.6	0.0647	0.0836
7:30	0.2	114	22	8.7	525.5	0.1201	0.1640
8:30	0.3	124	24	17.3	695.6	0.100	0.1443
9:30	0.6	190	28	85.6	833.5	0.1235	0.1578
10:30	1.2	295	31	292.6	929.7	0.2338	0.2440
11:30	1.5	416	34	497.6	977.7	0.1650	0.1741
12:30	1.9	551	34	639.3	974.1	0.1112	0.1230
13:30	2.2	466	34	384.3	919.4	0.0377	0.0410
14:30	3.4	356	35	334.1	817.0	0.0	0.0002
15:30	3.6	246	36	75.9	674.2	0.0	0.0
16:30	3.4	137	36	53.8	500.5	0.002	0.001
17:30	3.0	110	36	10.3	307.8	0.0	0.0
18:30	3.3	83	33	8.9	109.3	0.0	0.0
19:30	1.6	56	30	0.4	0.0	0.0	0.0

From Tilden and Seinfeld (1982).

San Bernardino County. The meteorological conditions and emission rates along the trajectory path are given in Table 9.2.

It is seen from Table 9.2 and Fig. 9.11 that, due to light and variable wind conditions, the air parcel trajectory remained near downtown Los Angeles until 10:00. It then moved eastward into the foothills of the San Gabriel mountains as the sea breeze became established. Thus, emission levels for the trajectory were high until 10:00 and then decreased rapidly. The mixing height, derived from temperature profiles, was 154 m at 5:00, decreased to 104 m as the trajectory moved closer to the ocean, and then increased rapidly at about 9:00. The maximum mixing height was 551 m at 12:30 but dropped substantially in the afternoon to levels below 100 m by 18:30.

Model simulation

Equations (9.7) and (9.8) were applied to simulate chemical species concentrations along the trajectory. Five cells ($N = 5$), at heights 21, 49, 70, 210, and 350 m, were employed in the calculations. The movement of the inversion-layer base was accounted for through the specification of the vertical variation of K_{zz}. The initial species concentrations for cells below the inversion base were obtained by interpolation of the hourly averaged concentrations reported by the 61 monitoring stations in operation. The concentrations above the inversion base were estimated based on observations of O_3 levels during episode conditions in Los Angeles. The emission rates, E_i, were developed by McRae et al. (1982a) and the reaction rates, R_i, were estimated by Falls et al. (1979).

The computed concentrations of species O_3, NO_x, and reactive hydrocarbons

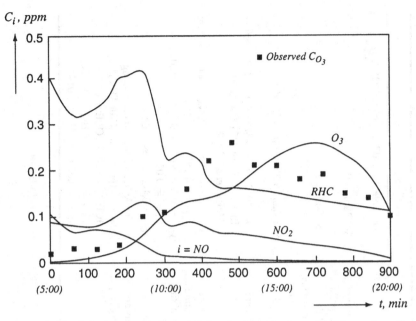

Figure 9.12. Calculated and observed chemical species concentrations along the trajectory shown in Fig. 9.11. From Tilden and Seinfeld (1982).

(RHCs) for the conditions in Table 9.2 are shown in Fig. 9.12, together with the measured concentration of O_3 along the trajectory. It is seen that the trends of the measured and simulated O_3 concentrations are very similar, although the model has been simplified significantly and uncertainties in model input parameters exist.

Let us now report the results of the global sensitivity analysis of model predictions with respect to variations of input parameters. The parameters selected for variations are listed in Table 9.3. Each parameter, including initial concentrations and emission rates, was varied independently and significantly (by ±50% of its nominal value). The global sensitivity values [defined by Eq. (9.21)] for NO, NO_2, RHCs, and O_3 for the ground-level cell are shown in Figs. 9.13 to 9.16, respectively.

Nitrogen oxide sensitivity

It is seen from Fig. 9.13 that in the first portion of the simulation ($t < 100$ min), NO is most sensitive to variations in the initial concentrations and mixing height. The large sensitivity of NO to variations in mixing height arises because the air above the inversion contains about 0.1 ppm of O_3 and negligible NO due to consumption by the reaction $NO + O_3 \rightarrow NO_2 + O_2$. Thus, increase in the mixing height leads to decrease in the ground-level NO concentration. The NO concentration is insensitive to variations in photolysis intensity, emission rates, and ambient temperature since the O_3 level is small.

In the middle portion of the simulation ($100 < t < 300$ min), the sensitivity of the NO concentration to initial conditions becomes small, while the important parameters

Table 9.3. Uncertain parameters considered in the sensitivity analysis

Type of Parameter	Parameter	Physical Significance	Nominal Value
Meteorological	Photolysis intensity	Determines the rate of photochemical activity	See Table 9.2 and McRae et al. (1982b)
	Mixing height	Determines the vertical extent of the air parcel containing the pollutants	See Table 9.2
	Ambient temperature	Controls the magnitude of the nonphotochemical rate constants	See Table 9.2
	Relative humidity[a]	Affects certain chemical reactions	
	Turbulent diffusivity	Determines the rate of vertical mixing of pollutants	See Table 9.2 and McRae et al. (1982a)
	Deposition velocity	The rate at which pollutants are removed at the ground surface	$NO = 0.12$; $O_3 = 0.381$; $NO_2 = 0.381$ m/min $RCHO = 0.60$; $HCHO = 0.60$ (McRae et al., 1982a)
Initial concentrations and concentrations aloft	Initial concentration below mixing height	Pollutant concentration near the ground at the start of the simulation	$y_{NO} = 0.104$; $y_{O_3} = 0.0$; $y_{NO_2} = 0.088$ ppm; $y_{THC} = 4.28$[b]
	Concentration and boundary conditions	Pollutants entrained when mixing height rises	$y_{NO} = 0.01$; $y_{O_3} = 0.0$; $y_{NO_2} = 0.01$ ppm; $y_{THC} = 0.05$
	Aloft	Upper boundary conditions	$y_{O_3} = 0.10$; others $= 0.0$
Emissions	Emission intensity	The rate at which pollutants are emitted into the trajectory air parcel	See Table 9.2. Hydrocarbon splitting factors[c] (ppmv/ppm C of THC) $HCHO = 0.0047$; $OLE = 0.0129$; $ARO = 0.0086$; $RCHO = 0.0020$; $ALK = 0.0502$; $C_2H_4 = 0.0142$.

From Tilden and Seinfeld (1982).

[a] To which the ground-level concentration is insensitive in preliminary calculations; thus, it is not included in the sensitivity analysis.

[b] y_{THC} denotes the total hydrocarbon concentration (ppm C).

[c] These splitting factors were derived from Los Angeles emission inventory data by McRae et al. (1982a).

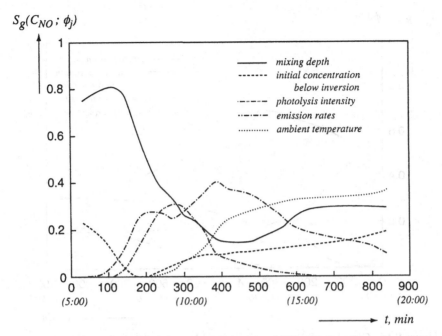

Figure 9.13. Global sensitivities of the ground-level NO concentration with respect to changes of various system parameters along the trajectory shown in Fig. 9.11. From Tilden and Seinfeld (1982).

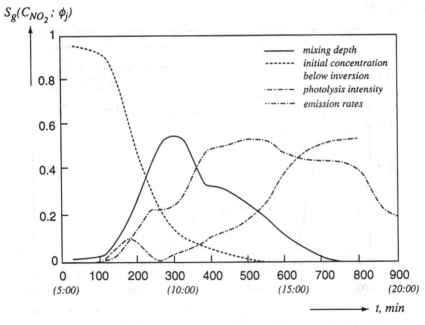

Figure 9.14. Global sensitivities of the ground-level NO₂ concentration with respect to changes of various system parameters along the trajectory shown in Fig. 9.11. From Tilden and Seinfeld (1982).

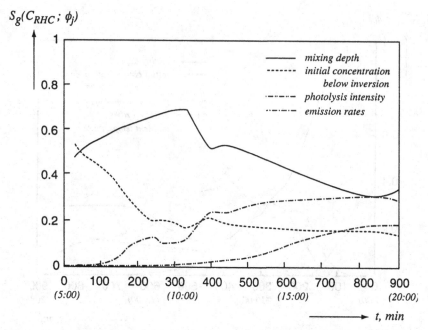

Figure 9.15. Global sensitivities of the ground-level reactive hydrocarbon concentration with respect to changes of various system parameters along the trajectory shown in Fig. 9.11. From Tilden and Seinfeld (1982).

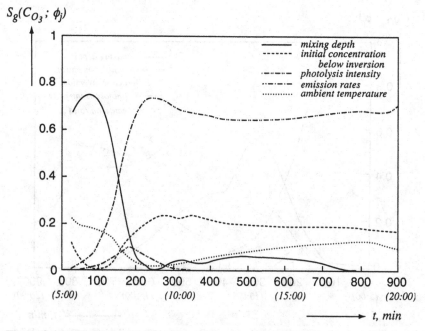

Figure 9.16. Global sensitivities of the ground-level O_3 concentration with respect to changes of various system parameters along the trajectory shown in Fig. 9.11. From Tilden and Seinfeld (1982).

are mixing height, emission rates, and photolysis intensity. In particular, the sensitivity to photolysis intensity reaches a local maximum close to 10:00, where from Table 9.2 the solar intensity increases rapidly and reaches near the highest level. In this stage, the concentrations of NO_2 and RHCs also reach their maximum, as shown in Fig. 9.12. However, starting in the late morning, turbulent mixing processes are enhanced; thus sensitivity to mixing height decreases progressively.

For $t > 400$ min, although the sensitivities of the ground-level NO concentration to ambient temperature, photolysis intensity, emission rates, and initial concentrations are relatively large, the results are of little practical importance since the calculated values of NO concentration in this period (see Fig. 9.12) are small.

Nitrogen dioxide sensitivity

The sensitivities of the ground-level NO_2 concentration to various input parameters are shown in Fig. 9.14. In the initial stage, the ground NO_2 concentration is sensitive only to the initial concentrations. The sensitivity to the mixing height, which is important for NO in this stage, now becomes small. This is because the solar intensity during this period being relatively small, the contribution of the $NO + O_3$ reaction is not meaningful; hence, increase in the mixing height does not influence the NO_2 level significantly.

However, soon after NO_2 reaches a local maximum in Fig. 9.12, the highest sensitivity occurs with respect to the mixing height, since from Table 9.2 the mixing height itself undergoes a large variation. For $t > 400$ min, the ground-level NO_2 concentration becomes the most sensitive to emission rates and photolysis intensity, owing to the high O_3 concentrations available, which promote the $NO + O_3$ reaction.

Reactive hydrocarbon sensitivity

The sensitivities of the RHCs are shown in Fig. 9.15. The RHCs are primary emission pollutants, but unlike NO_x, a large fraction of them (*e.g.*, alkanes and ethylene) exhibits slow photochemical reaction rates. Thus, the RHC sensitivity to photolysis intensity is generally small, except toward the end of the simulation, where both O_3 concentration and temperature are moderately high.

In the initial stage, the RHCs are most sensitive to variations in the mixing height and initial concentrations. In particular, the sensitivity to the mixing height reaches a maximum around 10:00 when the mixing height undergoes large variations (see Table 9.2). This arises because increase in the mixing height leads to a substantial decrease in the ground-level RHC concentrations, as clearly shown in Fig. 9.12. The RHCs also become sensitive to variations in emission rates in later portion of the simulation, since their concentration levels are moderate and the emission rates are very small.

Ozone sensitivity

As shown in Fig. 9.16, the large sensitivities of ozone in the initial stage of the simulation occur with respect to the mixing height and the initial concentrations, due to the large vertical gradient of O_3 concentration. Then the solar intensity increases rapidly, which initiates the photochemical processes that consume ozone. Thus, the ground-level O_3 concentration becomes most sensitive to the variation in the photolysis intensity, followed by initial conditions and ambient temperature. Note that the ozone sensitivity to the emission rates is small in later stages, since the O_3 concentration is relatively large.

Finally, it should be mentioned that the sensitivities of NO, NO_2, RHCs, and O_3 to variations in the relative humidity, vertical turbulent diffusivity K_{zz}, and deposition velocity $v_{i,d}$ were also investigated by Tilden and Seinfeld (1982). The results indicate that these parameters are not important in determining the ground-level species concentrations. Hence, based on the results of the above sensitivity analysis, it can be concluded that for photochemical air pollution, the most important parameters affecting the ground-level pollutant concentrations are mixing height, photolysis intensity, initial conditions, and emission rates. The relative importance of these parameters, however, varies during the course of the day, owing to interactions between atmospheric chemistry and meteorological conditions.

References

Carmichael, G. R., Peters, L. K., and Kitada, T. 1986. A second generation model for regional-scale transport/chemistry/deposition. *Atmos. Environ.* **20**, 173.

Cho, S.-Y., Carmichael, G. R., and Rabitz, H. 1987. Sensitivity analysis of the atmospheric reaction-diffusion equation. *Atmos. Environ.* **21**, 2589.

Cho, S.-Y., Carmichael, G. R., and Rabitz, H. 1988. Relationships between primary emissions and regional air quality and deposition in Eulerian models determined by sensitivity analysis. *Water Air Soil Pollut.* **40**, 9.

Cukier, R. I., Fortuin, C. M., Shuler, K. E., Petschek, A. G., and Schaibly, J. H. 1973. Study of the sensitivity of coupled reaction systems to uncertainties in rate coefficients. I. Theory. *J. Chem. Phys.* **59**, 3873.

Cukier, R. I., Schaibly, J. H., and Shuler, K. E. 1975. Study of the sensitivity of coupled reaction systems to uncertainties in rate coefficients. III. Analysis of the application. *J. Chem. Phys.* **63**, 1140.

Cukier, R. I., Levine, H. B., and Shuler, K. E. 1978. Nonlinear sensitivity analysis of multiparameter model systems. *J. Comp. Phys.* **26**, 1.

Dunker, A. M. 1980. The response of an atmospheric reaction-transport model to changes in input functions. *Atmos. Environ.* **14**, 671.

Dunker, A. M. 1981. Efficient calculation of sensitivity coefficients for complex atmospheric models. *Atmos. Environ.* **15**, 1155.

Falls, A. H., McRae, G. J., and Seinfeld, J. H. 1979. Sensitivity and uncertainty of reaction mechanisms for photochemical air pollution. *Int. J. Chem. Kinet.* **11**, 1137.

Gelinas, R. J., and Vajk, J. P. 1978. Systematic sensitivity analysis of air quality simulation models. *Final Report, EPA Contract No. 68-02-2942.* Pleasanton, CA: Science Applications.

Goodin, W. R., McRae, G. J., and Seinfeld, J. H. 1979. A comparison of interpolation methods for sparse data: application to wind and concentration fields. *J. Appl. Meteorol.* **18**, 761.

Hidy, G. M., and Mueller, P. K. 1976. The design of the sulfate regional experiment. *Report EC-125*, Vol. 1. Palo Alto, CA: Electric Power Research Institute.

Hsu, H. M., and Chang, J. S. 1987. On the Eulerian source-receptor relationship. *J. Atmos. Chem.* **5**, 103.

Kleinman, L. I. 1988. Evaluation of sulfur dioxide emission scenarios with a nonlinear atmospheric model. *Atmos. Environ.* **22**, 1209.

Koda, M., McRae, G. J., and Seinfeld, J. H. 1979a. Automatic sensitivity analysis of kinetic mechanisms. *Int. J. Chem. Kinet.* **11**, 427.

Koda, M., Dogru, A. H., and Seinfeld, J. H. 1979b. Sensitivity analysis of partial differential equations with applications to reaction and diffusion processes. *J. Comput. Phys.* **30**, 259.

Liu, M. K., and Seinfeld, J. H. 1975. On the validity of grid and trajectory models of urban air pollution. *Atmos. Environ.* **9**, 555.

McRae, G. J., Goodin, W. R., and Seinfeld, J. H. 1982a. Development of a second-generation mathematical model for urban air pollution – I. Model formulation. *Atmos. Environ.* **16**, 679.

McRae, G. J., Tilden, J. W., and Seinfeld, J. H. 1982b. Global sensitivity analysis – a computational implementation of the Fourier Amplitude Sensitivity Test (FAST). *Comput. Chem. Eng.* **6**, 15.

Mitchell, A. R. 1969. *Computational Methods in Partial Differential Equations.* New York: John Wiley.

Schaibly, J. H., and Shuler, K. E. 1973. Study of the sensitivity of coupled reaction systems to uncertainties in rate coefficients. II. Applications. *J. Chem. Phys.* **59**, 3879.

Seinfeld, J. H. 1986. *Atmospheric Chemistry and Physics of Air Pollution.* New York: John Wiley.

Tilden, J. W., and Seinfeld, J. H. 1982. Sensitivity analysis of a mathematical model for photochemical air pollution. *Atmos. Environ.* **16**, 1357.

10

Sensitivity Analysis in
Metabolic Processes

I N THE RECENT DECADES, growth in molecular biology has been explosive. The
details of molecular constituents and of chemical transformations in metabolism have
become increasingly clear, at least for a number of simpler organisms. An important
strategy for molecular biologists is to reduce a complex biochemical reacting system
to its elemental units, in order to explain it at the molecular level, and then to use this
knowledge to reconstruct it. However, cellular components exhibit large interactions
that are associative rather than additive. We still know relatively little about such interactions, what makes such an integrated system a living cell, and how a reconstructed
system will respond to variations in novel environments and to specific variations in
the metabolic pathway. To understand them, molecular biologists now recognize the
need for systematic methods that can be provided by mathematical analysis.

As an effective mathematical tool, sensitivity analysis has been widely applied
in biochemical reacting systems in order to understand the effects of variations in
activities of enzymes, in kinetic parameters, and in external modifiers on metabolic
processes. In other words, the objective is to understand how the function of an integrated biochemical system can be deduced from kinetic observations of its elemental
units. The applications of sensitivity analysis in biochemical reacting systems were
pioneered by Higgins (1963) and have spawned three basic theories: *biochemical systems theory* (Savageau, 1969a,b,c, 1971a,b, 1976), *metabolic control theory* (Kacser
and Burns, 1973; Heinrich and Rapoport, 1974a,b; Westerhoff and Chen, 1984; Fell
and Sauro, 1985; Reder, 1988; Giersch, 1988), and *flux-oriented theory* (Crabtree and
Newsholme, 1978, 1985, 1987). The most distinguishing difference (Fell, 1992) between these theories is the choice of the parameter that is varied for the determination
of sensitivities. In biochemical systems theory, the primary parameters for sensitivity
analysis are the rate constants for synthesis and degradation of metabolite pools. This
is similar to the sensitivity analysis of detailed kinetic schemes discussed in Chapter 8
and requires at least an approximate knowledge of the kinetic scheme and the reaction
kinetics. In metabolic control theory, enzyme concentrations (or activities) are usually
chosen as parameters for sensitivity analysis, and the response of a metabolic pathway

to an external modifier is derived from the resulting sensitivities. The main character-istic of metabolic control analysis is that it attempts, through appropriate assumptions, to perform sensitivity analysis using the minimum possible information about the re-action kinetics. The flux-oriented theory is intermediate between the other two, and the external modifiers are the primary parameters chosen for sensitivity analysis.

In this chapter we discuss two specific approaches to sensitivity analysis and illus-trate two examples of interest in biochemistry. The basic idea is to provide a flavor of the potential of these techniques in this field. The first, referred to here as the *general approach*, is that applied in previous chapters and can be classified in the framework of the biochemical systems theory mentioned above. This implies first solving simul-taneously the differential equations for mass balances and sensitivities of metabolite concentrations, and then evaluating the sensitivities of other quantities of interest (*e.g.*, fluxes) from the obtained results. The second approach is the *matrix method*, developed in the context of metabolic control theory, which requires only algebraic matrix operations. Besides their mathematical structure, the differences between them are that the general approach can be applied to either dynamic or steady-state condi-tions, without limitation on the reaction pathway and the number and types of system variables and parameters, while the matrix method is generally applicable only to steady-state conditions and its complexity grows exponentially as the complexity of reaction pathway increases. On the other hand, in the general approach, one needs to know the reaction mechanisms and kinetics along the metabolic pathway, while the matrix method requires less detailed information in this regard. It is expected that as the details of metabolic processes become clearer as time goes on, the general approach will be used more widely.

10.1 The General Approach to Sensitivity Analysis

10.1.1 Mathematical Framework

Mathematical models of metabolic systems generally assume that metabolites are homogeneously distributed over the enzymes that act on them, and hence internal dif-fusion processes are not involved. Let us consider a metabolic system whose detailed enzyme kinetics are known. It consists of n metabolite concentrations (dependent variables), m enzymatically catalyzed reactions, and s system input parameters. In this general approach, the s system input parameters may include all the possible physicochemical parameters: enzyme concentrations, reaction rate constants, initial conditions, external modifiers, transport coefficients through cell walls, *etc.* The mass balances of the metabolites may be represented by the general form (Hatzimanikatis *et al.*, 1996):

$$\frac{dx}{dt} = f[v(x, \phi), x, \phi] \tag{10.1}$$

with initial conditions (ICs)

$$x = x^i, \qquad v = v^i, \qquad \text{at } t = 0 \tag{10.2}$$

where x represents the vector of the n metabolite concentrations, f is a function vector determined by the mass balances, ϕ is the s-dimensional system input parameter vector, and v is the m-dimensional reaction rate vector. In principle, since v is a function of x and ϕ, one can express f as a function of x and ϕ only. However, in the practice of metabolic process modeling, f is often expressed also as an explicit function of v, as shown in Eq. (10.1). This arises because detailed expressions of the metabolic reaction rates v are generally not well known, due to lack of knowledge of the metabolic mechanisms, while the metabolic reaction rates may be obtained directly from experiments. In addition to metabolic reaction rates, the above mass balance equations may also include terms that account for other processes leading to changes in metabolite concentrations, such as dilution caused by increase in the biomass volume (Fredrickson, 1976) and transport through the cell walls.

When the metabolic process is at steady state, the metabolite concentrations become independent of time, and the above differential equation reduces to an algebraic equation

$$f[v(x, \phi), x, \phi] = 0 \tag{10.3}$$

It should be noted that a steady state for a metabolic system is merely a mathematical concept, which requires that the formation rate of each metabolite in the system be equal to its consumption rate, while in real metabolic systems only a quasi-steady state may occur when metabolite accumulation in time is small as compared to the other terms. If the detailed kinetics of the reaction pathway, v, are given, by solving Eq. (10.1) with ICs (10.2), one can obtain all the metabolite concentrations as functions of time or at steady state.

When analyzing metabolic processes, we are often interested in the effects of variations in system input parameters on metabolite concentrations, $i.e.$, to evaluate the sensitivities of the metabolite concentrations with respect to the input parameters. The normalized sensitivity of the ith metabolite concentration with respect to variations in the jth system input parameter is defined as

$$S_{i,j} = S(x_i; \phi_j) = \frac{\phi_j}{x_i} \cdot \left(\frac{\partial x_i}{\partial \phi_j} \right) \tag{10.4}$$

which in the context of metabolic control theory is often referred to as the *concentration control coefficient*, if the parameter ϕ_j is an enzyme concentration.

For metabolic systems, similar to the other reaction systems discussed in previous chapters, evaluation of the normalized sensitivities of metabolite concentrations requires solving the following sensitivity equations:

$$\frac{dS}{dt} = NE^T S + KS + N\Pi^T + \Lambda \tag{10.5}$$

which are obtained by direct differentiation of Eq. (10.1) with respect to system input parameters, where $N, K, \Lambda, S, E,$ and Π are matrices, defined by

$$N = \begin{bmatrix} n_{1,1} & \cdots & n_{1,m} \\ \vdots & \ddots & \vdots \\ n_{n,1} & \cdots & n_{n,m} \end{bmatrix}, \quad n_{i,j} = \frac{v_j}{x_i} \cdot \left(\frac{\partial f_i}{\partial v_j} \right) \tag{10.6}$$

$$K = \begin{bmatrix} \kappa_{1,1} & \cdots & \kappa_{1,n} \\ \vdots & \ddots & \vdots \\ \kappa_{n,1} & \cdots & \kappa_{n,n} \end{bmatrix}, \quad \kappa_{i,j} = \frac{x_j}{x_i} \cdot \left(\frac{\partial f_i}{\partial x_j} \right) \tag{10.7}$$

$$\Lambda = \begin{bmatrix} \lambda_{1,1} & \cdots & \lambda_{1,s} \\ \vdots & \ddots & \vdots \\ \lambda_{n,1} & \cdots & \lambda_{n,s} \end{bmatrix}, \quad \lambda_{i,j} = \frac{\phi_j}{x_i} \cdot \left(\frac{\partial f_i}{\partial \phi_j} \right) \tag{10.8}$$

$$S = \begin{bmatrix} S_{1,1} & \cdots & S_{1,s} \\ \vdots & \ddots & \vdots \\ S_{n,1} & \cdots & S_{n,s} \end{bmatrix}, \quad S_{i,j} = \frac{\phi_j}{x_i} \cdot \left(\frac{\partial x_i}{\partial \phi_j} \right) \tag{10.9}$$

$$E = \begin{bmatrix} \varepsilon_{1,1} & \cdots & \varepsilon_{1,m} \\ \vdots & \ddots & \vdots \\ \varepsilon_{n,1} & \cdots & \varepsilon_{n,m} \end{bmatrix}, \quad \varepsilon_{i,j} = \frac{x_j}{v_i} \cdot \left(\frac{\partial v_i}{\partial x_j} \right) \tag{10.10}$$

$$\Pi = \begin{bmatrix} \pi_{1,1} & \cdots & \pi_{1,m} \\ \vdots & \ddots & \vdots \\ \pi_{s,1} & \cdots & \pi_{s,m} \end{bmatrix}, \quad \pi_{i,j} = \frac{\phi_j}{v_i} \cdot \left(\frac{\partial v_i}{\partial \phi_j} \right) \tag{10.11}$$

and S is called the *concentration sensitivity matrix*. Matrices E and Π, in the framework of metabolic control theory (Reder, 1988; Schlosser and Bailey, 1990; Cornish-Bowden and Cardenas, 1990), are often referred to as *elasticity (coefficient) matrices*. In fact, as is readily seen from their definitions, the elements of matrices E and Π are the normalized sensitivities of the metabolic reaction rates v with respect to variations in the metabolite concentrations x and in the system input parameters ϕ, respectively. Moreover, if the reaction rate is expressed as a function of the metabolite concentrations using the power law, *i.e.*, $v_j \propto x_i^\beta$, which is the typical assumption in biochemical systems theory, the elasticity $\varepsilon_{i,j}$ becomes equal to β, the reaction order of the jth reaction with respect to the ith metabolite concentration.

Thus, for a metabolic system with known enzyme kinetics, by solving simultaneously the mass balances and sensitivity equations (10.1) and (10.5), one can obtain the values of all metabolite concentrations (x) and their sensitivities (S) with respect to variations in the system input parameters (ϕ).

In metabolic process control, in addition to metabolite concentrations, other quantities are also important. In particular, it is often required to investigate the sensitivities of fluxes through the reaction pathway or through global pathway segments with respect to variations in the system input parameters, since they are a measure of how one global metabolic output (flux) varies when one independent variable of the system (often an enzyme concentration) changes. Let us define a general r dimensional vector of the fluxes J that one desires to examine as

$$J = J[v(x, \phi), x, \phi] \tag{10.12}$$

Note that based on this definition, J could also be used to denote any other system quantity that can be expressed as a function of the reaction rates, the metabolite concentrations, and the input parameters, $i.e.$, quantities that can be evaluated from the solution of the model equation (10.1).

The normalized sensitivity of the ith flux J_i with respect to the jth system input parameter is defined as

$$C_{i,j} = C(J_i; \phi_j) = \frac{\phi_j}{J_i} \cdot \left(\frac{d J_i}{d \phi_j} \right) \tag{10.13}$$

which is often referred to as the *flux control coefficient* in metabolic control theory. The flux sensitivity equation for computing $C_{i,j}$ can be derived by directly differentiating Eq. (10.12), leading to

$$C = \Xi E^T S + H S + \Xi \Pi^T + \Theta \tag{10.14}$$

where matrices Ξ, H, Θ, and C are defined as

$$\Xi = \begin{bmatrix} \xi_{1,1} & \cdots & \xi_{1,m} \\ \vdots & \ddots & \vdots \\ \xi_{r,1} & \cdots & \xi_{r,m} \end{bmatrix}, \qquad \xi_{i,j} = \frac{v_j}{J_i} \cdot \left(\frac{\partial J_i}{\partial v_j} \right) \tag{10.15}$$

$$H = \begin{bmatrix} \eta_{1,1} & \cdots & \eta_{1,n} \\ \vdots & \ddots & \vdots \\ \eta_{r,1} & \cdots & \eta_{r,n} \end{bmatrix}, \qquad \eta_{i,j} = \frac{x_j}{J_i} \cdot \left(\frac{\partial J_i}{\partial x_j} \right) \tag{10.16}$$

$$\Theta = \begin{bmatrix} \theta_{1,1} & \cdots & \theta_{1,s} \\ \vdots & \ddots & \vdots \\ \theta_{r,1} & \cdots & \theta_{r,s} \end{bmatrix}, \qquad \theta_{i,j} = \frac{\phi_j}{J_i} \cdot \left(\frac{\partial J_i}{\partial \phi_j} \right) \tag{10.17}$$

$$C = \begin{bmatrix} C_{1,1} & \cdots & C_{1,s} \\ \vdots & \ddots & \vdots \\ C_{r,1} & \cdots & C_{r,s} \end{bmatrix}, \qquad C_{i,j} = \frac{\phi_j}{J_i} \cdot \left(\frac{d J_i}{d \phi_j} \right) \tag{10.18}$$

C is referred to as the *flux sensitivity matrix* and can be readily evaluated through Eq. (10.14) using the values of the metabolite concentrations, x, and their sensitivities, S, obtained from the solution of Eqs. (10.1) and (10.5), respectively.

Table 10.1. Yeast glycolytic pathway

$v_{in} = v_{in}^o - 0.5([G6P] - 2.7)$

$v_{HK} = 68.5 \left(\dfrac{0.00062}{[G^{in}][ATP]} + \dfrac{0.11}{[G^{in}]} + 1 \right)^{-1}$

$v_{POL} = 15 \left[\left(1 + \left(\dfrac{2}{[G6P]} \right)^{8.25} \right) \left(\dfrac{1.1}{[G6P][UDPG]} + \dfrac{1}{[UDPG]} + 1 \right) \right]^{-1}$

$V_{PFK} = 5283 \dfrac{[F6P][ATP]R}{R^2 + L_o L^2 T^2}$

$v_{GAPD} = 49.9 \left[1 + \dfrac{0.0025}{[G3P]} + \dfrac{0.18 S}{[NAD^+]} \left(1 + \dfrac{0.0025}{[G3P]} (1 + 3333[NADH]) \right) \right]^{-1}$

$v_{PK} = 68.8 \dfrac{[PEP][ADP](R_{PK} K_I + 0.2 L_{o,PK} L_{PK} T_{PK})}{\left(R_{PK}^2 + L_{o,PK} L_{PK}^2 T_{PK}^2 \right) (1 + 10^{pH - 8.02})}$

$v_{GOL} = 15 \dfrac{[FdP]}{24 + [FdP]}$

$S = 1 + 0.9091[AMP] + 0.6667[ADP] + 0.4[ATP]$

$R = 1 + [F6P] + 16.67[ATP] + 166.7[F6P][ATP]$

$T = 1 + 0.0005[F6P] + 16.67[ATP] + 0.0083[F6P][ATP]$

$L = \dfrac{1 + 0.76[AMP]}{1 + 40[AMP]} \quad L_o = \exp(4.17\ pH + 20.42) - 1658$

$R_{PK} = 1 + K_I[PEP] + 0.2[ADP] + 0.02 K_I[PEP][ADP]$

$T_{PK} = 1 + 0.02[PEP] + 0.2[ADP] + 0.004[PEP][ADP]$

$K_I = 17.92 + 10^{pH - 4.907} \quad L_{o,PK} = ab - a^2$

$a = 1.713 + 10^{pH - 6.306} \quad b = \dfrac{0.1192 + 10^{pH - 7.084}}{11.83 + 1.61 pH - 8.722 \sqrt{pH}}$

From Schlosser *et al.* (1994).
The concentration of UDPG was fixed at [UDPG] $= 0.7 \times 10^{-3}$ kmol/m^3.
The equilibrium step, G6P $\longleftrightarrow$ F6P, was described by [F6P] $= 0.3$[G6P].
The equilibrium step, G3P $\longleftrightarrow$ FdP, was described by [G3P] $= 0.01$[FdP].
For the adenylate kinase (AK) an equilibrium constant $q_{AK} = 1$ was considered and the total adenylate pool, [AN] $=$ [ATP] $+$ [ADP] $+$ [AMP], was treated as fixed parameters at a value [AN] $= 2.8 \times 10^{-3}$ kmol/m^3.
The ratio [NADH]/[NAD$^+$] and the sum [NADH] $+$ [NAD$^+$] were also treated as fixed at values 0.03×10^{-3} and 2.5×10^{-3} kmol/m^3, respectively.

10.1.2 A Case Study: The Yeast Glycolytic Pathway

The glycolytic pathway from yeast is now well understood, and is represented schematically in Fig. 10.1. The kinetic expressions for the various reactions involved are reported in Table 10.1 (Schlosser *et al.*, 1994). With this information, sensitivity analysis

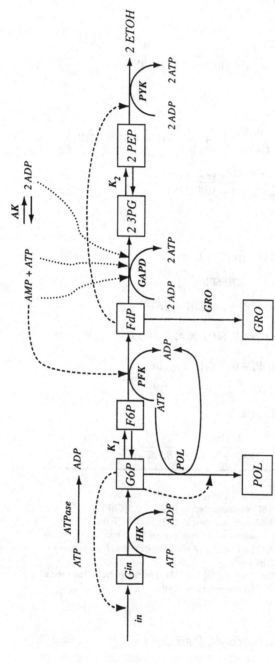

Figure 10.1. Anaerobic fermentation pathway of the yeast *Saccharomyces cerevisiae* with glucose as the sole carbon source, under nitrogen starvation. The solid arrows indicate reaction steps, broken arrows indicate activation, and dotted arrows indicate inhibition. **AK**, adenylate kinase; **ADP**, adenosine diphosphate; **AMP**, adenosine monophosphate; **ATP**, adenosine triphosphate; **ATPase**, net ATP consumption; **ETOH**, ethanol; **F6P**, fructose-6-phosphate; **FdP**, fructose-1,6-bisphosphate; **GAPD**, glyceraldehyde-3-phosphate dehydrogenase; **GRO**, glycerol production; **G6P**, glucose-6-phosphate; **HK**, hexokinase; **in**, glucose uptake; **K₁**, equilibrium step; **K₂**, equilibrium step; **PFK**, phosphofructokinase; **POL**, polysaccharide production; **PYK**, pyruvate kinase; **3PG**, 3-phosphoglycerate. From Hatzimanikatis and Bailey (1997).

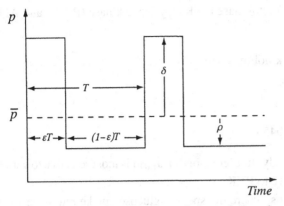

Figure 10.2. Pulsed periodic variation of the glucose uptake rate, p, as defined by Eq. (E.10.1). From Hatzimanikatis and Bailey (1997).

of the metabolite concentrations and production fluxes can be performed through the general approach discussed above. In the following, we report the results obtained by Hatzimanikatis and Bailey (1997), who analyzed the sensitivity of the ethanol flux J_{EtOH} (*i.e.*, the flux through the last reaction step in Fig. 10.1) with respect to changes in various enzyme concentrations. It should be noted that their analysis is based on an approximate (log) linear model, but their simulations indicate that it yields excellent agreement with the original nonlinear model. They considered two different operating modes for the system. In the first, the system is at steady state and the glucose uptake rate is constant and equal to $\overline{p}$. In the second, the glucose uptake rate is changed periodically in time according to the sequence of steps illustrated in Fig. 10.2, which can be represented by

$$p(t) = \overline{p} + \begin{cases} \delta, & t \in (jT, \{j + \varepsilon\}T) \\ \rho = \dfrac{\varepsilon\delta}{\varepsilon - 1}, & t \in (\{j + \varepsilon\}T, \{j + 1\}T) \end{cases} \qquad \text{(E10.1)}$$

where T is the period. The value of ρ is selected so that the average $\overline{p}$ equals the steady-state value above, *i.e.*,

$$\frac{1}{T}\int_0^T p(t)\,dt = \overline{p} + \frac{\delta\varepsilon T}{T} + \frac{\rho(1 - \varepsilon)T}{T} = \overline{p} \qquad \text{(E10.2)}$$

Let us first consider the sensitivity of the ethanol flux J_{EtOH} at the reference steady state. This can be evaluated by first solving Eqs. (10.1) and (10.5), which in this case are algebraic equations since their left-hand sides are set equal to zero. Next, using the obtained values of metabolite concentrations and their sensitivities, the ethanol flux J_{EtOH} and its sensitivities are evaluated through Eqs. (10.12) and (10.14). The values for the ethanol flux and its sensitivities with respect to variations in the concentrations of three enzymes, phosphofructokinase (PFK), pyruvate kinase (PYK), and ATPase, are

$$J_{EtOH} = 37.61 \times 10^{-3} \, \text{kmol/m}_{cell}^3/\text{min}$$

$$S(J_{EtOH}; PFK) = 0.3184$$

$$S(J_{EtOH}; PYK) = 0$$

$$S(J_{EtOH}; ATPase) = 0.0455$$

which indicate that at the steady state, ethanol flux J_{EtOH} is most sensitive to variations in PFK, while it is insensitive to PYK.

Under dynamic conditions, where the specific glucose uptake rate is varied periodically as described by Eq. (E10.1), in order to evaluate the sensitivities of ethanol flux J_{EtOH} with respect to variations in enzyme concentrations, we need to solve simultaneously the differential equations (10.1) and (10.5). Figure 10.3 shows the sensitivities of the average ethanol flux, $\bar{J}_{EtOH}$, with respect to variations in PFK, PYK, and ATPase, as functions of the period of glucose uptake variation, T. It may be seen that under dynamic conditions, the sensitivities of $\bar{J}_{EtOH}$ are substantially different from those at steady-state condition. For instance, the sensitivity of J_{EtOH} to PYK approaches zero at steady state, whereas, under dynamic conditions, $\bar{J}_{EtOH}$ becomes very sensitive to PYK, particularly for large values of the period, T.

Several other conclusions can be drawn from the results shown in Fig. 10.3. First, sensitivities generally do not change monotonically with the period, T. As can be seen, for the sensitivities of the ethanol flux $\bar{J}_{EtOH}$ to PFK and ATPase, there exists a range of T values where the sensitivities reach local maxima or minima. Second, it is interesting to note in Fig. 10.3c that, by selecting the values of period T and of the waveform parameters ε and γ, the sensitivity of $\bar{J}_{EtOH}$ to ATPase can be made either positive or negative. This observation is important, since it implies that similar changes in enzyme concentrations under certain conditions have a beneficial effect, whereas in other conditions the effect is counterproductive.

10.2 The Matrix Method from Metabolic Control Theory

It is probably fair to say that with the current state of knowledge, for most metabolic systems, information about reaction mechanisms and kinetics is not sufficient to perform sensitivity analysis using the general approach discussed above. In particular, steady-state metabolite concentrations are often implicitly defined as functions of enzyme concentrations by the steady-state condition of the system (Giersch, 1988), such that the sensitivity equations (10.5) and (10.14) cannot be derived by simply differentiating the mass balance equations (10.1) and (10.2). In these cases, we need a simpler, approximate method for sensitivity analysis that requires less information

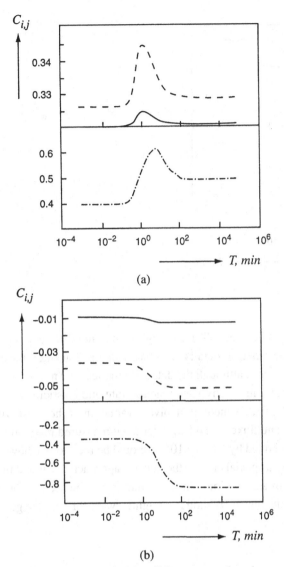

Figure 10.3. Sensitivities of the average ethanol production rate, $i = \bar{J}_{EtOH}$ with respect to various enzymes, $C_{i,j}$ as functions of the period, T, defined in Fig. 10.2, where (a) $j =$ phosphofructokinase, (b) $j =$ pyruvate kinase, and (c) $j =$ ATPase. Solid curves: $\varepsilon = 0.5$ and $\delta = 0.1$; broken curves: $\varepsilon = 0.5$ and $\delta = 0.2$; dashed-dotted curves: $\varepsilon = 0.8$ and $\delta = 0.1$. From Hatzimanikatis and Bailey (1997).

about the kinetics of the metabolic pathway. This leads to the development of the matrix method.

The matrix method has been developed and applied by several authors (Kacser and Burns, 1973; Heinrich and Rapoport, 1974a,b; Westerhoff and Chen, 1984; Fell

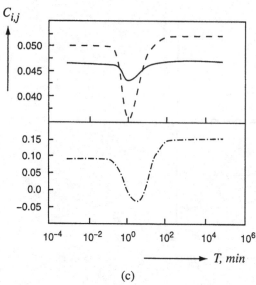

(c)

Figure 10.3. (cont.)

and Sauro, 1985; Reder, 1988; Giersch, 1988), although with minor differences in the terminology and in some mathematical details. The basic idea is that often for each reaction step (or pathway segment), although the detailed kinetics may not be known, one can measure experimentally its steady-state reaction rate and elasticity with respect to metabolites. Thus, the matrix method involves evaluation of the sensitivities of metabolite concentrations and fluxes, S and C, in terms of measured reaction rates and elasticities, *i.e.*, v and E defined by Eq. (10.10). It should be mentioned, however, that the matrix method is only a special case of the general approach discussed in the previous section under certain assumptions. In particular, it will be seen in the following that the derivation of the matrix method is straightforward through the general approach.

10.2.1 Model Framework

Assumptions

In its fundamental form, the matrix method is based on certain assumptions:

(1) The metabolic system is at steady state, so that the left-hand side of the mass balance equation (10.1) is set equal to zero.
(2) The system includes only the reaction terms, and no other physical processes such as transport.
(3) All reaction rates v are directly proportional to enzyme concentrations, and each enzyme affects only one reaction.
(4) The number of metabolites (n) does not exceed the number of enzymes (m).

(5) In the analysis, enzymes are treated as parameters rather than variables. Moreover, we analyze the sensitivities of the metabolite concentrations and fluxes with respect to variations in only the enzyme concentrations. The effects of other system input parameters are not investigated.

It is worth noting that some of the above assumptions can be relaxed, but at the expense of a significant increase in complexity of the method (*e.g.*, Acerenza *et al.*, 1989; Heinrich and Reder, 1991; Kacser and Sauro, 1990).

The algorithm

According to the above assumptions, in the following we restrict the system input parameter vector ϕ to include only the enzyme concentrations; hence ϕ becomes an m-dimensional vector. At steady-state conditions [assumption (1)], the mass balance equation (10.1) for the ith metabolite concentration x_i reduces to

$$f_i[\nu(x, \phi)] = \sum_{j=1}^{m} \alpha_{i,j} v_j = 0, \qquad i = 1, 2, \ldots, n \tag{10.19}$$

where $\alpha_{i,j}$ are coefficients indicating whether v_j is a source ($\alpha_{i,j} > 0$), a sink ($\alpha_{i,j} < 0$), or is irrelevant ($\alpha_{i,j} = 0$) for x_i. The above equation may be expressed in vector form as follows:

$$f = A[1 \quad \cdots \quad 1]^T = 0 \tag{10.20}$$

where A is an $n \times m$ matrix, given by

$$A = \begin{bmatrix} a_{1,1} & \cdots & a_{1,m} \\ \vdots & \ddots & \vdots \\ a_{n,1} & \cdots & a_{n,m} \end{bmatrix} = \begin{bmatrix} \alpha_{1,1} v_1 & \cdots & \alpha_{1,m} v_m \\ \vdots & \ddots & \vdots \\ \alpha_{n,1} v_1 & \cdots & \alpha_{n,m} v_m \end{bmatrix} \tag{10.21}$$

and its rank is assumed to equal n.

The sensitivity equation (10.5) in this case reduces to the following simple form (Hatzimanikatis, V., 1997, personal communication):

$$N E^T S + N = 0, \tag{10.22}$$

since from assumptions (2) and (3) we have

$$K = 0, \qquad \Pi = I, \qquad \Lambda = 0 \tag{10.23}$$

where I is the identity matrix of order m. Note that from Eqs. (10.6) and (10.10), matrices N and E have the same dimensions. Further, it can be shown based on the implicit function theorem that the inverse, $(N E^T)^{-1}$ exists; thus from Eq. (10.22) the concentration sensitivity matrix S is given by

$$S = -(N E^T)^{-1} N$$

or using matrix A

$$S = -(A E^T)^{-1} A \tag{10.24}$$

since in this case from Eqs. (10.6) and (10.21) we have

$$N = X^{-1} A \tag{10.25}$$

where

$$X = \text{diag}(x_1, x_2, \ldots, x_n) \tag{10.26}$$

Therefore, when the reaction rates v and the elasticity matrix E are given, the sensitivities of the metabolite concentrations with respect to variations in enzyme concentrations, S, can be evaluated through matrix operations, using Eq. (10.24). This is the essence of the matrix method.

In addition, if the vector J represents only fluxes of reaction steps, following the above procedure, we can also express the flux sensitivity matrix C using the measured reaction rates v and elasticity E. The final equation for C has the form

$$C = I - E^T (A E^T)^{-1} A \tag{10.27}$$

10.2.2 A Case Study: The Metabolic Pathway of Gluconeogenesis from Lactate

The pathway of gluconeogenesis from lactate, in terms of intermediate metabolite concentrations at different glucose production fluxes, fluxes of different segments of the pathway, and their interactions, is well established (Hue, 1981; Groen and Westerhoff, 1990). A simplified scheme of gluconeogenesis is shown in Fig. 10.4, where the reactions in the pathway have been grouped into four main segments: the first represents reactions between cytosolic pyruvate and phosphoenolpyruvate (PEP), the second is formed by pyruvate kinase, the third is composed of reactions between phosphoenolpyruvate and glyceraldehyde-3-phosphate (GAP), while the fourth includes reactions between glyceraldehyde-3-phosphate and glucose. For this simplified scheme, elasticities and fluxes for each of the four pathway segments have been measured experimentally and are reported in Table 10.2. Let us now compute the sensitivities of phosphoenolpyruvate and glyceraldehyde-3-phosphate with respect to variations in enzyme concentrations through the four segments.

It is seen from Fig. 10.4 that the concentration of PEP depends only on pathway segments 1, 2, and 3, while that of GAP on pathway segments 3 and 4. Thus, the mass balances for PEP and GAP are given by

$$\frac{dx_{\text{PEP}}}{dt} = f_{\text{PEP}} = v_1 - v_2 - v_3 \tag{E10.3}$$

$$\frac{dx_{\text{GAP}}}{dt} = f_{\text{GAP}} = v_3 - v_4 \tag{E10.4}$$

Table 10.2. Values of elasticities and fluxes of each pathway segment in Fig. 10.4 measured experimentally by Groen and Westerhoff (1990)

	Pathway segment			
	1	2	3	4
$\varepsilon_{j,\text{PEP}}$	−0.04	3.5	2.04	0
$\varepsilon_{j,\text{GAP}}$	0	0	−1.05	1.2
$J_j = v_j$	15.3	4.5	10.8	10.8

Operating conditions include saturated concentrations of lactate and pyruvate. Cells were perifused with concentration ratio of lactate and pyruvate equal to 10 in the presence of 0.1×10^{-3} kmol/m³ oleate.

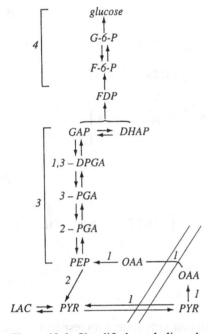

Figure 10.4. Simplified metabolic pathway of gluconeogenesis from lactate. **DHAP**, dihydroxyacetone phosphate; **FDP**, fructose-1,6-bisphosphate; **F-6-P**, fructose-6-phosphate; **GAP**, glyceraldehyde-3-phosphate; **G-6-P**, glucose-6-phosphate; **LAC**, lactate; **OAA**, oxaloacetate; **PEP**, phosphoenolpyruvate; **PYR**, pyruvate; **1,3-DPGA**, 1,3-bisphosphoglycerate; **2-PGA**, 2-phosphoglycerate; **3PGA**, 3-phosphoglycerate. From Groen and Westerhoff (1990).

or in vector form

$$\frac{d}{dt}\begin{bmatrix} x_{PEP} \\ x_{GAP} \end{bmatrix} = \begin{bmatrix} f_{PEP} \\ f_{GAP} \end{bmatrix} = A[1 \quad 1 \quad 1 \quad 1]^T \tag{E10.5}$$

where matrix A is given by

$$A = \begin{bmatrix} v_1 & -v_2 & -v_3 & 0 \\ 0 & 0 & v_3 & -v_4 \end{bmatrix} = \begin{bmatrix} 15.3 & -4.5 & -10.8 & 0 \\ 0 & 0 & 10.8 & -10.8 \end{bmatrix} \tag{E10.6}$$

From the values in Table 10.2, the corresponding elasticity matrix E can be constructed as

$$E = \begin{bmatrix} \varepsilon_{1,PEP} & \varepsilon_{2,PEP} & \varepsilon_{3,PEP} & \varepsilon_{4,PEP} \\ \varepsilon_{1,GAP} & \varepsilon_{2,GAP} & \varepsilon_{3,GAP} & \varepsilon_{4,GAP} \end{bmatrix} = \begin{bmatrix} -0.04 & 3.5 & 2.04 & 0 \\ 0 & 0 & -1.05 & 1.2 \end{bmatrix} \tag{E10.7}$$

so that

$$A E^T = \begin{bmatrix} 15.3 & -4.5 & -10.8 & 0 \\ 0 & 0 & 10.8 & -10.8 \end{bmatrix} \begin{bmatrix} -0.04 & 0 \\ 3.5 & 0 \\ 2.04 & -1.05 \\ 0 & 1.2 \end{bmatrix}$$

$$= \begin{bmatrix} -38.39 & 11.34 \\ 22.03 & -24.3 \end{bmatrix} \tag{E10.8}$$

The inverse, $(A E^T)^{-1}$, which is needed in Eq. (10.24) to compute the sensitivity matrix, is easily obtained as

$$(A E^T)^{-1} = \begin{bmatrix} -0.03558 & -0.01660 \\ -0.03225 & -0.0562 \end{bmatrix} \tag{E10.9}$$

Thus, the sensitivities of the PEP and GAP concentrations with respect to variations in the enzyme concentrations in the four pathway segments are given by

$$S = \begin{bmatrix} S_{PEP,1} & S_{PEP,2} & S_{PEP,3} & S_{PEP,4} \\ S_{GAP,1} & S_{GAP,2} & S_{GAP,3} & S_{GAP,4} \end{bmatrix} = -(A E^T)^{-1} A$$

$$= -\begin{bmatrix} -0.03558 & -0.01660 \\ -0.03225 & -0.0562 \end{bmatrix} \begin{bmatrix} 15.3 & -4.5 & -10.8 & 0 \\ 0 & 0 & 10.8 & -10.8 \end{bmatrix}$$

$$= \begin{bmatrix} 0.544 & -0.16 & -0.205 & -0.179 \\ 0.493 & -0.145 & 0.259 & -0.607 \end{bmatrix} \tag{E10.10}$$

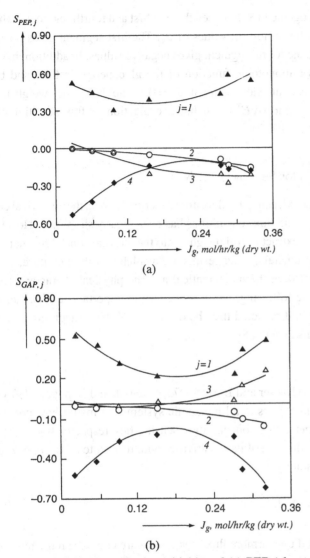

Figure 10.5. Normalized sensitivities of (a) **PEP** (phosphoenolpyruvate) and (b) **GAP** (glyceraldehyde-3-phosphate) with respect to variations in enzymes in the four gluconeogenic pathway segments shown in Fig. 10.4, as functions of the gluconeogenic flux, J_g. From Groen and Westerhoff (1990).

By repeating the above computations using elasticities and fluxes measured at different operating conditions, it is possible to obtain trends in the sensitivity values. An example (Groen and Westerhoff, 1990) is shown in Fig. 10.5, where the concentration sensitivities of phosphoenolpyruvate and glyceraldehyde-3-phosphate with respect to variations in enzyme concentrations in the four pathway segments are reported as a functions of the gluconeogenic flux. It is seen that the first and fourth segments exhibit the highest sensitivities for almost all values of J_g. The first segment gives positive sensitivity values, while the fourth segment gives negative values. In addition, neither of these sensitivities is a monotonic function of the gluconeogenic flux, and both exhibit their minimum absolute value at about $J_g = 0.18$ mol/hr/kg (dry weight).

It is clear that with E and $(A E^T)^{-1}$ at hand, evaluation of flux sensitivities is straightforward, using Eq. (10.27).

10.2.3 Some Useful Theorems for Sensitivity Analysis

Following the matrix method, using the algorithm described above, the numerical computation of concentration and flux sensitivities has been greatly simplified. However, it is generally not easy to extract useful insights into the enzyme control mechanisms from this procedure. Fortunately, other results are available in the literature, which provide a connection between the mathematical and the physical characteristics of metabolic systems. In the following, we introduce some important results, summarized as theorems. It should be noted that the derivation of these theorems was also based on the assumptions given in Section 10.2.1.

The summation theorem

The summation theorem (Kacser and Burns, 1973; Heinrich and Rapoport, 1974a,b) consists of two parts and applies to all metabolic systems, no matter how complex. They refer to flux and metabolite concentration sensitivities, respectively.

For a specific flux J_i, the sum of its sensitivities with respect to all the m enzymes in a metabolic system is unity, $i.e.$,

$$\sum_{j=1}^{m} C_{i,j} = 1, \qquad i = 1, 2, \ldots, r \tag{10.28}$$

This result signifies that if one or more flux sensitivities are large, the remaining ones must be small. In particular, for a linear reaction pathway with normal enzyme kinetics ($i.e.$, substrates activate and products inhibit), in which all flux sensitivities must be positive or zero, the upper limit of flux sensitivity with respect to any enzyme is 1, and it would occur only when flux sensitivities to all the other enzymes are zero. In such a case, the reaction step corresponding to that particular enzyme is the rate-controlling step. When the reaction pathway has branches and/or cycles, the flux sensitivities can be negative, even with normal kinetics, leaving open the possibility that some flux sensitivities have values substantially greater than 1 (Kacser, 1983).

For the sensitivities of the ith metabolite concentration with respect to all the enzymes, the corresponding summation relationship is

$$\sum_{j=1}^{m} S_{i,j} = 0 \qquad i = 1, 2, \ldots, n \tag{10.29}$$

It may be observed that the sensitivity values of PEP and GAP computed in the previous section, Eq. (E10.10), indeed satisfy this requirement.

The connectivity theorem

The connectivity theorem (Kacser and Burns, 1973) also consists of two parts, related to flux and metabolite concentration sensitivities. First, it states that the sensitivities of flux J_i with respect to the jth enzyme are related to the elasticities of the corresponding enzyme with respect to the kth metabolite by the relation

$$\sum_{j=1}^{m} C_{i,j} \cdot \varepsilon_{j,k} = 0 \tag{10.30a}$$

or in matrix form

$$\boldsymbol{C E}^T = \boldsymbol{0} \tag{10.30b}$$

This relation, which can be derived from Eq. (10.27), provides a route to understanding how enzyme kinetics affect the flux sensitivities. An important result of this theorem is that the ratio of the flux sensitivities for two adjacent enzymes is inversely proportional to the ratio of their elasticities with respect to their common intermediate metabolite. For example, consider the following simple linear pathway:

$$X_0 \xrightarrow{E_1} X_1 \xrightarrow{E_2} X_2 \xrightarrow{E_3} X_3, \ldots J_i$$

where X_k ($k = 0, 1, 2, 3$) are metabolites, E_j ($j = 1, 2, 3$) are enzymes, and J_i is the pathway flux. When Eq. (10.30) is applied to the metabolite X_1 common to enzymes E_1 and E_2, we have

$$C_{i,1}\varepsilon_{1,1} + C_{i,2}\varepsilon_{2,1} = 0 \tag{10.31}$$

which can be rewritten in the equivalent form

$$\frac{C_{i,1}}{C_{i,2}} = -\frac{\varepsilon_{2,1}}{\varepsilon_{1,1}} \tag{10.32}$$

Thus, based only on relationship (10.32), and without knowing anything else about the rest of the pathway, we can infer that a large elasticity value is always associated with a small corresponding value of flux sensitivity.

The corresponding connectivity theorem for metabolite concentration sensitivities is (Westerhoff and Chen, 1984)

$$\sum_{j=1}^{m} S_{i,j}\varepsilon_{j,k} = -\delta_{i,k} \tag{10.33a}$$

where $\delta_{i,k}$ is the Kronecker function, $i.e.$, $\delta_{i,k}$ equals 1 if $i = k$ and 0 if $i \neq k$. In matrix form, based on Eq. (10.24), this takes the form

$$\boldsymbol{S E}^T = -\boldsymbol{I} \tag{10.33b}$$

It should be noted that when the metabolites are involved in moiety-conserved cycles, the connectivity theorems (10.30) and (10.33) have to be modified for each of the metabolites containing the conserved group (Reder, 1988; Fell and Sauro, 1985; Hofmeyr *et al.*, 1986; Sauro *et al.*, 1987). In addition, in the case of linear pathways, the summation and connectivity theorems applied to all the metabolites provide exactly the number of simultaneous equations needed to evaluate the flux sensitivities with respect to all the enzymes in terms of elasticities. However, this is not the case for branched pathways or pathways involving certain types of cycles (*e.g.*, substrate cycles). In these cases, additional equations are required to correlate the flux sensitivities with the relative fluxes through different parts of the system, which are known as branching and substrate cycle theorems (Fell and Sauro, 1985; Reder, 1988).

Nomenclature

$a_{i,j}$ Element in matrix A, defined by Eq. (10.21)

A $n \times m$ matrix, defined by Eq. (10.21)

$C_{i,j}$ Normalized sensitivity of the ith flux with respect to the jth enzyme (or input parameter), defined by Eq. (10.13), also referred to as the flux control coefficient

C Flux sensitivity matrix, defined by Eq. (10.18)

E_j jth enzyme

E Elasticity matrix, defined by Eq. (10.10)

f Function vector, defined by Eq. (10.1) or (10.20)

H Matrix, defined by Eq. (10.16)

J Flux vector, defined by Eq. (10.12)

K Matrix, defined by Eq. (10.7)

m Number of enzymes or enzymatically catalyzed reactions

n Number of system dependent variables (metabolites)

$n_{i,j}$ Element in matrix N, defined by Eq. (10.6)

N Matrix, defined by Eq. (10.6)

s Number of system input parameters

$S_{i,j}$ Normalized sensitivity of the ith metabolite concentration with respect to the jth enzyme (or input parameter), defined by Eq. (10.4), also referred to as the concentration control coefficient

S Metabolite concentration sensitivity matrix, defined by Eq. (10.9)

v_j Reaction rate of the jth reaction step in the metabolic pathway

t Time

T Period

x Metabolite concentration vector

x_i Concentration of the ith metabolite

X_i ith metabolite

Greek Symbols

$\alpha_{i,j}$ Stoichiometric coefficient, defined by Eq. (10.19)

δ Wave-form parameter, defined by Eq. (E10.1)

$\delta_{i,k}$ Kronecker function, *i.e.*, $\delta_{i,k}$ equals 1 if $i = k$ and 0 if $i \neq k$

ε Wave-form parameter, defined by Eq. (E10.1)

$\varepsilon_{i,j}$ Element in matrix E, defined by Eq. (10.10)

$\eta_{i,j}$ Element in matrix H, defined by Eq. (10.16)

ϕ m-dimensional vector of enzyme concentrations or s-dimensional vector of system input parameters

ϕ_j jth element in vector ϕ

$\kappa_{i,j}$ Element in matrix K, defined by Eq. (10.7)

$\lambda_{i,j}$ Element in matrix Λ, defined in Eq. (10.8)

Λ Matrix, defined by Eq. (10.8)

$\pi_{i,j}$ Element in matrix Π, defined by Eq. (10.11)

Π Matrix, defined by Eq. (10.11)

$\theta_{i,j}$ Element in matrix Θ, defined by Eq. (10.17)

Θ Matrix, defined by Eq. (10.17)

ρ Wave-form parameter, defined by Eq. (E10.1)

$\xi_{i,j}$ Element in matrix Ξ, defined by Eq. (10.15)

Ξ Matrix, defined by Eq. (10.15)

Superscripts

i Initial condition

References

Acerenza, L., Sauro, H. M., and Kacser, H. 1989. Control analysis of time-dependent metabolic systems. *J. Theor. Biol.* **137**, 423.

Cornish-Bowden, A., and Cardenas, M. L. (eds.) 1990. *Control of Metabolic Processes* (NATO ASI Series A: Lifesciences, Vol. 190). New York: Plenum.

Crabtree, B., and Newsholme, E. A. 1978. Sensitivity of a near-equilibrium reaction in a metabolic pathway to changes in substrate concentration. *Eur. J. Biochem.* **89**, 19.

Crabtree, B., and Newsholme, E. A. 1985. A quantitative approach to metabolic control. *Curr. Top. Cell. Regul.* **25**, 21.

Crabtree, B., and Newsholme, E. A. 1987. The derivation and interpretation of control coefficients. *Biochem. J.* **247**, 113.

Fell, D. A. 1992. Metabolic control analysis: a survey of its theoretical and experimental development. *Biochem. J.* **286**, 313.

Fell, D. A., and Sauro, H. M. 1985. Metabolic control theory and its analysis: additional relationships between elasticities and control coefficients. *Eur. J. Biochem.* **148**, 555.

Fredrickson, A. G. 1976. Formulation of structured growth models. *Biotechnol. Bioeng.* **28**, 1481.

Giersch, C. 1988. Control analysis of biochemical pathways: a novel procedure for calculating control coefficients and an additional theorem for branched pathways. *J. Theor. Biol.* **134**, 451.

Groen, A. K., and Westerhoff, H. V. 1990. Modern control theories: a consumers' test. In *Control of Metabolic Processes*, A. Cornish-Bowden and M. L. Cardenas, eds., p. 101. New York: Plenum.

Hatzimanikatis, V. 1997. Personal communication.

Hatzimanikatis, V., and Bailey, J. E. 1997. Effects of spatiotemporal variations on metabolic control: approximate analysis using (log)linear kinetic models. *Biotechnol. Bioeng.* **54**, 91.

Hatzimanikatis, V., Floudas, C. A., and Bailey, J. E. 1996. Analysis and design of metabolic reaction networks via mixed-integer linear optimization. *A.I.Ch.E. J.* **42**, 1277.

Heinrich, R., and Rapoport, T. A. 1974a. A linear steady-state treatment of enzymatic chains. *Eur. J. Biochem.* **42**, 89.

Heinrich, R., and Rapoport, T. A. 1974b. A linear steady-state treatment of enzymatic chains: critique of the crossover theorem and a general procedure to identify interaction sites with an effector. *Eur. J. Biochem.* **42**, 97.

Heinrich, R., and Reder, C. 1991. Metabolic control analysis of relaxation processes. *J. Theor. Biol.* **151**, 343.

Higgins, J. 1963. Analysis of sequential reactions. *Ann. N.Y. Acad. Sci.* **108**, 305.

Hofmeyr, J.-H. S., Kacser, H., and Van der Merwe, K. J. 1986. Metabolic control analysis of moiety-conserved cycles. *Eur. J. Biochem.* **155**, 631.

Hue, L. 1981. Regulation of carbohydrate metabolism. *Adv. Enzymol.* **52**, 249.

Kacser, H. 1983. The control of enzyme systems *in vivo*: elasticity analysis of the steady state. *Biochem. Soc. Trans.* **11**, 35.

Kacser, H., and Burns, J. A. 1973. The control of flux. *Symp. Soc. Exp. Biol.* **27**, 65.

Kacser, H., and Sauro, H. M. 1990. Enzyme-enzyme interactions and control analysis: 2. The case of non-independence: heterologous associations. *Eur. J. Biochem.* **187**, 493.

Reder, C. 1988. Metabolic control theory: a structural approach. *J. Theor. Biol.* **135**, 175.

Sauro, H. M., Small, J. R., and Fell, D. A. 1987. Metabolic control and its analysis: extensions to the theory and matrix method. *Eur. J. Biochem.* **165**, 215.

Savageau, M. A. 1969a. Biochemical systems analysis: I. Some mathematical properties of the rate law for the component enzymatic reactions. *J. Theor. Biol.* **25**, 365.

Savageau, M. A. 1969b. Biochemical systems analysis: II. The steady-state solutions for an n-pool system using a power-law approximation. *J. Theor. Biol.* **25**, 370.

Savageau, M. A. 1969c. Biochemical systems analysis: III. Dynamic solutions using a power-law approximation. *J. Theor. Biol.* **26**, 215.

Savageau, M. A. 1971a. Concepts relating the behavior of biochemical systems to their underlying molecular properties. *Arch. Biochem. Biophys.* **145**, 612.

Savageau, M. A. 1971b. The behavior of intact biochemical control systems. *Curr. Top. Cell. Regul.* **6**, 63.

Savageau, M. A. 1976. *Biochemical Systems Analysis: A Study of Function and Design in Molecular Biology.* Reading, MA: Addison-Wesley.

Schlosser, P. M., and Bailey, J. E. 1990. An integrated modeling-experimental strategy for the analysis of metabolic pathways. *Math. Biosci.* **100**, 87.

Schlosser, P. M., Riedy, G., and Bailey, J. E. 1994. Ethanol production in baker's yeast: application of experimental perturbation techniques for model development and resultant changes in flux control analysis. *Biotechnol. Prog.* **10**, 141.

Westerhoff, H. V., and Chen, Y. 1984. How do enzyme activities control metabolite concentrations? An additional theorem in the theory of metabolic control. *Eur. J. Biochem.* **142**, 425.

Author Index

Subject Index